T0314047

The Everett Interpretation
of Quantum Mechanics

The Everett Interpretation of Quantum Mechanics

COLLECTED WORKS 1955–1980

WITH COMMENTARY

HUGH EVERETT III

Edited by JEFFREY A. BARRETT and PETER BYRNE

PRINCETON UNIVERSITY PRESS

PRINCETON & OXFORD

Library of Congress Cataloging-in-Publication Data

Everett, Hugh.
The Everett interpretation of quantum mechanics : collected works 1955–1980 with
commentary / Hugh Everett, III; editors, Jeffrey A. Barrett and Peter Byrne.
p. cm.
Includes bibliographical references and index.
ISBN 978-0-691-14507-5 (hardcover : acid-free paper) 1. Quantum theory.
2. Everett, Hugh.   I. Barrett, Jeffrey Alan.   II. Byrne, Peter, 1952–
III. Everett, Hugh. Selections.   IV. Title.
QC174.12.E96 2012
530.12–dc23
2011037956

British Library Cataloging-in-Publication Data is available

This book has been composed in Sabon LT Std

Printed on acid-free paper. ∞

Typeset by S R Nova Pvt Ltd, Bangalore, India
Printed in the United States of America

10 9 8 7 6 5 4 3 2 1

Reality resists imitation through a model.
  —Erwin Schrödinger, *The Present Situation in Quantum Mechanics* (1935)

Once we have granted that any physical theory is essentially only a model for the world of experience, we must renounce all hope of finding anything like "the correct theory."
  —Hugh Everett III, *The Theory of the Universal Wave Function* (1973)

CONTENTS

THIS VOLUME CONTAINS a collection of Hugh Everett III's work on pure wave mechanics and his related notes and correspondence. Several of these documents were unknown until quite recently and many are published here for the first time.

Our aim is to present the Everett papers in an accurate and readable form. We have corrected basic misspellings, typographical errors, and misidentifications in Everett's references without comment. For more significant changes or when there are notes or other salient features on the original manuscript, we provide footnotes that describe the changes and original text. We employ two levels of footnotes in this volume. The first level of (numbered) footnotes were a part of the original document. The second level of (lettered) footnotes contain textual notes, cross-references, and short discussions of the subject to make the text more accessible. We have left abbreviations where readability is not affected or where it is unclear how the abbreviations should be expanded. Digital scans of the original documents are available online at UCIspace. The original documents are now archived at the American Institute of Physics.

This volume also contains three introductory essays: a general introduction (by Barrett and Byrne), a biographical introduction (by Byrne), and a conceptual introduction (by Barrett). The introductory essays, which differ in style and approach, are intended to make Everett's writings more accessible and useful to readers with different professional backgrounds. The biographical introduction is intended to be accessible to the general reader. Barrett's conceptual introduction and explanatory footnotes are more technical in tone but also aim for accessibility.

The present volume reflects the cooperation and work of many friends and colleagues. Foremost, the project would have been impossible without the help of Mark Everett, who encouraged us to dig through the boxes of old papers in his basement. Jim Weatherall's industry and excellent editorial advice has been invaluable, especially in preparing the final manuscript.

We thank Samuel Fletcher and Thomas Barrett for their extensive careful work on this project. They did most of the LaTeX transcriptions from Peter Byrne's digital scans of Everett's papers, and much of the initial work in organizing, formatting, and copyediting the LaTeX transcriptions. We thank Brett Bevers, Christina Conroy, Porter Williams, and Andrew Bollhagen, who helped to transcribe the original documents. Brett Bevers also found the letter from Everett to Jaynes at the archive of Jaynes' papers at Washington University in St. Louis and helped to provide historical perspective on this letter. We thank Julie Shawvan for her excellent work on the index, and Ben Holmes at PUP for help all around. We thank Michelle

Light, Acting Head of Special Collections and Archives at UC Irvine, and her colleagues for their advice and support in setting up the permanent digital companion archive of the Everett source material at UCIspace@the Libraries. The permanent URL for this companion archive can be found at http://hdl.handle.net/10575/1060. This project has benefited from discussions with Craig Callender, Simon Saunders, David Wallace, Jason Hoelscher-Obermaier, Brian Skyrms, and Christian Wüthrich.

Finally, we would like to thank the editorial staff at Princeton University Press, which published Everett's theoretical work in 1973. The present volume and the UCIspace digital archive of Everett's papers were supported by UC Irvine and NSF grant No. SES-0924135.

Jeffrey A. Barrett
Irvine, California
December 2011

Peter Byrne
Petaluma, California
December 2011

PART I

Introduction

CHAPTER 1
_____

General Introduction

EVERETT AND HIS PROJECT

In July 2007, *Nature* celebrated the half-centenary of the "many worlds" interpretation of quantum mechanics with a splashy cover and a series of explanatory articles. That year, there were two international conferences dedicated to dissecting Hugh Everett III's claim that the universe is completely quantum mechanical.[1] Although the theorist had been dead for a quarter century, his controversial theory was alive and kicking.

First published in *Reviews of Modern Physics* in 1957 as "The 'Relative State' Formulation of Quantum Mechanics," the theory was not labeled "many worlds" until 1970, and then, not by Everett, but by his enthusiastic supporter, physicist Bryce S. DeWitt. Today, the Everett interpretation is one of a handful of contenders for explaining the structure of the quantum universe—whether or not its "branching" motion is interpreted as a metaphor for the linear evolution of the universal state or as modeling idealized or ontologically real worlds.

Everett was only 27 years old when he developed his theory, which would become his doctoral dissertation at Princeton. More interested in military game theory than theoretical physics, Everett never published another word on quantum mechanics. And yet his dissertation has stood the test of time and disbelief. Something in Everett's work has continued to resonate with physicists and philosophers alike so that, despite his many critics, three generations of researchers have returned to Everett's strange, counterintuitive theory, trying to find language to capture the quantum universe described mathematically by his pure wave mechanics.

This volume presents the two previously published versions of his theory, Everett's long and short theses, alongside a selected collection of his unpublished works and correspondence, which illuminate how Everett and his contemporaries struggled to answer questions that remain with us today.

Everett developed his interpretation of quantum mechanics, his relative-state formulation of pure wave mechanics, while a graduate student in physics at Princeton University. Matriculating in the fall of 1953, he began writing down his idea a year later. A detailed presentation of the theory,

_____

[1] July 2007 at University of Oxford, Oxford, England, and September 2007 at Perimeter Institute for Theoretical Physics, Waterloo, Canada.

the long thesis, was submitted by Everett to John Archibald Wheeler, his doctoral thesis advisor, in January 1956. It was circulated in April of that year to several prominent physicists, including Niels Bohr.[2]

The long thesis was Everett's earlier, more detailed formulation and discussion of his theory, whereas the short thesis was a highly redacted and refocused version of the long thesis, reworked under the direction of Wheeler to soften the force of Everett's attack on the orthodox Copenhagen interpretation.

The back story is that Wheeler had spent considerable effort in May 1956 trying to convince Niels Bohr and his colleagues at the Institute for Theoretical Physics in Copenhagen, Denmark, that Everett's work should not be taken as a fatal threat to their understanding of quantum mechanics. His efforts were in vain and, with his doctoral degree in limbo due to Wheeler's reluctance to accept his long thesis without a nod of approval from Bohr, Everett left Princeton and took a job outside of academics as a military operations researcher in Washington, D.C., in June 1956.

During the winter of 1957, he and Wheeler rewrote the long thesis, cutting about 75 percent of it, to make it, in Wheeler's phrase, "javelin proof."[3] Subsequently, Everett's doctoral thesis (1957a), the short thesis, was accepted in March 1957, and a nearly identical paper (1957b) was published by *Reviews of Modern Physics* in July of that year. Bryce S. DeWitt and Neill Graham (1973) later published an updated version of Everett's long thesis in their volume entitled *The Many-Worlds Interpretation of Quantum Mechanics*.[4]

Although Everett's notes and correspondence indicate that he continued to be interested in the conceptual problems of quantum mechanics and in the interpretation and reception of his model of pure wave mechanics, he did not play an active role in the public debates surrounding his theory in the 1970s. He died of a heart attack in 1982 without writing any further systematic presentation of it. For many years, his long and short theses remained the primary evidence for how he had intended his formulation of quantum mechanics to work.

---

[2] See the biographical introduction in this volume for a more detailed account of the circumstances surrounding Everett's development of his relative-state formulation of pure wave mechanics, especially starting on pg. 11.

[3] See pg. 212.

[4] The title of the long thesis submitted by Everett to Wheeler in January 1956 was "Quantum Mechanics by the Method of the Universal Wave Function." In April 1956, it was retitled, "Wave Mechanics Without Probability." After being edited in 1957, the approved dissertation (short thesis) was entitled, "On the Foundations of Quantum Mechanics." The short thesis was again retitled for publication in *Reviews of Modern Physics* in July 1957 as "'Relative State' Formulation of Quantum Mechanics." When the long thesis was published in 1973 in the DeWitt–Graham book, Everett settled on yet another title: "The Theory of the Universal Wave Function." Versions of the long thesis (pg. 72) and short thesis (pg. 173) are included in this volume.

In 2007, the investigative journalist Peter Byrne was invited by Everett's son, Mark Everett, to open a dozen cardboard boxes that had been stored for many years in his basement. The boxes contained numerous items of scientific interest, including correspondence about the theory with Niels Bohr, Norbert Wiener, Wheeler, and other prominent physicists. Hundreds of pages of handwritten and typed and retyped drafts of the long thesis document Everett's thought process as he formulated his theory from the fall of 1954 through the winter of 1956. Importantly, three "minipapers" give an overview of Everett's basic arguments as of September 1955. This newly discovered material helps to illuminate his previously published work, often in striking ways.

EVERETT'S TARGET: THE MEASUREMENT PROBLEM

In the long thesis, Everett directly attacked both the von Neumann–Dirac and the Copenhagen formulations of quantum mechanics. He held that neither orthodox formulation could adequately describe what happened to a physical system when it was measured. Everett believed that the standard von Neumann–Dirac collapse formulation of quantum mechanics, the version of the theory found in most textbooks, provided an incomplete and incoherent characterization of measurement and that Bohr's formulation of the theory, called the Copenhagen interpretation, was even worse since it simply stipulated that the process of measurement could not be understood quantum mechanically. Wheeler, as his thesis advisor, wanted Everett to present his controversial theory in a way that he believed would be more easily received by the physics community. This led to the much shorter thesis that Everett defended for his Ph.D. The short thesis still expressed dissatisfaction with the conventional formulations of quantum mechanics, but it now characterized their inadequacies less as fundamental conceptual flaws and more as roadblocks to applying quantum mechanics to field theories and cosmology.

The problem with the standard collapse theory, according to Everett, was that it required observers always to be treated as external to the system described by the theory, one consequence of which was that it could not be used to provide a consistent physical description of the universe as a whole since the universe contains observers. More specifically, the standard collapse theory has two dynamical laws: one says that physical systems evolve in a linear, deterministic way when not measured, and the other says that physical systems evolve in a nonlinear random way when measured. But since the standard theory does not say what constitutes a measurement, it is at best incomplete. And if one takes measuring devices and observers to be described by the deterministic linear law (and why shouldn't they be insofar as they are constructed of simpler systems that each follow the linear deterministic law?), then the collapse theory is logically inconsistent. This is the notorious quantum measurement problem for the standard textbook formulation of quantum mechanics.

Everett was not alone in his dissatisfaction with the prevailing interpretation of quantum mechanics. Other notable discontents included Erwin Schrödinger, Albert Einstein, Boris Podolsky, Nathan Rosen, and David Bohm. Indeed, Bohm, who left Princeton just before Everett arrived,[5] had devised a deterministic "hidden-variable" formulation of quantum mechanics that addressed the quantum measurement problem and made the same empirical predictions as the conventional formulations for those experiments where they made coherent predictions at all. Everett, however, believed that his simpler approach rendered Bohm's hidden variables "superfluous."[6]

Everett tackled the measurement problem by promoting what he called "pure wave mechanics."[7] His formalism characterized the physical state of the universe with a "universal wave function," which describes a superposition of possible classical states that evolves in a perfectly continuous and linear way. This is the simplest possible formulation of quantum mechanics, said Everett, because it entirely avoids the quantum measurement problem, and, unlike most other formulations of quantum mechanics, it can be put in a form that is compatible with the constraints of general relativity. In this sense, it provides an ideal quantum mechanical foundation for modern field theories. Everett's theory is consequently one of the most popular formulations of quantum mechanics among both physicists and philosophers.

Going further than previous critics of the standard collapse postulate, Everett's proposed solution to the measurement problem was to drop the random nonlinear dynamics from the standard collapse theory and take the resulting pure wave mechanics, governed by the time-dependent Schrödinger equation alone, as a complete physical theory. His goal was to deduce the empirical predictions of the standard collapse theory as the subjective experiences of observers who are themselves treated as physical systems described by the theory. He referred to pure wave mechanics with the interpretive apparatus provided by his fundamental principle of the relativity of quantum states as the relative-state formulation of quantum mechanics. It is, however, unclear precisely how Everett intended for the relative-state formulation to be understood. There is agreement among those who study Everett's interpretation of quantum mechanics that his

---

[5] Bohm's contract was not renewed by Princeton after he took the Fifth Amendment while testifying before Congress to the communist-hunting House Un-American Activities Committee.

[6] See Everett's discussion of Bohmian mechanics in the long thesis (pg. 153).

[7] See pgs. 65, 77, 178–80, and 196, for examples of how Everett characterized pure wave mechanics and his relative-state interpretation of it.

interpretation requires interpretation, and many people have attempted to explain exactly what he had in mind. Indeed, it is fair to say that most no-collapse interpretations of quantum mechanics have at one time or another either been directly attributed to Everett or suggested as charitable reconstructions.[8]

That said, the various many-worlds formulations of quantum mechanics have proven to be the most popular reconstructions of Everett's theory. This way of understanding the relative-state formulation is largely due to Bryce DeWitt's energetic promotion in the early 1970s of what he called the EWG theory, for Everett, Wheeler, and DeWitt's student Neill Graham. Whereas Everett himself never mentioned many worlds or parallel universes in either version of his thesis, DeWitt's interpretation of Everett so captured people's imagination that it remains the most popular understanding of Everett's theory.[9] Nonetheless, a half century after the theory was first published, much work continues to be done to formulate a clear and compelling many-worlds interpretation of pure wave mechanics. The most recent many-worlds interpretations characterize worlds as emergent entities that are roughly individuated by decoherence considerations.[10]

In the end, Everett's remarkable achievement was in providing a compelling case that pure wave mechanics alone constitutes a complete and accurate physical theory and makes the same empirical predictions as the standard collapse theory. According to him, the quantum measurement problem was simply a misunderstanding generated by unnecessarily adding a postulate that measurement is special to a theory that works without that postulate. Although most researchers believe that Everett was not entirely successful in deriving the standard quantum mechanical predictions from the mathematics of pure wave mechanics alone, he got close enough to motivate many others to try filling in the details in his project. Because of the simplicity of the mathematical formalism, its universal scope, and its other theoretical virtues, the stakes are high in understanding Everett's theory and in finding an acceptable interpretation of it.

But in the 1950s at Bohr's Institute for Theoretical Physics in Copenhagen, saying what Everett said was considered "heresy" (Leon Rosenfeld)[11] and "theology" (Alexander Stern).[12] Wheeler (who was researching a theory of quantum gravity) had tried to convince Bohr and his colleagues that the "relative state" model was a theoretical advance, but he ran into a phalanx of closed minds. In 1959 Everett and Bohr met in Copenhagen to discuss the controversial theory, which removed the epistemological barrier that

---

[8] See, for examples, the interpretations discussed in the conceptual introduction (pg. 37).

[9] See the conceptual introduction (pg. 41) for further discussion of DeWitt's splitting-worlds formulation of Everett's theory.

[10] See the discussion of the emergent-worlds formulation (pg. 45).

[11] Rosenfeld to Bell, 11/30/71, in Byrne (2010, pg. 316).

[12] Stern to Wheeler, 5/20/56; in this volume (pg. 215).

Bohr and his fellow travelers had erected between the overlapping realms of microscopic and macroscopic events. But Bohr dismissed Everett's work, and eventually, so did Wheeler.

History has been more accepting of Everett's theory than his contemporaries were. Shortly after the issue of *Nature* dedicated to the "many worlds" interpretation, the British Broadcasting Corporation and *NOVA* aired an award-winning television program, "Parallel Worlds, Parallel Lives," which is about the theory and, also, Everett's sad relationship with his rock singer son, Mark. But Everett was not around to take pleasure in these events. Of all of the late scientist's immediate family, only his son, the family's sole survivor, witnessed the world paying homage to the strange, brilliant, revolutionary idea widely known as the "many worlds" interpretation of quantum mechanics.

Jeffrey A. Barrett
Irvine, California
December 2011

Peter Byrne
Petaluma, California
December 2011

CHAPTER 2

Biographical Introduction

BASEMENT TREASURE

In the spring of 2007, the rock musician Mark Everett and the journalist Peter Byrne descended into the basement of the songwriter's house in Los Angeles. One wall was lined with wooden shelves holding the family saga, two dozen cardboard boxes bursting with photographs and memorabilia and paper trails though time. They opened up his father's boxes, which had been gathering dust since they had been packed a quarter century earlier.

The musty cartons held old textbooks, physics and operations research papers, stacks of letters, used airplane tickets, cancelled checks to liquor stores, crumpled hotel receipts, a cheesy Super 8 pornographic film, and a scrap of paper on which the physicist had parodied the standard ontological proof of the existence of God in the predicate calculus. Several boxes were stuffed with thousands of sheets of yellow legal paper covered with algorithms variously designed to radar-track ballistic missiles bearing nuclear warheads or to outwit the housing and stock markets. Other boxes held artwork made by the kids for Father's Day and Christmas. And beneath the childish art were letters from some of the most renowned quantum physicists and philosophers of the midcentury: John Wheeler, Norbert Wiener, Phillip Frank, Niels Bohr, Henry Margenau, H. J. Groenewold, and others.

Chief among the finds in the basement were successive versions of handwritten drafts of Everett's dissertation, along with his research materials and notes. There was a series of short, typed papers in which he summarized his main ideas, including "Probability in Wave Mechanics," written for Wheeler and Bohr in the fall of 1955.[1] In this outline of his theory, written in ordinary language, he compared his splitting, branching quantum states to splitting amoebas and human observers, much to his thesis advisor's displeasure. In a note, Wheeler urged Everett to eschew metaphors of splitting macroscopic objects "because of parts subject to mystical misinterpretations by too many unskilled readers."[2]

---

[1] This is one of the three minipapers reproduced in this volume (chapter 6, pg. 64).
[2] Wheeler to Everett, 09/21/55; in this volume (pg. 71).

As the papers were sifted and sorted, it became clear that even though Everett never wrote another word of quantum physics, he had followed the rehabilitation of his theory with great interest, though galled by what he perceived as the failure of his intellectual peers, including John Bell, Norbert Wiener, and Bryce DeWitt, to comprehend the character of the probability measure at the core of his theory. Here is what he scribbled next to DeWitt's statement that Everett's probability derivation was not satisfying: "Goddamit, you don't see it".[3]

In the decades since Everett recorded that particular disappointment, physicists and philosophers around the world have been hard at work trying to improve upon his formal and linguistic arguments. But in this volume we present only documents pertaining to how Everett and the contemporaries with whom he corresponded viewed his daring move to follow the linearity of the Schrödinger equation to its logical end—only to discover a purely quantum mechanical universe that can be considered to contain macroscopic superpositions of all objects, including copies of mice, cannonballs, and human observers, all carving out (in some sense) separate historical trajectories inside a global superposition that Everett termed the "universal wave function."

LIFE OF EVERETT: THE SHORT STORY

Hugh Everett III was born on Armistice Day, November 11, 1930, in Washington, D.C., to a military-minded father and a bohemian mother. He was raised in suburban Bethesda, Maryland, and spent most of his life in the metropolitan area of the capital city.

Hugh Everett, Jr. (his father), held a bachelor's degree in civil engineering, a master's degree in patent law, and a doctorate in juridical science. During World War II, he served the general staff on the European front as a logistics expert. In the 1960s, he engineered the construction of fallout shelters for top secret military installations in the capital region. A heavy drinker and pipe smoker, Colonel Everett, 77, died of lung cancer in 1980.

Katharine Kennedy Everett (his mother) abandoned a teaching career to concentrate on writing during the Great Depression. National newspapers and magazines regularly published her pulp fiction (including some science fiction) and her floridly phrased poetry (penned with a feminist perspective). At the time of her death in 1962 from breast cancer, she was active in the nuclear disarmament movement, while remaining proud of her son's career in "rocket science" and his "cosmic" security clearance. She did not know, however, that his job entailed designing software to operate the nuclear war fighting machine.

---

[3] See Everett's handwritten notes on DeWitt's paper in this volume (pg. 280).

The Everetts divorced, not amicably, when their son was five years old, and the split scarred him psychologically. For much of his adult life, he suffered from depression; he lacked empathy for others; he shied away from emotional intimacy. He viewed his business and social and familial interactions as utility-maximizing games, often treating people callously as he attempted to calculate the cost–benefits of relationships.

Despite his love of pure reason (in the form of mathematics and computer programming), Everett was addicted to alcohol, food, tobacco, and sex. Obese and compulsive, he refused to visit medical or psychiatric professionals; he cheated on his devoted wife, Nancy; and he neglected his emotionally needy children, Mark and Elizabeth. He loved fine wines, French haute cuisine, and week-long parties on cruise ships. Needing enough disposable cash to subsidize his habits, he was constantly inventing potentially marketable ideas in operations research and computer science, but he consistently failed to follow through on implementing promising business ventures. He died despondent, nearly bankrupt, and drunk.[4]

Personal and business failings aside, Everett is best remembered as a radiantly intelligent mathematician, physicist, game theorist, and pioneer in the science of electronic computation. In his midtwenties, during his first year as a graduate student at Princeton University, he wrote one of the seminal papers in the theory of games ("Recursive Games") before devising his counterintuitive solution to the measurement problem in quantum mechanics (which he initially approached from a game theoretical perspective). And after the publication of his quantum theory was met with either silence or outright rejection by members of the physics establishment, he immersed himself in military operations research at the Pentagon, working for the top secret Weapons System Evaluation Group (WSEG), which was operated by the nonprofit think tank, Institute for Defense Analyses. Sadly, despite Wheeler's perennial urging, Everett eschewed further research in quantum mechanics, even after his theory began to be publicly recognized as compelling and viable in the early 1970s.

ORIGINS OF THE THEORY[5]

A lifelong atheist, Everett attended a military, Catholic high school, St. John's. He excelled at science and math but failed his cadet drill classes. At Catholic University, his mathematics professor considered him to be the most brilliant undergraduate he had ever encountered. A chemistry major, he plowed through classes in advanced mathematics, game theory, and,

---

[4] See Byrne (2010) for further biographical details.

[5] The following history is drawn from Everett's papers and interviews with his former colleagues as documented in Byrne (2010).

as required, theology, reportedly driving his Jesuit philosophy professor to despair by making a logical argument against the existence of God. A science fiction fan, he was briefly attracted to L. Ron Hubbard's "science of Dianetics." Ever the technician, he used a strobe light to photograph sporting events, selling the action photos to newspapers. And in 1953, he worked for the summer as an associate mathematician at an operations research lab operated by Johns Hopkins University in Silver Spring, Maryland.

In the fall of 1953, Everett entered Princeton University as a doctoral student in physics. He took a course in electromagnetism, a seminar in algebra, and introductory quantum mechanics, with Robert H. Dicke. He studied (in German) John von Neumann's classic textbook, *Mathematical Foundations of Quantum Mechanics* (1932), and David Bohm's *Quantum Theory* (1951). But it was game theory that captivated him during his first year.

He regularly attended weekly seminars on game theory in Fine Hall, hosted by Albert Tucker and Harold Kuhn, who also organized a series of formal conferences attended by Everett that featured the illuminati of the craft, including John von Neumann (Institute for Advanced Study, Princeton); Oskar Morgenstern (professor of economics at Princeton University); John Forbes Nash (Massachusetts Institute of Technology); and Lloyd. S. Shapley (RAND Corporation).

At the 1955 game theory conference, Everett presented "Recursive Games," a paper on military tactics written during his first year at Princeton. Kuhn, who mentored Everett, as well as Nash (later to win the 1994 Nobel Prize in Economic Science for his work on equilibriums in cooperative games), considered Everett's paper extraordinary. It devised a method for determining a payoff point in games allowing infinitely many moves. It was first published in *Annals of Mathematics* in July 1957. And in 1997, Kuhn republished it in his book, *Classics in Game Theory*, alongside Nash's seminal paper on game equilibriums.

But in the fall of 1954, Everett was hard at work researching and writing his dissertation. He took only one class: Methods of Mathematical Physics, with Eugene Wigner, a philosophically inclined physicist. In Wigner's class, he came face to face with the mathematical contradiction between the continuous, linear evolution of the state of a quantum system as governed by the Schrödinger wave equation and the discontinuous, nonlinear collapse dynamics that the standard theory says occurs whenever one measures the system. The threat of inconsistency is real here since the standard collapse theory does not say when this discontinuous dynamics occurs except to indicate that it happens whenever a measurement occurs. But since measuring devices are themselves constructed of systems that the theory describes as obeying the continuous linear dynamics, the composite system consisting of the measuring device and the system being measured should also evolve in a continuous, linear way.

The standard collapse theory was formulated in the 1930s by John von Neumann and Paul Dirac. They added the collapse dynamics to the usual quantum linear dynamics to guarantee that a single, definite measurement result is randomly generated whenever a measurement is made. The problem is that the linear dynamics predicts that physical systems are typically in states where they do not have any definite familiar properties. Hence, von Neumann and Dirac postulated that measured systems randomly "jump" to states where they have whatever definite property is being measured whenever they are observed (and we do not know why). This postulate became known as the reduction or collapse of the wave function. Before a measurement occurs, the wave function describing a physical system in a quantum superposition represents a situation in which many different measurement results are possible but typically none are actual. One might think of the wave function as spread out over the different results that the measurement could yield. But as soon as some property of the system is measured in such a way as to create a record of the measurement outcome, the wave function instantly jumps, collapses, or reduces to a state where it represents a situation in which precisely one measurement result is realized. This is the case whether the record is captured inside a cloud chamber or a particle accelerator or is trapped as a bit inside a digital computer or as a chemical trace in a human brain. But Everett held that the standard collapse theory was entirely unacceptable since, rather than explaining how an experiment yields a definite result by appealing to the usual linear quantum dynamics, it simply stipulated that measurements yield definite results at just the right times and with just the right quantum statistics.

Nor was Everett impressed by the Copenhagen interpretation, which is attributed mainly to Niels Bohr and Werner Heisenberg. The Copenhagen interpretation deals with measurement by positing one set of physical laws (classical) for the macroscopic realm and another set of laws (fundamentally unknowable) for the microscopic realm. It declares that knowledge about what goes on inside the quantum realm is forever inaccessible and can never be reliably pictured. Rather, we must describe this world through the distorting filter of the language of classical physics. Everett thought that this dichotomy between quantum and classical description, which Bohr held to be a necessity, was a "philosophic monstrosity" (pg. 255). Ultimately, Everett believed that one should drop the collapse postulate and treat the entire universe as quantum mechanical and see what happened if one included the observer in the wave function—as opposed to arbitrarily treating the observer as classical and placing him (and his measurement device and the rest of the universe) outside the wave function of the quantum object observed.

Everett was not the first physicist to think in these terms. In 1935, the inventor of wave mechanics, Erwin Schrödinger, had found wave function collapse to be an unnatural suspension of the laws governing the "orderly course of natural events."[6] And in 1952, Schrödinger remarked, "[I]t would seem that, according to the quantum theorist, Nature is prevented from rapid jellification only by our perceiving or observing it. And I wonder that he is not afraid, when he puts a ten-pound note into his drawer in the evening, he might find it dissolved in the morning, because he has not kept watching it."[7]

We do not know if Everett feared jellification, but he was clearly inspired by his teacher, Wigner, who also questioned the standard interpretation. In 1961, Wigner published "Remarks on the Mind-Body Problem," which purported to solve the measurement problem by postulating that it is observation by human consciousnesses that collapses wave functions. Wigner held that to be consistent the standard theory of quantum mechanics needs to tell us precisely when collapses occur, and they must occur precisely when a scientist becomes conscious that the system he is observing has determinate properties to be observed. What could be responsible for this jump at precisely this instant? Aha! said Wigner: It is the mind of the scientist that collapses the wave function of the object that he observes into the single result recorded by the scientist. Human consciousness, in other words, creates physical reality, posited Wigner. One may wonder, however, whether this proposal improves much on the original collapse postulate.

Wigner did not publish this interpretation until after Everett wrote his theory.[8] But there is little doubt that he had been discussing the problem with his students for many years. In an early handwritten draft of his dissertation, Everett made a drawing of the dilemma facing Wigner's Friend.[9]

And in another handwritten draft, Everett made a drawing of his own solution to the problem of infinite regression.

In the spring of 1955, Everett passed his general examinations and received a master's degree in physics. By that time, he had been writing his doctoral thesis for half a year. And he had been thinking about it for at least a year.

The catalytic insight leading to Everett's theory may have occurred on or about April 14, 1954, when Albert Einstein gave the last lecture of his life (he died a year later) to a class on general relativity taught by Wheeler. It is believed that Everett attended this lecture. He certainly knew that Einstein had remarked that day that quantum mechanics was true, as far as it went, but that it did not fully describe the quantum world. Speaking of the standard interpretation of quantum mechanics, which privileges

---

[6] Schrödinger (1983, pg. 160). See also Byrne (2010, pg. 100).

[7] Schrödinger (1995, pgs. 19–20). See also Byrne (2010, pg. 101).

[8] Wigner (1961). See also Byrne (2010, pg. 409).

[9] See the discussion of Everett's version of the Wigner's Friend story in the conceptual introduction (pg. 30).

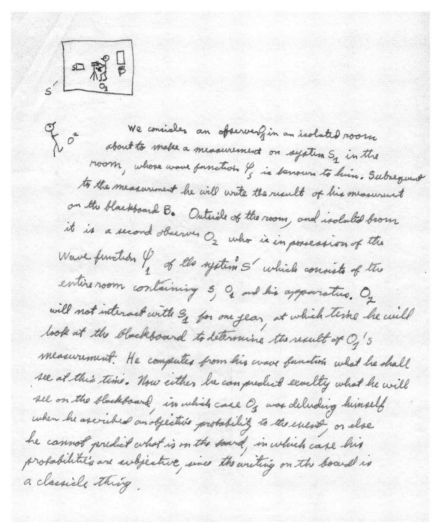

We consider an observer O_1 in an isolated room about to make a measurement on system S_1 in the room, whose wave function ψ_S is known to him. Subsequent to the measurement he will write the result of his measurement on the blackboard B. Outside of the room, and isolated from it is a second observer O_2 who is in possession of the Wave function ψ_1 of the system S' which consists of the entire room containing S, O_1 and his apparatus. O_2 will not interact with S_1 for one year, at which time he will look at the blackboard to determine the result of O_1's measurement. He computes from his wave function what he shall see at this time. Now either he can predict exactly what he will see on the blackboard, in which case O_1 was deluding himself when he ascribed an objective probability to the event, or else he cannot predict what is on the board, in which case his probabilities are subjective, since the writing on the board is a classical thing.

Figure 2.1. Everett's drawing of Wigner's Friend's dilemma. The Hugh Everett Papers, courtesy of Mark Everett.

the role of an external observer in the making of a quantum mechanical measurement, Einstein (echoing Schrödinger) said, "It is difficult to believe that the description is complete. It seems to make the world quite nebulous unless somebody, like a mouse, is looking at it."[10]

In the fall of 1954, Bohr was in residence at the Institute for Advanced Study located near Princeton University; he conferred privately with

[10] Byrne (2010, pg. 132); Tauber (1979, pg. 187).

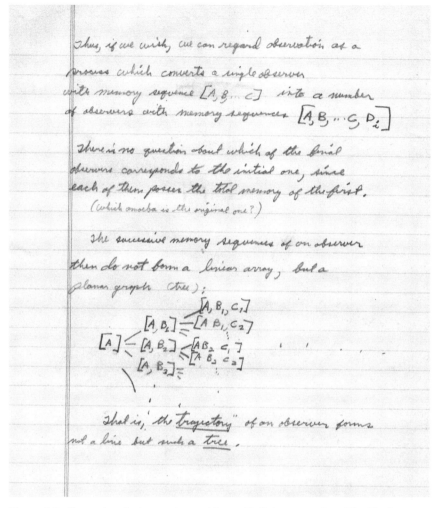

Figure 2.2. Everett's solution to the problem of infinite regression. The Hugh Everett Papers, courtesy of Mark Everett.

Wheeler, Everett, and Charles Misner, Everett's classmate. Everett attended a public lecture at the graduate college, in which Bohr declared that, because of his philosophy of "complementarity," finding a single measurement result (where more than one was possible) was not a problem.[11] Everett found Bohr's view to be preposterous. He talked about his doubts with Misner and Bohr's young assistant, Aage Petersen. More than two decades later, Everett recalled[12] that his theory had crystallized during a drunken

[11] Pais (1991, pg. 435). See also Byrne (2010, pg. 88).
[12] See the transcript of the Everett–Misner tape in this volume in chapter 23 (pg. 299).

conversation in his room with Petersen and Misner (whose thesis advisor was Wheeler). Be that as it may, Everett spent the next year researching and refining and writing down the argument of his dissertation in pencil on dozens of yellow legal pads, under Wheeler's supervision.

To Split or Not To Split

In mid-1955, Everett began dating Nancy Gore (whom he married in 1956). She was an excellent typist, and in the late summer of 1955 she typed up three extracts from Everett's handwritten drafts. He gave these "minipapers" to Wheeler for comment early in the fall semester.

It appears that Wheeler had intended to show the minipapers to Bohr. Wheeler revered Bohr as his mentor, and he wanted input from him because Everett's thesis directly attacked the complementarity model. And although Bohr was not an overt advocate of wave collapse, since he did not view measurement as a problem, he did not dispute the collapse postulate's well-established, pragmatic role in physics. And it was commonly considered that the Copenhagen interpretation embraced wave function collapse insofar as it accounted for measurement interactions at all. Although Wheeler found much to like in Everett's thesis, he was not eager to incur the displeasure of the icon of quantum physics.

After receiving Wheeler's written and verbal comments on the minipapers, Everett excised his most evocative metaphor, the splitting amoebas, although he kept superposed cannonballs and splitting observer and mouse states. He submitted the typed long thesis to Wheeler in January 1956. Titled "Quantum Mechanics by the Method of the Universal Wave Function," it was composed of six chapters and two appendices (one appendix of mathematical proofs, a second appendix speaking to questions of interpretative method). Folders found in the basement of Everett's son show that each chapter and appendix went through two or three and sometimes four revisions. The equations in the revisions changed as the author refined the formalism over time. Paragraphs and pages and subsections were edited and reedited until thickets of handwritten edits required rewriting the whole chapter by hand.

After January 1956, Everett made a few more corrections (most likely at Wheeler's strong suggestion), and in April 1956 a final version—renamed "Wave Mechanics Without Probability"—was mailed to Bohr in Denmark.

A detailed account of what happened next is presented in Part III of the current volume. The upshot of the affair was that Bohr and the colleagues he assigned to examine Everett's work, including Petersen, saw no merit in it. In fact, those who contacted Everett and Wheeler on Bohr's behalf disagreed with the premise that a measurement problem exists. And they found Everett's proposed solution, modeled as a nondenumerable infinity of "equally real," branching subsystems of a completely quantum mechanical universal wave function to be absurd and wrong on its face because it did not agree with Bohr's philosophy of complementarity. Adding insult to heresy, in the eyes of the Copenhagenists, the upstart thesis treated as credible (if not preferred) a range of (nonlinear) "hidden variables" interpretations as proposed by Bohm, Einstein, and Wiener, as well as the stochastic interpretation proposed by the German theoretical physicist Friedrich Bopp. Everett believed that his own (linear) interpretation of quantum mechanics was superior to those nonstandard theories, but he did not discount them, as did Bohr and his colleagues, who strove to protect Bohr's scientific and philosophical legacy as if it was the answer for all time.

Wheeler fought hard to promote the formal argument of Everett's theory in Copenhagen, but after his mentor rejected it, as expressed at length in a letter from Alexander Stern, he insisted that Everett tone down his language and make his argument more palatable to those who objected to its philosophical implications.[13] It is worth noting that despite his aversion to the epistemological barrier that Bohr's interpretation inserted between the microscopic and macroscopic realms, Everett viewed complementarity as subsumed by his theory, not as destroyed by it. His notes contain several comment of this type: "Complementarity contained in general form in present scheme."[14]

The Everett interpretation was briefly discussed at the Conference on the Role of Gravitation in Physics held at the University of North Carolina, Chapel Hill, in January 1957. This meeting was organized by Wheeler, Bryce S. DeWitt, and Cecile DeWitt-Morette. Much of the conference was devoted to the task of quantizing gravity, especially by using the path integral method invented by Richard Feynman, who attended.[15]

Toward the end of the last day, Wheeler suggested that Everett's conceptualization of a universal wave function was useful to the task at hand. The universal wave function idea was attractive to Wheeler because, unlike the standard collapse interpretation, it did not mandate that observations take place from a vantage point outside the quantum system observed; quite the contrary, it presumed that wave functions must include observers. This was particularly useful for creating a theory of quantum gravity for the universe, since, obviously, an observer cannot stand outside the universe.

Cecile DeWitt later reported that Feynman did not like Everett's proposal. According to her synopsis of the conference proceedings, Feynman said, "The concept of a 'universal wave function' has serious difficulties. This is so since the function must contain amplitudes for all possible worlds

---

[13] The exchange among Stern, Wheeler, and Everett is included in this volume in chapter 12 (pg. 214).

[14] See Byrne (2010, pg. 142).

[15] See Byrne (2010, pg. 180). The conference report is available as DeWitt (1957).

depending upon all quantum mechanical possibilities in the past and thus one is forced to believe in the equal reality of an infinity of possible worlds."[16] Feynman may have been one of the first physicists to publicly critique Everett's theory because it led to an extravaganza of possible worlds, but he was not the last to do so.

Shortly after the conference, Wheeler sat down with Everett and they cut, condensed, and rewrote the dissertation. They eliminated three-quarters of the original material, including the entire chapter called "Probability, Information, and Correlation." They excised the direct criticisms of Bohr, as well as references to macroscopically splitting objects. The argument was recast, not as a solution to the measurement problem, but as a method of facilitating Wheeler's project of developing a theory of quantum gravity that might mesh with general relativity. It was retitled "On the Foundations of Quantum Mechanics."

Princeton awarded Everett his Ph.D. in physics in April 1957. By that time, he had been doing top secret operations research for a year at the Pentagon's WSEG. According to the terms of his contract, he needed his doctorate to keep the job; he had no choice but to acquiesce to Wheeler's editorial demands since, if he did not comply, he was not going to receive his degree. That said, Wheeler may very well have done him a professional favor by insisting that controversial language elements be removed, leaving the logical consistency of pure wave mechanics to speak for itself.

Slightly revised, Everett's dissertation was published in the July 1957 edition of *Reviews of Modern Physics* as "'Relative State' Formulation of Quantum Theory." An entire section of the journal (guest-edited by DeWitt) was devoted to the proceedings of the Chapel Hill conference on gravity. (Everett had not attended the conference, but Wheeler insisted that his thesis be published in the proceedings.) At the last minute, while proofreading the article for publication, Everett added a footnote that explained why the "splitting process" was not observable. The article appeared alongside a companion piece written by Wheeler that extolled his student's theoretical breakthrough, comparing it to the paradigm shifts in physics initiated by Einstein, Newton, and Maxwell.

OPERATIONS RESEARCH

After the publication of Everett's article, the world of physics remained largely silent (at least in public) about his startling conjecture. Wheeler kept urging Everett to travel to Copenhagen and meet with Bohr to thrash out their disagreements. But Everett was busy calculating kill ratios for

---

[16] See DeWitt (1957, pg. 149), Byrne (2010, pg. 182), and the discussion in the conceptual introduction (pg. 39).

radioactive fallout from nuclear conflagrations[17] and designing software for the ultra top secret National Security Agency and inventing nuclear war simulation programs for the Joint Chiefs of Staff.

As the chief mathematician and computer engineer at WSEG, Everett was instrumental in creating "Report 50," a top-secret investigation detailing the U.S. military's offensive capabilities in a nuclear war (and its astonishing lack of defensive capabilities). The findings of the (now partially declassified) report convinced President John F. Kennedy and his Secretary of Defense, Robert McNamara, to try to steer the nuclear war plan away from the prevailing scenario of massive first strike and retaliation (charmingly termed "wargasm" by Everett's colleague and friend, Herman Kahn) toward a more flexible mode of destruction based upon deploying a triad of hydrogen bomb-bearing bombers, submarines, and intercontinental ballistic missiles. The report, which Everett personally briefed to McNamara in early 1961, was used as the basic blueprint for weaponizing the concept of mutual assured destruction.

Everett also led the WSEG computational team involved in designing the targeting programs for the new single integrated operational plan (SIOP), which became operational in 1962. The concept of a SIOP was sold to political leaders as a computerized design capable of generating a flexible response based upon political exigencies, but it turned out not to be very flexible. It was only capable of firing off a series of five nested and increasingly violent attack modes, which would have pulverized Russia, China, and Eastern Europe to varying degrees, rained fallout everywhere, and caused nuclear winter.

In 1959 (while visiting Bohr in Copenhagen), Everett invented the generalized Lagrange multiplier method, also known as the Everett algorithm or "magic multipliers." A creation of the information age, the algorithm solves complex, nonlinear optimization problems with sampling techniques. It enables the calculation of ranges of consequences or prices for making real-world decisions to expend specific amounts of a resource to overcome a constraint. Problems of this type include maximizing the efficiency of assigning targets to hydrogen bombs, scheduling just-in-time manufacturing runs, allocating bus routes to most efficiently desegregate school systems, and projecting results of funding specific foreign and domestic policies: all tasks that Everett performed for the military–industrial complex.

The magic multiplier was the cornerstone of his career at the Pentagon, and, after 1964, at his own think tank, Lambda Corporation in Arlington, Virginia, where he continued working for military intelligence and designing

---

[17] A declassified version of this fallout study was made public in 1959. Three years later, Linus Pauling was awarded the Nobel Peace Prize for promoting nuclear disarmament. In his acceptance speech, he applauded Everett and his WSEG collaborator, George E. Pugh, for projecting the globally disastrous effects of fallout. They had determined that even a modestly sized exchange of nuclear warheads would poison all life on Earth.

computerized war games for the Joint Chiefs of Staff, as well as antiballistic missile systems based on Bayesian inference methods (none of which worked since they are relatively easy to overcome). He also invented compression algorithms for the first generation of relational databases. His algorithms categorized and tagged information for retrieval without having to search the entire memory mechanism for each byte. These "attribute value" algorithms were instantly applied to business and governmental tasks by American Management Systems, a private company set up by Everett's former bosses in McNamara's office. The company eventually made billions of dollars, but Everett died almost broke.

THE THEORY MATURES

Although operations research was his passion, theoretical quantum mechanics was not entirely ignored by Everett. In the spring of 1959, he discussed his theory several times with Bohr at his institute in Copenhagen; but the great physicist mumbled incomprehensibly while lighting and relighting his pipe. The meeting was, according to Everett, "a hell of a . . . doomed from the beginning."[18]

In 1962, Everett was invited to explain his theory of pure wave mechanics (the linear evolution of quantum states without the collapse postulate) to a private gathering of physics luminaries at Xavier University in Cincinnati, Ohio. The panel included Eugene Wigner, Boris Podolsky, Nathan Rosen, P.A.M. Dirac, and Wendell H. Furry. Stumped by the measurement problem, some panel members were attracted by the logic of the relative-state formulation, despite the "parallel worlds" of "science fiction" implicit in Everett's solution.[19]

In 1967, DeWitt published a paper in *Physical Review*, "Quantum Theory of Gravity," which modeled the universe as a system of branching worlds. DeWitt claimed (as did Everett, but for different reasons) that the quantum mechanical probability postulate was derivable from the formalism of quantum mechanics. (However, DeWitt later backtracked on that claim, which had been based upon counting the number of branching worlds, a method that does not work for a non-denumerable infinity of worlds.)

In 1970, DeWitt wrote a "deliberately sensational" article[20] in *Physics Today* that propelled Everett's theory into the mainstream of physics with sentences such as: "[E]very quantum transition taking place on every star,

---

[18] See the transcript of Misner's and Everett's discussion in this volume (pg. 307).

[19] See Podolsky's comments to Everett in Chapter 19 (pg. 273).

[20] DeWitt used these words to describe his 1970 article in his referee report on Ben-Dov (1990).

in every galaxy, in every remote corner of the universe is splitting our local world on earth into myriads of copies of itself. Here is schizophrenia with a vengeance." In a referee report for the *American Journal of Physics* (which was not made public until 2009), he wrote that this extravagant metaphor was scientifically "sloppy." But it is often quoted to characterize DeWitt's interpretation of Everett's theory as a realm of many "worlds," as opposed to the more prosaic "relative states."[21]

In the early 1970s, the physicist-historian Max Jammer interviewed DeWitt while researching his soon-to-be classic text, *The Philosophy of Quantum Mechanics* (1974). He had never heard of Everett's theory until DeWitt explained it to him. In the last chapter of his book, Jammer detailed the "many worlds" theory, describing it as "one of the most daring and most ambitious theories ever constructed in the history of science" (Jammer, 1974, pg. 134). Correspondence between Everett and Jammer is included in this volume.[22]

In 1973, DeWitt and his graduate student, Neill Graham, edited a collection of Everett's works that included Everett's previously unpublished long thesis (chapter 8 in this book) and the *Reviews of Modern Physics* articles written by Everett (the short thesis, chapter 9) and Wheeler (the assessment of the theory, chapter 10). The volume included papers on Everett's theory by DeWitt and Graham and a paper by Leon N Cooper and Deborah Van Vechten that proposed a similar theory of a noncollapsing wave function that includes the observer but uses the path integral method (they were unaware of Everett's theory when they wrote it).

Princeton University Press published the DeWitt–Graham volume as *The Many-Worlds Interpretation of Quantum Mechanics* (1973). Its main feature was the inaugural publication of the long thesis of April 1956, which DeWitt said he had not known existed until Everett sent him one of his last remaining copies. But Everett had mentioned the existence of the long thesis at the very end of his May 31, 1957, letter to DeWitt: "With respect to your 'minor' criticisms, most of them are explicitly dealt with in the original work from which the article was condensed. I hope, sometime soon, to revise it and make it available, as it contains a much fuller discussion of the various points, as well as a discussion of the present alternative formulations of quantum mechanics. *It is just impossible to do full justice to the subject in so brief an article as the one you read.*"[23]

For publication by the Princeton Press, Everett made selected revisions and retitled the long thesis, "The Theory of the Universal Wave Function." It was DeWitt who attached the appellation "many worlds" to the title of

---

[21] See Byrne (2010, pg. 308).
[22] See chapter 22.
[23] Everett to DeWitt, 5/31/57, emphasis added (pg. 256).

the book, and, consequently, to the theory.[24] And DeWitt's interpretation of Everett's interpretation of quantum mechanics caught on. It echoed the splitting amoeba-with-a-shared-memory metaphor that had so dismayed Wheeler in Everett's 1955 minipaper "Probability in Wave Mechanics."[25]

Of course, not all proponents of the relative-state theory endorse DeWitt's splitting-worlds interpretation of Everett, which changed over the years, reflecting related developments in quantum theory, such as decoherence considerations. Some theorists shy away from strongly positing the existence of separate, physically noncommunicating, ontologically present worlds while they model a universal state composed of "overlapping" worlds. Others take the splits literally in arguing for either acceptance or rejection of the "Everettian" approach.[26]

For his part, Everett was more circumspect than DeWitt's "sloppy" *Physics Today* description of the ontology of the many worlds, resting his case on his oft-repeated statement that the branching wave functions were all "equally real." Everett viewed the branches as macroscopically separate and pragmatically irreversible (from our point of view), without discounting that reversible interference effects operate at a fine-grained level. Nor is he on record as confirming or discounting that his theory supported the existence of countless copies of splitting observers and servomechanisms in a system of "many worlds." Rather, as we point out in the second section of this introduction, although Everett made the case for pure wave mechanics, he was open to considering the viability of various interpretations of quantum mechanics and various interpretations of his own theory (as long as they were physical and not purely mentalist). And, currently, that range of interpretations goes from a purely relative-state formulation (that needs no preferred basis)[27] to a system of splitting observers correlating to their environments on sets of separating, decohering, historically unique trajectories within the global superposition described by a universal wave function. That said, evidence of macroscopic superpositions would certainly sway the issue, as would confirmation of wave collapse (however that could be demonstrated).

DeWitt eventually softened his view on the necessity of a splitting universe occurring with every atomic interaction when he broadly embraced

---

[24] There is no record that Everett ever publicly objected to the "many worlds" label. He did, however, make the point of distinguishing his choice of language from DeWitt's in his personal correspondence with Lévy-Leblond (pg. 313).

[25] See chapter 6 (pg. 69) for Everett's amoeba story in this minipaper.

[26] See Saunders et al. (2010) for an extensive recent discussion of matters pertaining to the "Everettian" canon.

[27] A "basis" is a set of "vectors" in the mathematical space in which quantum mechanical states are written that provides a way of representing all possible physical states. There is an infinite number of possible bases and no clear physical reason why any one of them should be preferred for representing states.

the decoherence approaches proposed by Dieter Zeh, Wojciech Zurek, James B. Hartle, Murray Gell-Mann, and others beginning around 1970. Cosmologists and decoherence theorists have used Everett's model of a completely quantum mechanical universe as a springboard to formulate a variety of interpretations. We shall leave it up to the reader to determine to what degree, if any, these approaches reflect Everett's intentions, language, and formalism.

In 1976, to Everett's delight, the science fiction magazine *Analogue* published a four-page article on the "Everett–Wheeler" interpretation of quantum mechanics. It was based largely on DeWitt's *Physics Today* article. He sent copies of the article to friends, including Wheeler. A year later, Everett was invited by DeWitt and Wheeler to make a presentation on his theory at the University of Texas in Austin. There he met David Deutsch, a young graduate student under DeWitt who was to embrace a strong version of Everett's theory in his search for a theory of quantum computation.

In the history of physics, Everett's theory stands out as an example of how difficult it can be to translate a logically consistent formalism into explanatory language. That has not stopped a wide range of scientists and philosophers from building upon Everett's core insight that the evolution of quantum systems can be described without resorting to collapse of the wave function or epistemological partitions between the microscopic and macroscopic realms. Everettians and their opponents have written scores of academic papers debating the finer points of the theory as they build up and tear down theoretical structures ranging from many worlds to many minds, and at the center of these debates is the nature of probability.

In the second appendix of his long thesis, Everett points out that since his theory could not make predictions differentiating it from the standard collapse, hidden variables, or stochastic theories, it was largely a matter of taste how one interprets the quantum mechanical equations.[28] For himself, he was convinced of its empirical correctness (as far as explaining the subjective experience of quantum measurement), and he did not believe that any model was capable of fully capturing "reality." He observed, "Once we have granted that any physical theory is essentially only a model for the world of experience, we must renounce all hope of finding anything like '*the* correct theory.' There is nothing which prevents any number of quite distinct models from being in correspondence with experience (i.e., all 'correct'), and furthermore no way of ever verifying that any model is completely correct, simply because the totality of all experience is never accessible to us" (chapter 8, pg. 170).

---

[28] The second appendix starts in chapter 8 (pg. 168). See also Everett's letter to DeWitt for a discussion of such matters of taste in chapter 16 (pg. 254).

Leaving it to the reader to determine for herself the degree to which Everett's theory conforms to physical reality, we present Everett's theory in its original and unabridged format.

Peter Byrne
Petaluma, California
December 2011

CHAPTER 3

Conceptual Introduction

Everett held that the two orthodox options for understanding quantum mechanics, the standard von Neumann–Dirac collapse formulation and Bohr's Copenhagen interpretation, encountered the quantum measurement problem, and were hence ultimately untenable.

The quantum measurement problem arises in the standard collapse theory from a conflict between its two dynamical laws. One law says that the state of a system evolves in a deterministic, linear way when no measurement is made of the system. The other law says that the state of a system evolves in a stochastic, nonlinear way when the system is measured. But the theory does not say what constitutes a measurement, and, consequently, does not specify when each of the two laws applies. Everett thus argued, in the context of what has since been called a Wigner's Friend story, that the theory is inconsistent if one considers multiple observers and seeks to describe the observers themselves quantum mechanically.

Everett took the Copenhagen interpretation to be even less satisfactory. This formulation of the theory maintains consistency by simply stipulating that observers never be treated quantum mechanically. Although Everett agreed that this was consistent, he held that such a solution was not only ad hoc but precluded from the outset the possibility of an explanation of the classical behavior of macroscopic systems from the quantum mechanical behavior of their parts. The thought was that since the Copenhagen formulation of quantum mechanics maintained consistency only by brute-force stipulation, its consistency was not an honest virtue. Since observers are constructed of simpler systems, which the theory tells us behave quantum mechanically, Everett thought it reasonable to require an explanation for the emergence of the apparent classical behavior of macroscopic systems like observers and their measurement records.

Everett believed that he was able to derive an account of determinate measurement records by considering how an observer would become correlated to the system he is observing if one supposed that the observer and his object system evolved quantum mechanically in accord with the standard linear dynamics alone. More specifically, Everett's proposal was to adopt pure wave mechanics, a theory that treated every physical system in precisely the same quantum mechanical way, including observers themselves, then to deduce determinate measurement records that exhibit the standard quantum statistics. This would show that the theory was both consistent and empirically faithful, where empirical faithfulness is a condition closely related to empirical adequacy but perhaps somewhat weaker than the traditional notion. Everett called his interpretation of pure wave mechanics the *relative-state formulation of quantum mechanics*. It has, however, never been entirely clear how to best interpret Everett's interpretation of quantum mechanics.

The two main problems in interpreting Everett's relative-state formulation are (1) *the determinate record problem*, explaining the sense in which observers, treated purely quantum mechanically, might be taken as having determinate records at the end of a measurement, and (2) *the probability problem*, explaining how the standard quantum statistics might be taken to arise in a theory that is fully deterministic and apparently involves no special uncertainty regarding postmeasurement states.

How one takes Everett to have solved these two problems depends on how one interprets Everett's relative-state formulation of quantum mechanics itself. There have been many interpretations of Everett's theory, but easily the most popular have been those in the tradition of Bryce DeWitt's splitting-worlds or parallel-universes interpretation. Although significant progress has been made over the past several decades, it remains unclear how one might best address these problems on a splitting-worlds interpretation of Everett.

There is reason to believe, however, that Everett himself understood such interpretational issues to involve nothing more than one's choice of language for describing the correlations represented by the universal wave function. Moreover, it seems that Everett himself did not care much what language one used as long as one ended up with a description of the model that was consistent and revealed the empirical faithfulness of pure wave mechanics. He took the theory to be empirically faithful if one could find representations of observers' determinate measurement records and the standard quantum statistics in the correlation model of pure wave mechanics, and he believed that he accomplished this in his thesis.

Although one might argue that Everett's project was in fact successful by his lights, one might want somewhat more from a successful physical theory than he got. This desire for a richer explanation of experience than Everett provided drives continued research in pure wave mechanics.

THE QUANTUM MEASUREMENT PROBLEM

Everett clearly distinguished between two main orthodox options for understanding quantum mechanics: the standard von Neumann–Dirac collapse formulation and Bohr's Copenhagen interpretation. He primarily focused on the former textbook version. As he reported to his friend Aage

Petersen,

> the particular difficulties with quantum mechanics that are discussed in my [thesis] have mostly to do with the more common (at least in this country) form of quantum theory, as expressed for example by von Neumann, and not so much with the Bohr (Copenhagen) interpretation" (pg. 239).

That both the long and short theses are largely concerned with problems encountered by the von Neuman–Dirac formulation, however, was not because Everett preferred the Copenhagen interpretation; rather, Everett throught each of the orthodox formulations of quantum mechanics encountered different but similarly fatal problems.

Everett characterized the standard von Neumann–Dirac collapse formulation of quantum mechanics, which he referred to as the *external* or *popular interpretation*, as involving the following principles:[1]

1. Representation of states: The state of a physical system S is represented by a vector ψ_S of unit length in a Hilbert space $\mathcal{H}$.
2. Representation of observables: Every physical observable O is represented by a Hermitian operator $\hat{O}$ on $\mathcal{H}$, and every Hermitian operator on $\mathcal{H}$ corresponds to some complete observable. Equivalently, physical properties and quantities are represented by the *orthogonal* vectors that describe states where the system determinately has or does not have the property or for which the system determinately possesses the various possible values of the quantity.
3. Interpretation of states: A system S has a determinate value for observable O if and only if it is in an eigenstate of O: that is, S has a determinate value for O if and only if $\hat{O}\psi_S = \lambda\psi_S$, where $\hat{O}$ is the Hermitian operator corresponding to O, ψ_S is the vector representing the state of S, and the eigenvalue λ is a real number. In this case, one would with certainty get the result λ if one measured O of S. That is, a system S determinately has a physical property $O = \lambda$ if and only if $\hat{O}$ operating on ψ_S, the vector representing S's state, yields $\lambda\psi_S$. In this case, we say that S is in an eigenstate of O with eigenvalue λ. Similarly, S determinately does not have property $O = \lambda$ if and only if $\hat{O}$ operating on ψ_S yields $\gamma\psi_S$ and $\gamma \neq \lambda$.
4. Dynamical laws:

   I. Linear dynamics: *If no measurement is made* on a physical system, it will evolve in a deterministic linear way. More specifically, if the state of S is given by $\psi(t_0)_S$ at time t_0, then its state at a time t will be given

---

[1] For examples of Everett's presentation of the theory see (pgs. 58, 73 and 175). *Rules 1, 2, 4I*, and *4II* are explicit in Everett's presentation of the theory in his thesis. *Rule 3* is implicit in both how he assigns properties to absolute states and in when he claims there are no absolute determinate properties or states. *Rule 5* is also implicit in his use of the standard collapse theory.

by $\hat{U}(t_0, t_1)\psi(t_0)_S$, where $\hat{U}$ is a unitary operator on $\mathcal{H}$ that depends only on the energy properties of S.

II. Nonlinear collapse dynamics: *If a measurement is made*, the system S will randomly, instantaneously, and nonlinearly jump to a state where it either determinately has or determinately does not have the property being measured, with the probability of each possible postmeasurement state determined by the system's initial state. More specifically, if a measurement is made of the system S, it will randomly, instantaneously, and nonlinearly jump to an eigenstate of the observable being measured such that if the initial state is given by ψ_S and if ϕ_S is an eigenstate of O, then the probability of S jumping to ϕ_S when O is measured is equal to $|\psi_S\phi_S|^2$ (the square of the magnitude of the projection of the premeasurement state ψ_S onto the eigenstate ϕ_S). Equivalently, when a measurement is made, S instantaneously and randomly jumps from the initial superposition on the left to the eigenstate of the observable being measured on the right

$$\psi = \sum_k c_k \phi_k \longrightarrow \phi_j \qquad (3.1)$$

with probability $|c_j|^2$.

5. Composition of systems and properties: If system S_1 is represented by an element ϕ_{S_1} of vector space $\mathcal{H}_1$ and S_2 by an element ψ_{S_2} of $\mathcal{H}_2$, then the composite system $S_1 + S_2$ is represented by an element $\phi_{S_1} \otimes \psi_{S_2}$ of $\mathcal{H}_1 \otimes \mathcal{H}_2$. Similarly, if property P_1 of system S is represented by orthogonal elements of $\mathcal{H}_1$ and an independent, i.e., a quantum mechanically compatible, property P_2 of S by orthogonal elements of $\mathcal{H}_2$, then both properties can be simultaneously represented by orthogonal elements of $\mathcal{H}_1 \otimes \mathcal{H}_2$. This rule tells one how to combine the vector space representations of different physical systems or of different properties or quantities into a representation that simultaneously describes each system, property, or quantity.

The notorious quantum measurement problem arises in the standard theory from a conflict between the deterministic linear dynamics expressed by *rule 4I* and the stochastic nonlinear dynamics expressed by *rule 4II*. The upshot, according to Everett, is that "[i]n its unrestricted form this view can lead to paradoxes ... and is therefore untenable" (pg. 152). He described the inconsistency of the theory in the context of what is now thought of as a Wigner's Friend story.[2]

---

[2] Everett presented his versions of the story some years before Wigner himself told this story it in print. Whereas Everett used the Wigner's Friend story to show what is wrong with standard collapse theory, Wigner used the story to argue that consciousness causes collapses in the standard theory (Wigner, 1961). For Everett's discussions see (pgs. 176 and 73). See also pg. 15 for his picture of the experiment.

In the first of his two versions of the Wigner's Friend story, Everett allows for observer B to measure a quantum mechanical observable of his friend observer A. Everett then explains why one cannot consistently apply *rule 4I* and *rule 4II* as they stand.

> The question of the consistency of the scheme arises if one contemplates regarding the observer and his object-system as a single (composite) physical system. Indeed, the situation becomes quite paradoxical if we allow for the existence of more than one observer. Let us consider the case of one observer A, who is performing measurements upon a system S, the totality $(A + S)$ in turn forming the object-system for another observer, B.
>
> If we are to deny the possibility of B's use of a quantum mechanical description (wave function obeying wave equation [*rule 4I*]) for $A + S$, then we must be supplied with some alternative description for systems which contain observers (or measuring apparatus). Furthermore, we would have to have a criterion for telling precisely what type of systems would have the preferred positions of "measuring apparatus" or "observer" and be subject to the alternate description. Such a criterion is probably not capable of rigorous formulation.
>
> On the other hand, if we do allow B to give a quantum description to $A+S$, by assigning a state function ψ^{A+S}, then, so long as B does not interact with $A+S$, its state changes causally according to [*rule 4I*], *even though A may be performing measurements upon S*. From B's point of view, nothing resembling [*rule 4II*] can occur (there are no discontinuities), and the question of the validity of A's use of [*rule 4II*] is raised. That is, *apparently* either A is incorrect in assuming [*rule 4II*], with its probabilistic implications, to apply to his measurements, or else B's state function, with its purely causal character, is an inadequate description of what is happening to $A + S$ (pgs. 73–74).

Insofar as Everett has shown that the state of the composite system $A + S$ must depend on who is describing it, one gets a straightforward logical contradiction if one supposes that quantum state assignments are objective and nonperspectival, something that Everett took to be part of the standard understanding of quantum mechanical states on the orthodox view.

To further illustrate the problem, Everett asked his reader to consider a second version of the Wigner's Friend story, a story that he characterized as an "amusing, but *extremely hypothetical* drama" (pg. 74). Assume again that A is about to perform a measurement upon a system S and that after performing his measurement he will record the result in a notebook. Suppose that A knows the state function of S, and that it is not an eigenstate of the measurement he is about to perform. A then believes that the outcome of his measurement is undetermined and that the process is correctly described by the stochastic *rule 4II*. Observer B, in the meantime, is in possession of the state function of the entire room, including the composite system $A + S$. B computes the state function of the room for one week in the future according to the deterministic *rule 4I*. After a week, *B* still knows the state function of the room, which he believes to be a complete description of the room and its contents. *B* then opens the door to the room and looks at *A*'s notebook. As Everett tells it,

> Having observed the notebook entry, he turns to *A* and informs him in a patronizing manner that since his (*B*'s) wave function just prior to his entry into the room, which he knows to have been a complete description of the room and its contents, had non-zero amplitude over other than the present result of the measurement, the result must have been decided only when *B* entered the room, so that *A*, his notebook entry, and his memory about what occurred one week ago had no independent objective existence until the intervention by *B* (pgs. 74–75).

In each version of the story, if the standard quantum state is taken as complete, the central point for Everett is that *A* and *B* cannot both be right in their state attributions. If *A* in fact got a single determinate result to his measurement of *S* as stipulated by the collapse theory and if the quantum state is complete, then this result must be represented in the quantum state that *B* assigns to the composite system *A* + *S*, but the state that *B* calculates from *rule 4I* does not select any particular result as the one that was in fact realized and hence cannot be complete. In short, assuming state completeness and that *A* gets a single determinate measurement result, both of which are supposed on the standard collapse formulation, yields a straightforward contradiction.

What makes the story extremely hypothetical is that, due to decoherence effects between observer *A* and his environment and the associated technological difficulty in ever actually performing an appropriate measurement, it would be virtually impossible for any observer ever to determine the state function of a macroscopic composite system like *A* + *S* in practice. That Everett took such a hypothetical story to pose a problem for the standard theory, however, makes it clear that he took the measurement problem to be a conceptual, not a practical, problem. That the experiment would be virtually impossible to perform did not excuse quantum mechanics from providing a coherent description of what would happen if it were performed. Everett wanted a formulation of quantum mechanics that both made the right empirical predictions for the experiments we can perform and provided a consistent description of what would happen in those experiments we will perhaps never perform. He wanted a consistent theory that could be taken as providing a complete model of all physical interactions whatsoever.

The measurement problem for the standard collapse theory might be summarized then as follows. Since *rule 4I* and *rule 4II* typically predict different physical states, the standard von Neumann–Dirac formulation must provide disjoint conditions for when to use each to be logically consistent. The theory says to use *rule 4I* all of the time *except when a measurement is made*, but since it does not tell us what constitutes a *measurement*, the theory is at best incomplete. Furthermore, if one supposes that observers and their measuring devices are constructed from simpler systems that each obey the linear deterministic dynamics *rule 4I*, as Everett supposed and as experimental evidence suggests, then the standard theory predicts that composite systems consisting of observers and their measuring devices must also evolve in the linear deterministic way described by *rule 4I*. But if this is so, then nothing like the random, discontinuous evolution described by *rule 4II* can ever occur, and the theory is logically inconsistent insofar as it requires such an evolution on measurement. Regarding the standard von Neumann–Dirac formulation of quantum mechanics, Everett concluded that "[i]t is now clear that the interpretation of quantum mechanics with which we began is untenable if we are to consider a universe containing more than one observer" and "[w]e must therefore seek a suitable modification of this scheme, or an entirely different system of interpretation" (pg. 75).

Although he did not believe that it encountered this sort of contradiction, what proponents took as the more philosophically sophisticated Copenhagen formulation of quantum mechanics was, for Everett, even less satisfactory. He characterized it as "the interpretation developed by Bohr" where

> [t]he ψ function is not regarded as an objective description of a physical system (i.e., it is in no sense a conceptual model), but is regarded as merely a mathematical artifice which enables one to make statistical predictions, albeit the best predictions which it is possible to make. This interpretation in fact denies the very possibility of a single conceptual model applicable to the quantum realm, and asserts that the totality of phenomena can only be understood by the use of different, mutually exclusive (i.e., "complementary") models in different situations. All statements about microscopic phenomena are regarded as meaningless unless accompanied by a complete description (classical) of an experimental arrangement. (pgs. 152–53)

According to the Copenhagen interpretation (1) the understandable physical content of quantum mechanics is only what can be described using the ordinary language of classical mechanics and (2) a physical system only has a quantum mechanical state relative to a specification of a classical experimental arrangement designed to measure a particular property of the system and a specification of how the system was prepared and how the result of the experiment will be recorded. Different classical specifications of the experimental arrangement typically provide, in Bohr's language, *complementary* quantum mechanical descriptions of the system being observed. That such complementary descriptions in terms of quantum mechanical states may appear to be mutually incompatible is not a problem because they are descriptions in terms of quantum mechanical states, that is, in terms of *superpositions* of classical states, and hence not themselves proper objects of understanding, a role that the Copenhagen view held could only be played by an ordinary classical state. One uses the quantum mechanical description associated with a particular experimental arrangement to make statistical predictions for that arrangement using the Born rule, *rule 4II*. When one has another prediction to make, one starts afresh in assigning complementary states to the associated systems relative to the new experimental setup without requiring any dynamical explanation that ties the quantum mechanical states one used in the past to the quantum mechanical states one will use for one's next prediction. Indeed, to require such an explanation would be to pretend that it was possible to describe the quantum mechanical world directly without the aid of classical concepts (Bohr, 1949).

Such a formulation of quantum mechanics did not fit at all well with Everett's desire for a consistent theory that could provide a complete model of all physical processes whatsoever. As he wrote Petersen:

> The Bohr interpretation is to me even more unsatisfactory, and on quite different grounds. Primarily my main objections are the complete reliance on classical physics from the outset (which precludes *even in principle* any deduction at all of classical physics from quantum mechanics, as well as any adequate study of measuring processes), and the strange duality of adhering to a "reality" concept for macroscopic physics and denying the same for the microcosm. (pg. 239)

Everett echoed this sentiment in a letter to Bryce DeWitt, written the same day: "the Copenhagen interpretation is hopelessly incomplete because of its a priori reliance on classical physics (excluding *in principle* any deduction of classical physics from quantum theory, or any adequate investigation of the measuring process)." Less diplomatically, Everett reported to DeWitt that the Copenhagen interpretation is a "philosophical monstrosity" (pg. 255). In the earlier long version of his thesis, Everett concluded that "[w]hile undoubtedly safe from contradiction, due to its extreme conservatism, [the Copenhagen formulation] is perhaps overcautious" (pg. 153).

So, for Everett, the standard von Neumann–Dirac collapse formulation of quantum mechanics led to a straightforward contradiction if one supposed that measuring devices are physical systems like any other and the Copenhagen interpretation avoided inconsistency by simply stipulating that measurement interactions cannot be understood quantum mechanically.

Hence, neither orthodox option allowed for a coherent quantum mechanical picture of the process of measurement or of the classical behavior of physical systems. This was the quantum measurement problem as Everett understood it.

EVERETT'S PROPOSED RESOLUTION

Everett's proposal for solving the measurement problem was to treat measurement interactions in precisely the same way as all other physical interactions. In particular, he proposed dropping the stochastic collapse dynamics, *rule 4II*, from the standard theory and taking the resulting deterministic pure wave mechanics as providing a complete and accurate model of all physical systems. He then intended to "deduce the probabilistic assertions of [*rule 4II*] as *subjective* appearances . . . thus placing the theory in correspondence with experience." Consequently, "[w]e are then led to the novel situation in which the formal theory is objectively continuous and causal, while subjectively discontinuous and probabilistic"(pg. 77). To make inferences about the experiences of observers, Everett needed a model of observers as physical systems that might be treated within pure wave mechanics.

Everett's model for observers was simple. An ideal observer was a physical system with memory registers whose states might become perfectly correlated to the states of object systems over the course of measurement interactions. Given the linearity of the dynamics of pure wave mechanics, if such an observer M begins in a ready-to-make-a-measurement state $\psi[\text{"ready"}]_M$ and measures the observable O of system S, with eigenstates ϕ_S^i, then M's memory becomes correlated to S's state as follows:

$$\psi[\text{"ready"}]_M \sum_i a_i \phi_S^i \longrightarrow \sum_i a_i \psi[\text{"}i\text{"}]_M \phi_S^i \qquad (3.2)$$

And repeated measurements simply lead to more complicated entangled superpositions, each term of which, when written in the determinate-record basis as above, describes M as having recorded a different sequence of measurement results.

The next step was to seek an interpretation of such final absolute states.[3] To this end, Everett introduced the notion of *relative states*, which he reported "will play a central role in our interpretation of pure wave

---

[3] The absolute state of an unentangled system is the state given by the universal wave function. Insofar as the states of physical subsystems are typically entangled, they typically fail to have absolute states and only have relative states.

mechanics" (pg. 99). This new type of state provided Everett with an expression of his principle of the fundamental relativity of states:

> There does not, in general, exist anything like a single state for one subsystem of a composite system. Subsystems do not possess states that are independent of the states of the remainder of the system, so that the subsystem states are generally correlated with one another. One can arbitrarily choose a state for one subsystem, and be led to the relative state for the remainder. Thus we are faced with a fundamental *relativity of states*, which is implied by the formalism of composite systems. It is meaningless to ask the absolute state of a subsystem—one can only ask the state relative to a given state of the remainder of the subsystem. (pg. 180)

Everett held, in agreement with the standard interpretation of quantum states, that subsystems of entangled composite systems, systems like M and S above, do not possess absolute physical properties or even absolute states to call their own. With his new type of state, however, he added to the standard interpretation that while M and S do not possess absolute states, such subsystems possess *relative states* that are determined by the correlation structure characterized by, in the most general case, the absolute state of the universe.[4] If one arbitrarily chooses, as Everett put it, to assign the state ϕ_S^j to S, then the correlation structure assigns the relative state $\psi["j"]_M$, where result "j" is recorded by M. If, however, one chooses to assign a different state ϕ_S^k, $k \neq j$, to S, then the correlation structure assigns the relative state $\psi["k"]_M$ where result "k" is recorded by M. Or if one chooses another way of individuating the subsystems, one involving three subsystems, for example, where one subsystem contains parts of both S and M, then chooses to assign a state to one of these subsystems, one would typically end up with relative states that look quite different.

Insofar as Everett insisted that one can arbitrarily choose a state for one subsystem and be led to the relative state for the remainder of a composite system, he held that the specification of relative states does not require stipulating any ultimately preferred basis.[5] Just as no particular relative state for an observer has the special status of being the relative state that is in fact realized, there is no preferred decomposition of the composite state into a preferred set of relative states. The upshot is that correlation structure determined by the universal state characterizes a rich collection of relative states; some of these states are recognizable as states describing systems with quasi-classical properties, some may even be taken as describing observers with determinate measurement records, but most have no classical

---

[4] More specifically, the correlation structure is characterized by the universal wave function and all the ways that one might decompose it, given one's choice of basis and choice of how to individuate physical subsystems. In his unpublished notes, Everett referred to each such decomposition as a "cross section" of the universal wave function. See, for example, pg. 66.

[5] See Everett's later notes (fn. gf on pg. 227 and fn. la on pg. 287) for examples of his view that branching is always relative to one's choice of basis.

analogue. In short, relative states gave Everett a convenient way of talking about the correlation structure determined by the universal state in the context of pure wave mechanics. The correlation structure might be thought of as characterized by the complex-valued correlations that obtain between the properties of physical systems on all possible individuations of physical systems given the universal state.

It is in such relative states that Everett found a representation of determinate experience. Given the linearity of the dynamics of pure wave mechanics,

> [i]t is then an inescapable consequence that after the interaction has taken place there will not, generally, exist a single observer state. There will, however, be a superposition of the composite system states, each element of which contains a definite observer state and a definite relative object-system state. Furthermore, as we shall see, *each* of these relative object-system states will be, approximately, the eigenstates of the observation corresponding to the value obtained by the observer which is described by the same element of the superposition. Thus, each element of the resulting superposition describes an observer who perceived a definite and generally different result, and to whom it appears that the object-system state has been transformed into the corresponding eigenstate. In this sense the usual assertions of [the collapse of the state on measurement] appear to hold on a subjective level to each observer described by an element of the superposition. (pg. 78).[6]

A particular determinate experience then is explained by there being a relative observer record that describes the observer as having that particular relative experience (or in Everett's words the "relative phenomena" (pg. 158). When this relative record is the result of a reliable measurement, one where a strong correlation was produced between the state of the object system and the physical record of the result, it is associated with a corresponding relative state for the observed system. Furthermore, concerning the standard quantum statistics, Everett took the square of the coefficients associated with each of the relative states to provide a measure of the typicality of branches. He then showed that on the linear dynamics relative records in typical branches exhibit the standard quantum statistics.[7] Everett took this notion to explain observed quantum statistics, and he consequently took pure wave mechanics to make the same predictions as the standard collapse formulation. Taken together, Everett held that pure wave

---

[6] See also pg. 121.

[7] This theorem requires Everett's norm-squared notion of typicality. It does not hold if one takes a typical property to be one that occurs in most branches *by count* when the state is written in the determinate record basis. See pgs. 122, 187, 273, and 295 and the subsequent discussions for Everett's presentation of this result and Barrett (1999, limiting properties of the bare theory) for a detailed discussion of this issue.

mechanics made the standard statistical predictions for typical determinate relative observer records.

To summarize, Everett believed that pure wave mechanics provided a complete and accurate physical theory, and he took the particular relative measurement records, as represented in the appropriate elements or branches of the absolute state, to explain our particular experience. Furthermore, given the sense of typicality that Everett proposed as the most natural in the context of pure wave mechanics, he showed that sequences of measurement records in typical elements, branches, or relative states exhibit the standard quantum statistics. Many readers of Everett, however, have wanted more than this.

INTERPRETATIONS OF EVERETT

Although Everett believed he had shown that pure wave mechanics was a fully adequate physical theory, there has been a long tradition of being dissatisfied with where he left his project. The two main problems in interpreting Everett's relative-state formulation are (1) *the determinate record problem*, explaining why and the precise sense in which observers might be taken as having determinate measurement records, and (2) *the probability problem*, explaining exactly how quantum statistics are to be understood in a theory that is fully deterministic and apparently involves no special uncertainty. Of course, to take these as problems is to deny that what Everett himself said fully addresses these issues.[8] Many supporters of Everett's have consequently sought to provide additional explanations where they felt that Everett's own were lacking. This has typically involved adding new postulates, metaphysical assumptions, or other interpretational apparatus to pure wave mechanics.

Each of the following has been proposed as a strategy for reconstructing Everett's theory and filling in the details that he omitted: the bare theory[9],

---

[8] One might, for example, want to know why finding our experience on a particular branch explains our experience, especially when our experience on other branches is typically incompatible with what we in fact observe. Furthermore, one might wonder why one should expect the standard quantum statistics from the fact that a typical branch, in Everett's sense of typical, would exhibit the standard quantum statistics, especially since most branches in the determinate record basis, in the usual counting sense of most, typically do not exhibit the standard quantum statistics. Resolving the determinate record problem in the context of Everett's interpretation may also require one to address what has been called in the literature the preferred basis problem, depending on one's approach to interpreting Everett.

[9] In this interpretation, one supposes that Everett intended to drop the collapse dynamics but to keep the standard eigenvalue–eigenstate link and take the resulting pure wave mechanics as complete with no additional interpretive apparatus. Versions of this option have been presented and discussed by Geroch (1984), Albert and Loewer (1988), Albert (1992), and Barrett (1999). See Bub et al. (1998) for criticism of this formulation.

splitting worlds[10], decohering histories[11], relative facts[12], single mind and many minds[13], many threads[14], and emergent worlds.[15] While this is not an exhaustive list, nor are the features of these options mutually exclusive, each of these proposals provides a different context for addressing the two main interpretational problems facing pure wave mechanics. While some are relatively far afield from how Everett himself characterized his project, each

[10] This is DeWitt's original many-worlds interpretation of Everett (DeWitt, 1970). In this interpretation one supposes that Everett meant to understand branches as parallel, splitting worlds. The thought is that one writes the universal state in terms of a physically preferred basis that makes measurement records determinate in each term of the superposition. One then interprets each element of the superposition as a physical world in the state described by that term. This view has been discussed extensively in the literature and may be thought of as having been developed and extended by the decohering histories and emergent-worlds interpretations. Criticisms are discussed in Bell (1987, pg. 284), Dowker and Kent (1996), and Barrett (1999).

[11] Gell-Mann and Hartle (1990) understood Everett's theory as one that describes many, mutually decohering histories. This approach has also been used to address some of the problems faced by DeWitt's original splitting-worlds formulation. This view is sometimes called many histories or consistent histories. It shares with emergent worlds a focus on decoherence considerations that might be taken to hold for all practical purposes. See Dowker and Kent (1996) for a critical discussion of this view.

[12] This interpretation is arguably the approach that fits best with how Everett describes his principle of the fundamental relativity of states. It involves simply denying that there are typically any absolute matters of fact about the properties of physical systems or the records, experiences, and beliefs of observers. The idea is to treat branches of the superposition as indexical akin to time. See Saunders (1995, 1996, 1998) for an example of how this might work.

[13] Everett held that on his formulation of quantum mechanics "the formal theory is objectively continuous and causal, while subjectively discontinuous and probabilistic" (pg. 134). Albert and Loewer (1988) have sought to capture this feature in their many-minds theory by distinguishing between the time evolution of an observer's physical state, which is continuous and causal, and the evolution of an observer's mental state, which is discontinuous and probabilistic. This amounts to a sort of hidden variable theory, where the observer's mental state is always determinate. One must then provide a dynamic for how the mental state evolves. See Barrett (1999, 2008) for a discussion of this view.

[14] Instead of adding a nonphysical hidden variable that represents mental states directly as suggested by Albert and Loewer (1988), one might add a physical variable that one supposes determines mental states. If one chooses position, one ends up with a hidden variable theory like Bohmian mechanics. One might think of this as a many-worlds or many-histories theory where each initial specification of positions for each particle corresponds to a world. The features of that world at each time are determined by the hidden variable dynamics. See Barrett (1999) for a discussion of this strategy.

[15] This strategy is closely related to consistent histories. One postulates a physically real universal wave function. One uses decoherence considerations to select a quasi-classical preferred basis at a specified level of description in which to represent the absolute, universal state. This prevents one from having to specify an ad hoc preferred basis. Worlds are then represented by emergent structure in the universal state. This might be thought of as a development of DeWitt's splitting-worlds interpretation. See the discussion of the Saunders–Wallace interpretation later in this chapter and Albert (2010), Kent (2010), Maudlin (2010) and Wallace (2010a, 2011).

proposal has relative virtues. In most cases, they involve adding something interpretational to the basic formalism of pure wave mechanics. When they do, the pragmatic question is, precisely what is being added and whether the explanatory virtues one gets from the addition are worth the conceptual costs involved.

The many-worlds interpretations warrant further discussion because of their historical importance. They also illustrate the sort of metaphysical assumptions that interpreters have been willing to add to pure wave mechanics to achieve a richer interpretation than Everett himself proposed in the long and short theses.[16]

Splitting Worlds

Although Bryce DeWitt was initially skeptical of Everett's project, he eventually became one of Everett's strongest supporters.[17] DeWitt's (1970) splitting-worlds interpretation is easily the most popular understanding of Everett's theory and is largely responsible for Everett's fame outside of academics. Indeed, for many, DeWitt's interpretation of Everett has almost entirely eclipsed what Everett himself said about this theory, and equivocation between the two is commonplace. For his part, Everett did not use the language of splitting-worlds, many-worlds, or parallel universes in his presentation of his theory in either the long or short thesis. Everett was, however, willing to talk in terms of many worlds when others suggested using such language.[18]

---

[16] Everett continued to endorse what he said in the long and short theses throughout his life. See for example pg. 315. In contrast, both Wheeler's and DeWitt's views on the significance and the proper interpretation of Everett's formulation of quantum mechanics changed significantly over time.

[17] See DeWitt's letter to Everett's adviser John Wheeler (pg. 242) and Everett's reply to DeWitt (pg. 252). DeWitt's splitting-worlds interpretation of Everett itself evolved over time. And although DeWitt later apologized for his initial presentation of Everett on the grounds that it was too sensational, he also maintained that he had not misinterpreted Everett. Furthermore, in DeWitt (1988), an unpublished referee report that he wrote for Yoav Ben-Dov's paper "Everett's Theory and the 'Many Worlds' Interpretation" for the *American Journal of Physics*, DeWitt said that Everett had never indicated to him that he had misrepresented his views. In later correspondence with Lévy-Leblond, however, Everett both referred to DeWitt as his Boswell and sought to distinguish his position from DeWitt's. See Everett's reply to Lévy-Leblond (pg. 313) and the accompanying draft version. Everett also believed that DeWitt had failed to properly understand his derivation of quantum probabilities in pure wave mechanics (pg. 280).

[18] The most salient example is the Xavier conference (pgs. 273–74) exchange described below, but see also the letter to Raub (pg. 316). Everett also used the analogy of splitting amoebas to describe the branching of the state on measurement in an early paper for Wheeler (pg. 69). Although this does not involve worlds, it might suggest a splitting process. On the other hand, Everett's reply to Lévy-Leblond's letter suggests that he did not approve of DeWitt's choice of language (pg. 313). It is also worth noting that whereras Everett made

Although DeWitt was the first to present Everett's theory in print as involving a metaphysical commitment to many worlds, two earlier mentions of worlds in connection with Everett are known. The earliest occurred at a conference on quantum gravity organized by John Wheeler in January 1957, where Richard Feynman is reported to have said

> The concept of a "universal wave function" has serious difficulties. This is so since the function must contain amplitudes for all possible worlds depending upon all quantum mechanical possibilities in the past and thus one is forced to believe in the equal reality of an infinity of possible worlds. DeWitt (1957); quoted in Byrne (2010, pg. 182)

The other pre-DeWitt mention of worlds was at the Xavier conference on the conceptual foundations of quantum mechanics in 1962 and is noteworthy since Everett himself was involved in the discussion.

When Everett's theory was first mentioned at the Xavier conference, Boris Podolsky commented, "Oh yes, I remember now what it is about—it's a picture about parallel times, parallel universes, and each time one gets a given result he chooses which one of the universes he belongs to, but the other universes continue to exist" (pg. 271). After a preliminary discussion of the theory, the participants decided that they should hear from Everett himself, so he was invited to fly to Cincinnati to join the conference the next day. When he had the chance to present his theory, Everett explained it in a way that agreed with his description in the long and short theses:

> The picture I have is something like this: Imagine an observer making a sequence of observations on a number of, let's say, originally identical object systems. At the end of this sequence there is a large superposition of states, each element of which contains the observer as having recorded a particular definite sequence of results of observation. I identify a single element as what we think of as an experience, but still hold that it is tenable to assert that all of the elements simultaneously coexist. In any single element of the final superposition after all these measurements, you have a state which describes the observer as having observed a quite definite and apparently random sequence of events. Of course, it's a different sequence of events in each element of the superposition. In fact, if one takes a very large series of experiments, in a certain sense one can assert that for almost all of the elements of the final superposition the frequencies of the results of measurements will be in accord with what one predicts from the ordinary picture of quantum mechanics. That is very briefly it. (pg. 273)[19]

---

substantial changes to the long thesis for publication in the DeWitt and Graham anthology, he did not change his words to better accord with DeWitt's talk of splitting universes or worlds. See also the discussion of splitting in the biographical introduction to this volume (pg. 17).

[19] This is a bit misleading. It is typically not true that almost all elements *by count* will exhibit the standard quantum statistics; rather, almost all elements will exhibit the standard quantum statistics in the norm-squared coefficient measure. See Everett's discussion of typicality in this

Podolsky then suggested that Everett change his descriptive language:

> Perhaps it might be a little clearer to most people if you put it in a different way. Somehow or other we have here the parallel times or parallel worlds that science fiction likes to talk about so much. (pg. 273)

To which Everett replied:

> Yes, it's a consequence of the superposition principle that each separate element of the superposition will obey the same laws independent of the presence or absence of one another. Hence, why insist on having a certain selection of one of the elements as being real and all of the others somehow mysteriously vanishing. (pg. 274)

Later in the exchange Podolsky said "It looks like we would have a non-denumerable infinity of worlds" and Everett replied "Yes" (pg. 274). When Abner Shimony asked whether Everett associated awareness with each term in the superposition, Everett replied "Each individual branch looks like a perfectly respectable world where definite things have happened" (pg. 275). By conference policy Everett had the chance to revise the transcription of his own comments and, while he made several minor changes, he kept his use of the term *world* in this last comment.

While Everett was willing to talk in terms of worlds informally, he never used this language in the long or short thesis.[20] Rather, both of his published presentations of pure wave mechanics use the language of elements, branches, and relative states.[21]

DeWitt, however, seems to have believed that Everett had always meant for each element or branch in a preferred decomposition of the absolute state to be understood as a complete and metaphysically real copy of the physical world. Indeed, DeWitt (1970) attributed this view to Everett, Everett's thesis advisor John Wheeler, and DeWitt's graduate student Neill Graham. In his popular presentation of the theory, DeWitt consequently called it the EWG interpretation, and he presented the metaphysical commitment to splitting worlds as its central feature.

DeWitt introduced the EWG commitment to many physical worlds in the context of the Schrödinger's cat thought experiment.

---

second sense in the long thesis (pg. 123) and Barrett (1999, limiting properties of the bare theory) for further discussion.

[20] Everett (1973) is a slightly revised version of the original long version of Everett's Ph.D. thesis circulated in 1956 (reprinted as Chapter 8 of this volume). Everett (1957a) is the much shorter official version of his thesis as revised and redacted by Everett and Wheeler to soften the direct attack of the long thesis on the Copenhagen interpretation (reprinted as Chapter 8 of this volume). This is the version that Everett defended for his Ph.D.

[21] See Barrett (2011a) and Barrett (2011b) for a discussion of Everett's ambivalence concerning talk of worlds. See Lévy-Leblond's letter and Everett's reply (pg. 313) and his comments on DeWitt's understanding of the derivation of probability (pgs. 280–81) for examples of how Everett distinguished between his and DeWitt's positions.

The animal [is] trapped in a room together with a Geiger counter and a hammer, which, upon discharge of the counter, smashes a flask of prussic acid. The counter contains a trace of radioactive material—just enough that in one hour there is a 50% chance one of the nuclei will decay and therefore an equal chance the cat will be poisoned. At the end of the hour the total wave function for the system will have a form in which the living cat and the dead cat are mixed in equal portions. Schrödinger felt that the wave mechanics that led to this paradox presented an unacceptable description of reality. However, Everett, Wheeler and Graham's interpretation of quantum mechanics pictures the cats as inhabiting two simultaneous, noninteracting, but equally real worlds. (DeWitt, 1970, pg. 31)

DeWitt took this to follow from "the mathematical formalism of quantum mechanics as it stands without adding anything to it." More specifically, DeWitt claimed that EWG had proven a metatheorem that the mathematical formalism of pure wave mechanics interprets itself in terms of many physically real worlds.

Without drawing on any external metaphysics or mathematics other than the standard rules of logic, EWG are able, from these postulates, to prove the following metatheorem: The mathematical formalism of the quantum theory is capable of yielding its own interpretation. (DeWitt, 1970, pg. 3)

DeWitt gave Everett credit for the metatheorem, Wheeler credit for encouraging Everett, and Neill Graham credit for clarifying the metatheorem. He reported that

[t]he obstacle to taking such a lofty view of things, of course, is that it forces us to believe in the reality of all the simultaneous worlds represented in the superposition [...] in each of which the measurement has yielded a different outcome. Nevertheless, this is precisely what EWG would have us believe. According to them the real universe is faithfully represented by a state vector similar to that [above] but of vastly greater complexity. This universe is constantly splitting into a stupendous number of branches, all resulting from the measurement like interactions between its myriads of components. Moreover, every quantum transition taking place on every star, in every galaxy, in every remote corner of the universe is splitting our local world on earth into myriads of copies of itself. (DeWitt, 1970, pg. 33)

Concerning this vast and nonlocal splitting of physical worlds, DeWitt famously reflected:

I can recall vividly the shock I experienced on first encountering this multiworld concept. The idea of 10^{100+} slightly imperfect copies of oneself all constantly splitting into further copies, which ultimately become unrecognizable, is not easy to reconcile with common sense. Here is schizophrenia with a vengeance. ... Here we must surely protest. We do not split in two, let alone into 10^{100+}! To this EWG reply: To the extent that we can be regarded simply as automata and hence on a par with ordinary measuring apparatuses, the laws of quantum mechanics do not allow us to feel the splits. (DeWitt, 1970, pg. 33)[22]

And, on behalf of Everett, Wheeler, and Graham, DeWitt concluded:

Finally, the EWG interpretation of quantum mechanics has an important contribution to make to the philosophy of science. By showing that formalism alone is sufficient to generate interpretation, it has breathed new life into the old idea of a direct correspondence between formalism and reality. The reality implied here is admittedly bizarre. To anyone who is awestruck by the vastness of the presently known universe, the view from where Everett, Wheeler and Graham sit is truly impressive. (DeWitt, 1970, pg. 35)

One of the virtues of DeWitt's interpretation of Everett is that it provides a relatively straightforward explanation for why an observer gets a determinate outcome to a measurement in an entangled state like that described in expression (3.2) above: there simply is a different *physical copy* of the observer for each element in the absolute entangled superposition, when written in the observer's determinate-record basis, and each physical copy gets a perfectly ordinary result *in her copy world*. For DeWitt, quantum statistics were then to be explained by the fact that the measurement records in most worlds would exhibit the standard quantum statistics, as his student Graham had argued.[23]

When DeWitt and Graham subsequently included both versions of Everett's thesis with their own work in the 1973 Princeton University Press anthology *The Many-Worlds Interpretation of Quantum Mechanics*, the language of worlds from the title of the collection stuck to Everett's interpretation of quantum mechanics. But although DeWitt's understanding of the theory may have agreed with that of Graham, there is good reason to suppose that it did not agree well with either Wheeler's or Everett's

---

[22] Although 10^{100+} is a big number, DeWitt believed that there were only a finite number of worlds because of the finite precision of our measurements (DeWitt, 1971, pg. 42). For his part, Everett took there to be an uncountably infinite number of branches depending on the decomposition of the state one considers. See pg. 274 for this view.

[23] DeWitt, at least initially, and Graham held that the notion of "typical" that should matter here is the *numerical count* of the copy worlds; hence, Graham postulated that worlds split so that the number of copy worlds associated with a particular term in the absolute state would be proportional to the norm squared of the coefficient on that term. Graham's intuitions concerning the sort of typicality required here were in part a result of his understanding, with DeWitt, that branches were properly interpreted as causally independent physical worlds. The intuition was that since there were in fact multiple physical copies of the world, the only notion of "typical" that should matter physically was the notion that tracked the features of *most* worlds. Everett was deeply frustrated by this misunderstanding and the related thought that his initial formulation of the theory with his notion of typicality—which had nothing to do with the number of DeWitt worlds—was somehow mistaken or incomplete. See for example Everett's note on the DeWitt paper (pg. 281).

understanding. For his part, Wheeler directly attributed the many worlds interpretation of Everett to DeWitt.

> Bryce DeWitt, my friend at Chapel Hill, chose to call the Everett interpretation the "many worlds" interpretation, and DeWitt's terminology is now common among physicists (although I don't like it). The idea has entered into the general public consciousness through the idea of "parallel universes." Although I have coined catchy phrases myself to try to make an idea memorable, in this case, I opted for a cautious, conservative term. "Many worlds" and "parallel universes" were more than I could swallow. (Wheeler and Ford, 1998, pgs. 269–70)

Wheeler preferred the more conservative "relative state" language for describing Everett's theory, the language that Everett in fact used consistently throughout both the earlier long and later short versions of his thesis. The choice of language one used to describe Everett's theory was not simply a matter of convention for Wheeler. Wheeler believed that DeWitt's popular many-worlds language represented an "oversimplified way" of understanding Everett's theory (Wheeler and Ford, 1998, pg. 269). And for his part, Everett later reported that DeWitt's talk of worlds was not his own language.[24]

The extent to which pure wave mechanics might be taken to provide its own metaphysical interpretation is subtle both historically and conceptually. Everett believed that pure wave mechanics was capable of yielding its own interpretation in that it somehow allowed one to deduce the empirical predictions of the standard collapse theory as experiences of observers who are themselves treated as physical systems in the theory. The difficult interpretational question, however, concerns exactly how this was supposed to work. DeWitt seems to have believed that the mathematical formalism of pure wave mechanics alone entailed the existence of many splitting worlds and that the determinate physical states of these worlds then explained why one should expect the experiences predicted by the standard collapse theory. But it cannot work quite like this.

Because, as a point of logic, purely mathematical postulates entail only purely mathematical theorems, one cannot deduce any metaphysical commitments whatsoever regarding the physical world from the mathematical formalism of pure wave mechanics alone. Consequently, pure wave mechanics can only entail the sort of metaphysical commitments that DeWitt envisioned if a proper statement of the theory is already taken to involve interpretational principles that go beyond the bare mathematical formalism. The right question then is whether and to what extent pure wave mechanics might be taken to provide its own interpretation by dint of interpretational principles that one might properly understand as a natural part of the theory. That said, if DeWitt meant to suggest that Everett had

---

[24] See chapter 24 (pg. 313).

proven a metatheorem that pure wave mechanics, even properly conceived, involves a commitment to the existence of metaphysically real splitting worlds, then it is unclear what he could have had in mind. On even a very broad understanding of what might count as such a metatheorem, there is nothing answering to DeWitt's description in either the long or short version of Everett's thesis. Of course, that Everett never claimed to have deduced the existence of splitting worlds from a version of pure wave mechanics does not mean that it is impossible to do so. But whether and to what extent such a deduction might be possible depends on what metaphysical assumptions one takes to be properly included in a full statement of pure wave mechanics.[25]

Decoherence and Emergent Worlds

Insofar as the mathematical formalism of pure wave mechanics itself does not involve any metaphysical commitment to the existence of any particular physical entities, showing how it might be possible to deduce the existence of many worlds from pure wave mechanics involves finding the most conservative or plausible metaphysical extension of pure wave mechanics that might be taken to entail their existence. Relatively recently, it has been suggested that a particular sort of realism regarding the global quantum mechanical state is sufficient to allow one to deduce the existence of many splitting worlds as emergent entities from decoherence considerations in the context of pure wave mechanics. Although this line of argument goes beyond what Everett himself proposed, it does represent a natural development of DeWitt's interpretation of Everett, and it furthers Everett's project insofar as it remains in the service of his goal of getting as much as possible from pure wave mechanics while adding as little interpretational apparatus as possible.

---

[25] Everett did take himself to have proven an important result about pure wave mechanics, but it does not have anything in particular to do with the metaphysics of worlds. Rather, Everett believed that he could derive the standard quantum probabilities from pure wave mechanics alone. More specifically, he believed that he could show, and arguably did, that there was a unique measure of typicality that satisfied a small handful of constraints and such that most branches in that measure that described measurement records at all would describe records that exhibited the standard quantum statistics. Indeed, there is good reason to suppose that this was precisely what he meant when he said that he could deduce the standard predictions of quantum mechanics from pure wave mechanics. In any case, his characterization of "typicality" in pure wave mechanics and his associated limiting results for unbounded sequences of measurements were what Everett took to be his most significant accomplishments in connection with pure wave mechanics. See pgs. 122, 187–89, 273, and 295 for Everett's discussions of his notion of typicality in pure wave mechanics. The limiting results are described in the short thesis (pg. 193) and in more detail in the long thesis (pg. 127). See also Barrett (1999, limiting properties of the bare theory) for a discussion of the limiting results in the context of the bare theory formulation of pure wave mechanics.

The thought of those who wish to add these considerations is that since interference effects are the manifestly nonclassical behaviors of quantum mechanical systems, decoherence considerations might explain the emergence of classical, or quasi-classical, properties of quantum mechanical systems by explaining why interference effects are unobservable. In the context of Everett's pure wave mechanics, these quasi-classical properties might then be taken to be properties of physical systems in alternative emergent quasi-classical worlds.

Identifying the Everett–Wheeler formulation with DeWitt's popular interpretation, Wojciech Zurek described the problem of explaining classicality in pure wave mechanics in a 1991 article for *Physics Today*.

> The many-worlds interpretation was developed in the 1950s by Hugh Everett III with the encouragement of John Archibald Wheeler. In this interpretation all of the universe is described by quantum theory. Superpositions evolve forever according to the Schrodinger equation. Each time a suitable interaction takes place between any two quantum systems, the wavefunction of the universe splits, so that it develops ever more "branches".
>
> Everett's work was initially almost unnoticed. It was taken out of mothballs over a decade later by Bryce DeWitt, who managed—in part, through his *Physics Today* article (September 1970, page 30)—to upgrade its status from virtually unknown to very controversial. The many-worlds interpretation is a natural choice for quantum cosmology, which describes the whole universe by means of a state vector. There is nothing more macroscopic than the universe. It can have no *a priori* classical subsystems. There can be no observer "on the outside." In this context, classicality has to be an *emergent* property of the selected observables or systems. (Zurek, 1991, pg. 37)

Zurek then identified decoherence as the mechanism for the emergence of classicality in pure wave mechanics. Zurek gave credit to H. Dieter Zeh (1970) and to Murray Gell-Mann and James Hartle (1990) for the idea of using decoherence to individuate branches in Everett.

Zeh (1970) was the first to suggest adding decoherence considerations to Everett's pure wave mechanics to explain the dynamical stability of relative states. Decoherence occurs when a dynamical variable of a system in a superposition of classically well-defined states becomes strongly correlated with the system's environment or another internal degree of freedom. This correlation destroys the interference effects one might have observed that involve only the original system and the correlated variable. Consequently, the expectation values of the variable can be thought of as behaving classically. In this sense, a decohering interaction between a system and its environment determines what decompositions of the system's state provide branches that one might interpret as quasi-classical worlds. Insofar as branches may evolve to exhibit such worldlike behavior, one might take the worlds themselves to be emergent from the structure of the absolute state of the system and its environment. More concretely, Zeh argued that because each branch would interact differently with its environment, the "two world components" representing a superposition of the left- and right-handed orientations of a sugar molecule, for example, "would behave practically independently after they had been prepared" and that the classical handedness of the sugar would thus be dynamically stable (Zeh, 1970, pg. 346).

Zurek took Gell-Mann and Hartle to have added three crucial conceptual ingredients to Everett's formulation of quantum mechanics: "the notion of sets of alternative coarse grained histories of a quantum system, the decoherence of the histories in a set, and their approximate determinism near the effectively classical limit" (Zurek, 1991, pg. 42). For their part, Gell-Mann and Hartle took Everett's formulation of quantum mechanics to be suggestive but incomplete.

> It did not adequately explain the origin of the classical domain or the meaning of "branching" that replaced the notion of measurement. It was a theory of "many worlds" (what we would rather call "many histories"), but it did not sufficiently explain how these were defined or how they arose. Also Everett's discussion suggests that a probability formula is somehow not needed in quantum mechanics, even though a "measure" is introduced that, in the end, amounts to the same thing (Gell-Mann and Hartle, 1990, pg. 430).

They saw their many histories interpretation then as "an attempt at extension, clarification, and completion of the Everett interpretation" with the ultimate goal of finding a "coherent formulation of quantum mechanics for science as a whole, including cosmology" (Gell-Mann and Hartle, 1990, pg. 430). And to this end, they appealed to decoherence considerations as a way to individuate quantum histories. But it was arguably Zeh's (1970) and Zurek's (1991) idea to understand decoherence as the mechanism for the *physical emergence* of quasi-classical worlds from the objective global state. In any case, Gell-Mann, Hartle, Zeh, and Zurek recognized that they were adding interpretive tools to Everett's theory.[26] Although Everett did not explicitly appeal to decoherence considerations to account for the dynamical stability of branches or to select a preferred basis, such considerations do address the problem of what branches should count as worlds in DeWitt's interpretation of Everett, and they thus provide the foundation of the recent emergent-worlds interpretations of Everett.

---

[26] Zeh reported that "Everett's relative state interpretation is ambiguous...since the dynamical stability conditions [as provided by decoherence considerations] are not considered" (Zeh, 1970, pg. 347).

Perhaps the best developed of the recent emergent-world formulations of Everett is the one presented by David Wallace and Simon Saunders.[27] One of its virtues is that it provides a clear example of the sort of move required to deduce worlds, or any other metaphysically rich interpretation, from a version of pure wave mechanics.

Although Wallace has taken Everett's theory to be just pure wave mechanics, his understanding of this theory involves significantly more than just a statement of the mathematical formalism. In particular, pure wave mechanics, properly conceived, involves a metaphysical commitment that, at the most fundamental level, "the quantum state is all there is." Furthermore, the quantum state as represented by the absolute wave function is thought of as physically real in the same sense as a field might be taken as physically real in classical field theory. Worlds in this view are to be understood as emergent entities. More specifically, worlds are "mutually dynamically isolated structures instantiated within the quantum state, which are structurally and dynamically 'quasiclassical' " where the "existence of these 'worlds' is established by decoherence theory" (Wallace, 2010a, pgs. 69–70). The thought in the Saunders–Wallace view then is to identify worlds in terms of approximate emergent substructures of the correlation structure exhibited by the quantum state. Since the quantum state is physically real, worlds are also physically real inasmuch as they are identified with structures of the quantum state. In characterizing the emergence of worlds, the theory does some of its own interpretational work, but telling the story of emergent worlds here requires a prior commitment to an appropriate sort of state realism.[28]

In contrast with DeWitt's reading of Everett, then, Wallace does not insist that one somehow gets the *metaphysics* of splitting worlds from the mathematical formalism of pure wave mechanics alone. Rather, as one would expect, to get worlds as physically real emergent entities, one must first stipulate the more basic ontology from which they emerge. And here this means adopting a version of quantum state realism where the universal

---

[27] Although this general view has many proponents, it is fair to call it the Saunders–Wallace interpretation since no one has done more to clarify the details. See Saunders et al. (2010) for a collection of papers concerning this view and potential criticisms. Wallace takes the emergent-worlds interpretation to be identical with pure wave mechanics and consequently to involve no additions to the theory. One might perhaps better understand the emergent-worlds interpretation as pure wave mechanics and decoherence considerations together with a metaphysical interpretation of the theory in light of such considerations. See especially Wallace (2010a and 2010b).

[28] Following a suggestion by David Deutsch (1999), probabilities on the Saunders–Wallace view are taken to be recovered by arguing that in a universe described by pure wave mechanics a rational agent would act as if the Born rule (*rule 4II*) obtained. See also Wallace (2003 and 2010b). For his part, Wallace currently understands this as one way that one might attempt to get probabilities in the theory rather than as an integral part of his interpretation of Everett. In any case, there is no analogous argument in Everett.

wave function represents the sort of physical entity whose decohering substructures might be identified as physically real worlds. Rather than deriving this metaphysical commitment from the mathematical formalism, it is simply assumed as a primitive postulate of the theory, properly conceived, that the quantum state is all there is and that it is the sort of entity that might explain the existence of physically real emergent worlds by exhibiting an appropriate structure under the dynamics. The point here is that although it may be possible to argue that pure wave mechanics somehow yields its own interpretation, such a strategy is a significantly richer understanding of the theory than is provided by the mathematical formalism alone—it would require an understanding of pure wave mechanics that already includes significant metaphysical or other interpretational commitments.

This approach gets right Everett's desire to take some version of pure wave mechanics as a complete and accurate physical theory, and it fits well with his sense that the wave function alone determines all physical facts, but it does not mesh as well with other features of Everett's description of his theory. Although there is much to say on each of these points, there is good reason to suppose that (1) Everett's understanding of pure wave mechanics did not ultimately involve any special metaphysical commitment to the existence of the absolute quantum state or to real splitting worlds, (2) Everett did not appeal to decoherence considerations to individuate branches or to select a physically preferred basis and that it was important to his understanding of branches that they might be individuated with respect to any basis whatsoever, and (3) it was also important to Everett's understanding of branches that they are never dynamically isolated and consequently might always, at least in principle, interfere with each other.[29] Rather, as discussed above, Everett thought that a correlation between the observer and his object system was itself sufficient to explain the determinacy of the measurement result since it determined a relative state for the observer.[30] Furthermore, Everett explained quasi-classical behavior of relative systems on branches, when it occurs, as resulting from the low dispersion of systems correlated to massive systems. The argument went as follows:

> [a]ny general state can at any instant be analyzed into a *superposition* of states each of which ... represent[s] the bodies with fairly well defined positions and

---

[29] Although such an experiment would typically be extremely difficult to perform for macroscopic systems, this is the sense in which branches were operationally real for Everett: "It is ... improper to attribute any less validity or 'reality' to any element of a superposition than any other element, due to [the] ever present possibility of obtaining interference effects between the elements. All elements of a superposition must be regarded as simultaneously existing" (pg. 150). Moreover, as suggested in fn. 28 on pg. 48, Everett's understanding of probability in pure wave mechanics did not involve considerations of rationality. Rather, Everett explained probability by arguing that there was a particularly natural sense of "typical" such that a typical branch should be expected to exhibit the usual quantum statistics.

[30] For Everett's explanation of this point see, for example, pgs. 121 and 186–87.

momenta. Each of these states then propagates approximately according to classical laws, so that the general state can be viewed as a superposition of quasi-classical states propagating according to nearly classical trajectories. In other words, if the masses are large or the time short, there will be strong correlations between the initial (approximate) positions and momenta and those at a later time, with the dependence being given approximately by classical mechanics. (pgs. 136–37)[31]

Everett relied on the low dispersion of states involving massive systems under the linear dynamics and hence understood the conditions of quasiclassicality to involve large masses or short times. Finding classical appearances in the correlation structure described by pure wave mechanics would have been easier had he explicitly appealed to decoherence considerations. This would have provided a rough way to individuate quasi-classical branches that was relatively independent of the masses of the systems involved and that might become increasingly sharp over longer times.[32]

While the explanations provided by the emergent-worlds interpretation arguably go well beyond Everett's presentation of pure wave mechanics, the emergent-worlds interpretation is clearly in the Everett tradition and is certainly worth serious consideration on its own merits. It is easy to imagine that Everett would have appreciated the continued attention to his work. It is unclear, however, that he would have considered the extensions, clarifications, and completions offered by DeWitt, Graham, Zeh, Gell-Mann, Hartle, Zurek, Saunders, Wallace, and others as necessary to his theory.

ON THE FAITHFUL INTERPRETATION OF EVERETT

The historical question of how to best reconstruct Everett's understanding of pure wave mechanics may never be fully settled, but Everett himself always maintained that his interpretation was complete as he had described it in the long and short theses. Indeed, there is good reason to suppose that Everett's project was in fact successful by his own standards and without any further interpretation. Everett himself may have wanted something rather different from a physical theory than what his interpreters have required.

DeWitt and Everett began their correspondence long before DeWitt gave his popular presentation of Everett's theory. For his part, DeWitt was

---

[31] See also Everett's discussion of classical laws as correlation laws in pure wave mechanics in chapter 8 (pg. 158).

[32] Perhaps the closest Everett got to this was in a comment in his handwritten notes for the long thesis, where he indicates that he knows that the more degrees of freedom a systems has, the harder it is to observe interference effects.

initially very critical of pure wave mechanics on the grounds that he never experienced more than one branch of the global state, never felt himself branching, and was hence certain that he simply did not branch (pg. 246). The problem DeWitt explained to Wheeler was that Everett's model was *too rich*:

> I do agree that the scheme which Everett sets up is beautifully consistent; that any single one of the states $\psi^O[\alpha_i^1, \alpha_i^2 \ldots]$, when separated from the superposition which makes up the total or "universal" state vector $\psi^{S_1+S_2+\cdots+O}$, gives an excellent representation of a typical memory configuration, with no causal or logical contradictions, and with "built-in" statistical features. The whole state vector $\psi^{S_1+S_2+\cdots+O}$, however, is simply too rich in content, by vast orders of magnitude, to serve as a representation of the physical world. It contains all possible branches in it at the same time. In the real physical world we must be content with just one branch. Everett's world and the real physical world are therefore not isomorphic. (pgs. 246–47)

Wheeler passed DeWitt's comments to Everett, and Everett began his reply to DeWitt by summarizing his understanding of the proper cognitive status of physical theories.

> First, I must say a few words to clarify my conception of the nature and purpose of physical theories in general. To me, any physical theory is a logical construct (model), consisting of symbols and rules for their manipulation, *some* of whose elements are associated with elements of the perceived world. If this association is an isomorphism (or at least a homomorphism) we can speak of the theory as correct, or as faithful. The fundamental requirements of any theory are logical consistency and correctness in this sense.
>
> However, there is no reason why there cannot be any number of different theories satisfying these requirements, and further (somewhat arbitrary) criteria such as usefulness, simplicity, comprehensiveness, pictorability, etc., must be resorted to in such cases. There can be no question of which theory is "true" or "real"—the best that one can do is reject those theories which are *not* isomorphic to sense experience. (pg. 253)

In the second appendix to the long thesis, Everett had already argued that "[o]nce we have granted that any physical theory is essentially only a model for the world of experience, we must renounce all hope of finding anything like '*the* correct theory' " (pg. 170). Consequently, since pure wave mechanics is manifestly logically consistent, Everett just needed to argue for empirical faithfulness.[33]

---

[33] The relative-state interpretation of pure wave mechanics is consistent inasmuch as it is modeled by the correlation structure determined by the evolution of the universal wave function.

A crucial point in deciding on a theory is that one does *not* accept or reject the theory on the basis of whether the basic world picture it presents is compatible with everyday experience. Rather, one accepts or rejects on the basis of whether or not the *experience which is predicted by the theory* is in accord with actual experience. (pgs. 253–54)

And, Everett argued, the relative-state formulation of quantum mechanics

is in full accord with our experience (at least insofar as ordinary quantum mechanics is) ... just because it *is* possible to show that no observer would ever be aware of any "branching," which is alien to our experience as you point out. (pg. 254)

Although Everett did not explain why one would never experience branching in his letter to DeWitt, the argument is easy to reconstruct from how Everett modeled an observer's experience. A relative observer's experience is fully represented by his relative physical memory sequence, and, as Everett demonstrates, such relative memory sequences simply do not contain records of branching events. Indeed, as discussed earlier, Everett shows that a typical relative memory sequence, in his particular sense of typical, exhibits the standard quantum statistics. Hence,

[t]he theory *is* isomorphic with experience when one takes the trouble to see what the theory itself says our experience will be. Little more can be asked of it without exposing a naked philosophic prejudice of one kind or another. (pg. 255)

In the second appendix to the long thesis, Everett had explained that "[t]he word homomorphism would be technically more correct, since there may not be a one-one correspondence between the model and the external world" (pg. 169). The map is a homomorphism because (1) there may be elements of the theory that do not directly correspond to experience and (2) a particular theory may not seek to explain all of experience. Case (1) is particularly important here: Everett considered the surplus experiential structure represented in the various branches of the absolute state to be explanatorily harmless. Everett responded to DeWitt that

[f]rom the viewpoint of the theory, all elements of a superposition (all "branches") are "actual," none any more "real" than another. It is completely unnecessary to suppose that after an observation somehow one element of the final superposition is selected to be awarded with a mysterious quality called "reality" and the others condemned to oblivion. We can be more charitable and allow the others to coexist—they won't cause any trouble anyway because all the separate elements of the superposition ("branches") individually obey the wave equation with complete indifference to the presence or absence ("actuality" or not) of any other elements. (pgs. 254–55)

Pure wave mechanics then is empirically faithful because our actual experience is represented by relative measurement records on a branch that is associated with a measure of typicality determined by pure wave mechanics alone. In this view, Everett's solution to the determinate record problem consists of noting that the values of determinate measurement records for those measurements we take ourselves to have performed are represented as relative measurement records in the correlation structure described by pure wave mechanics. Indeed, they are isomorphic to a substructure of the correlation structure that models the theory.[34] Similarly, his solution to the probability problem might be understood to consist of noting that there is in fact a measure of typicality over relative measurement records that can be determined from the correlation structure alone and that covaries with the standard quantum mechanical expectations.[35]

The upshot is that since this is sufficient to show that pure wave mechanics is empirically faithful, and since pure wave mechanics is manifestly consistent, if empirical faithfulness and consistency constitute sufficient conditions for acceptability, Everett has shown that pure wave mechanics is an acceptable physical theory. The fact that it is also simple and comprehensive and provides a readily picturable model for all possible physical interactions distinguishes it from the less able competition.

In this sense, one might take Everett himself to have explained both determinate measurement records and the standard quantum statistics while avoiding what he would almost certainly have taken to be unnecessary metaphysical complications, given his commitments to empiricism.[36] Here

[34] The picture here is something like this. One begins with the absolute universal wavefunction. Then one chooses a decomposition that produces branches whose relative states characterize determinate measurement records. The theory is empirically faithful in the most basic sense if one finds a branch whose records correspond to one's own experience. The other aspect of empirical faithfulness for Everett was that in pure wave mechanics the coefficient of the branch containing one's experience covaries with one's empirically warranted quantum expectations. See, for example, pgs. 121 and 186–87 for descriptions of the role that Everett understood the correlation structure to play in the account of determinate experience in pure wave mechanics.

[35] See pgs. 78 and pg. 121 for examples of Everett's explanation of how determinate measurement outcomes are represented in terms of relative states and pg. 100 for an example of how he understands conditional expectations to be represented by the norm-squared measure over relative states. See also pgs. 122, 187–89, 273, and 295 for Everett's explanation of why the norm-squared measure is a particularly natural measure of typicality in pure wave mechanics. He argues is that it is the only probability measure that satisfies a small handful of conditions that Everett took to be both natural and analogous to conditions found in classical statistical mechanics. Of course, the norm-squared measure is ultimately appropriate since it allows Everett to represent within pure wave mechanics those conditional expectations that are in fact supported by observed quantum statistics.

[36] A realist might insist on somehow reifying the entire correlation structure in terms of worlds or minds, but one would want an argument that one got increased richness in explanation that somehow justified the fancy metaphysics. For his part, Everett might have

metaphysical claims are replaced by various ways in which one might picture and describe the correlation structure that models the theory. With respect to language one uses to describe the model of pure wave mechanics, Everett was both ecumenical and nondogmatic. Insofar as Everett took his project to be one of finding a representation of our experience in a comprehensive model of pure wave mechanics, he might be taken to have fully succeeded on his own terms without the help of further interpretation. The remaining question is just whether one should want more than what Everett himself wanted, and if so, what, more precisely?[37]

Although it is suggestive to read Everett as having succeeded on his own terms, there may never be an interpretation of Everett's formulation of quantum mechanics that fully meshes with everything he said. However, this is, perhaps, not a bad thing. Everett's work has proven to be of enduring value in part because it has continued to inspire the ingenuity of others in the service of making sense of pure wave mechanics.

Jeffrey A. Barrett
Irvine, California
December 2011

---

understood such a move as at most simply providing another way, in addition to the language of relative states, of describing the correlation structure characterized by pure wave mechanics.

[37] Although the notion of faithfulness is weaker than standard notions of empirical adequacy, in judging the empirical adequacy of a physical theory one always has some degree of freedom in choosing what aspect of the theory's model should correspond to empirical evidence, how it should correspond, and how the empirical evidence should itself be represented. See van Fraassen (2008) for a recent extended discussion of this point. In this sense, Everett's notion of empirical faithfulness is a variety of empirical adequacy. Furthermore, one might indeed desire more from a satisfactory formulation of quantum mechanics, but we have arguably found nothing so far that is clearly preferable to pure wave mechanics understood as empirically faithful. See Barrett (2010, 2011a and 2011b).

PART II

The Evolution of the Thesis

CHAPTER 4

Minipaper: Objective versus Subjective Probability (1955)

This minipaper presents an early formulation of the quantum mea-surement problem in terms of a conflict between subjective and objective probabilities. Everett ultimately provided a much clearer formulation of the measurement problem in terms of a conflict between different state attributions in the long thesis. The structure of the experimental setup, however, is similar. There are two versions of this minipaper. One, handwritten, contains the drawing (pg. 15) of the Wigner's Friend experiment. The second version (presented here) was typed after being edited and condensed from the handwritten draft.[a]

Everett

Objective vs Subjective probability.[b]

Since the root of the controversy over the interpretation of the formalism of quantum mechanics lies in the interpretation of the probabilities given by the formalism, we must devote some time to discussing these interpretations. There are basically two types of probabilities, which may be called subjective and objective probabilities, respectively. A subjective probability refers to an estimate by a particular observer which is based upon incomplete information, and as such is not a property of the system being observed, but only of the state of information of the observer. An objective probability on the other hand is regarded as an intrinsic property of a system, i.e., to what might be called "really" random processes. To illustrate, we consider the following experiment: A deck of cards is shuffled, and one card is selected and placed face down upon the top of a table. An observer, A, is asked whether or not that card is the ace of spades, whereupon he would probably reply that the "probability" that it is the ace of spaces is 1/52. This probability would be a subjective probability, because it clearly refers to the state of information of the observer, and not

---

[a] See also the discussion of the minipapers in the biographical introduction (chapter 2, pg. 17). A scan of the earlier handwritten version can be found online at UCIspace (pg. xi).

[b] Everett's copy of this document had handwritten marginal notes from his advisor John Wheeler. These are included here as notes with minimal editing in their approximate location.

to the system, namely the card, which is in actuality either the ace, or not the ace, and not a probability mixture. Nevertheless, A's statement that the probability is 1/52 has meaning, since if the experiment were to be repeated a large number of times, and he were to state each time that the card on the table is the ace of spades, then he would be correct roughly 1/52 of the time. However, if there were another observer, B, present, who is informed each time of the color of the card, he would answer differently, either 0 or 1/26 depending upon the color, and he would also be correct in the same sense that A was. Still another observer, who caught a glimpse of the card each time it was placed on the table would have yet another answer, in which no probability was present at all. This illustrates that subjective probabilities are *not* invariant from observer to observer, due to differing states of information.

Since an *objective* probability is conceived to be a property of a system, and hence independent of states of information, it must be invariant from one observer to the next. That is if two observers[c] ascribe different probabilities to some aspect of the same system,[d] then *at least one of these probabilities is subjective*! One possible method of arriving at a probability which satisfies this criterion of invariance is as a limit of subjective probabilities as information becomes more and more perfect. This obviously satisfies the criterion, but can never be verified in practice. We cannot go further towards a positive definition of objective probabilities, but we *can* investigate the consistency of ascribing objective probabilities to events in certain situations, with regard to the above negative criterion. In particular we wish to investigate the limitations of interpreting the probabilities of quantum mechanics as objective probabilities, which are imposed by this criterion. We shall refer to this view as the objective interpretation of quantum mechanics.

As a starting point we might investigate the consequences of the following set of postulates:[e]

1. Every physical system s possesses a state function, ψ_s, which gives the *objective* probabilities of the results of any measurement which might be performed upon the system.[f]

2. The state function ψ_s of an isolated system changes causally with time as long as the system remains isolated.

We shall now show that the above postulates are inconsistent by consideration of the following situation: We suppose that we have a system,

---

[c] Wheeler boxes the word "observers" and writes in the margin: "distinguish observation and observer".

[d] Wheeler writes in the margin: "either/or taken too strongly".

[e] Postulates 1 and 2 roughly correspond to von Neumann's (1955) Processes 1 and 2, respectively. See the conceptual introduction (chapter 3, pgs. 28–29) for a discussion of von Neumann's postulates.

[f] Wheeler writes in the margin: "by whom".

S_1, and a measuring device M_1, for measuring some property of S_1, which is connected to a recording device which will record the results of the measurement at a classical level, such as the position of a relay arm, and we assume that the measuring device is arranged to make the measurement automatically at some time. We further assume that the entire system, S_2, consisting of S_1, M_1, and the recording device, is isolated from any external interactions.

Now, the system S_1 possesses a state function ψ_1 which gives the objective probability of the results of the measurement. (We shall also assume that ψ_1 is not an eigenstate of the measurement, so that we shall have a probability which is neither 0 nor 1. But if we now consider the total system S_2, before the measurement, it also possesses a state function ψ_2. Furthermore, ψ_2 at any later time is strictly determined by our initial ψ_2 so long as there are no outside interactions, so that in particular ψ_2 for some time after the measurement has taken place is strictly determined by its value before the measurement. We now consider what this later ψ_2 may say about the configuration of the recording device. If it gives a probability mixture over the configurations, then clearly these probabilities are of the *subjective*[g] type, since they refer to something which actually exists in a pure state, because in reality the configuration of the recording device has already been determined. (Just as in the case of the man and the card). On the other hand if this later ψ_2 gives the exact configuration of the recording device, then clearly the outcome of the measurement was determined before it took place, since the later ψ_2 was strictly determined by the earlier, in which case the probability given by ψ_1 was not objective. So that we see that in any case at least one of the probabilities was subjective, and the postulates are untenable.

There are several ways of removing this inconsistency, such as the following modifications:

1. Not every physical system possesses a state function, i.e., that even in principle quantum mechanics cannot describe the process of measurement itself. This is somewhat repugnant since it leads to an artificial dichotomy of the universe into ordinary phenomena, and measurements.

2. The wave function of an isolated system is *not* always causally determined, but may suffer abrupt discontinuities from mixed states into probability mixtures of pure states, corresponding to an internal phenomenon which is regarded as a measurement. This is quite tenable, but leaves entirely unknown what is to be regarded as such a measurement, so that for this interpretation no formalism has been developed to give the points of discontinuity.[h]

---

[g] Wheeler writes in the margin: "Orders of mag *of time* Amplif'n means not pure".

[h] Although Everett's language of pure and mixed states is at least somewhat nonstandard, the thought behind this option is clear enough. He is describing a theory that provides

3. The probabilities occurring in quantum mechanics are *not* objective. That is, they correspond to our ignorance of some hidden parameters. Under such an interpretation the inconsistency resolves easily, since the outside wave function possesses more information than the internal one, such as phase factors, etc., for the interaction, so that it leads to a causal description.

conditions under which a stochastic collapse of the quantum mechanical state would occur. The standard collapse theory might qualify if it were completed in an appropriate manner.

CHAPTER 5

Minipaper: Quantitative Measure of Correlation (1955)

*This minipaper lays some groundwork for the application of infor-
mation theory as a measure of the degree of quantum correlation or
entanglement. From very early in the project, Everett thought that
Shannon information theory (Shannon and Weaver, 1949) would provide insight into pure wave mechanics. Although statistical measures,
such as standard deviation, would have sufficed, Everett developed
information theoretic measures to represent such things as the width
of distributions and the degree of correlation between the states of
multiple systems. He thought that the standard quantum uncertainty
relations were best expressed in an information theoretic way. There
are two versions of this minipaper, a handwritten one drawn from the
main body of the evolving draft thesis and the slightly different typed
version given to Wheeler that is presented here.[i]*

Prof. Wheeler   H. Everett

Quantitative Measure of Correlation

If one makes the statement that two variables, X and Y, are correlated,
what is basically meant is that one learns something about one variable
when he is told the value of the other. By utilizing some of the concepts
of information theory this notion can be made to yield a natural and
satisfactory quantitative measure of the correlation of two variables. In
information theory one defines the quantity of information contained in a
probability distribution for a random variable X, with density $P(x)$, to be:

$$I_X = \int P(x) \log P(x)\, dx,$$

a quantity which agrees quite closely with our intuitive ideas about
information.[j]

---

[i] See also the discussion of the minipapers in the biographical introduction (chapter 2
pg. 17). A scan of the earlier handwritten version of this minipaper can be found online at
UCIspace (pg. xi).

[j] This is the standard definition of information from Shannon information theory applied
here to a continuous distribution. See Shannon and Weaver (1949).

Suppose now that we are given a joint distribution $P(x, y)$ over two random variables X and Y, for which we seek a measure of the correlation between the two variables. Let us focus our attention upon the variable X. If we are not informed of the value of Y, then the probability distribution of X (marginal or a priori distribution) is $P(x) = \int P(x, y)\,dy$, and our information about X is given by $I_X = \int P(x) \log P(x)\,dx$. However, if we are now told that Y has the value y, the probability distribution for X changes to the *conditional* distribution $P_y(x) = P(x, y)/P(y)$, with information $I_X^y = \int P_y(x) \log P_y(x)\,dx$. According to what has been said we wish the correlation to measure how much we learn about X by being informed of the value of y, that is, the change in information. However, since this change may depend upon the particular value of Y which we are told, the natural thing to do is to consider the *expected* change in information about X, given that we are to be told the value of Y. This quantity we shall call the correlation of X with Y, denoted by $C(X, Y)$. Thus:

$$C(X, Y) = \int P(y) \left[I_X^y - I_X \right] dy$$

It turns out that this quantity is also the expected change in information about Y given that we are to be informed of the value of X, so that the correlation is symmetric in X and Y, and it is in fact equal to the information of the joint distribution less the sum of the information for the two a priori distributions:

$$
\begin{aligned}
C(X, Y) &= I_{XY} - I_X - I_Y \\
&= \int P(x, y) \log P(x, y)\,dx\,dy - \int P(x) \log P(x)\,dx \\
&\quad - \int P(y) \log P(y)\,dy \\
&= \int \int P(x, y) \log \frac{P(x, y)}{P(x)P(y)}\,dx\,dy
\end{aligned}
$$

It has the further property that it is zero if and only if the two variables are independent, and is otherwise strictly positive, ranging to $+\infty$ in the case of a functional dependence of x on y (perfect correlation).

It is in several respects superior to the usual correlation coefficient of statistics, which can be zero even when the variables are not independent, and which can assume both positive and negative values. A negative correlation can give, after all, quite as much information as a positive correlation.

Finally, it is invariant to changes of scale, i.e., if $z = ax$ then $C(Z, Y) = C(X, Y)$, so that if we decided to measure x in meters instead of inches, for example, the correlation with Y would be unchanged, so that it is in some sense an absolute measure.[k]

These notions can be easily extended to distributions over more than two variables by introducing further "correlation numbers," a process which leads to a simple and elegant algebra for these quantities.

Extending to wave mechanics, we define the correlation between observables X and Y, with wave function in x, y representation $\psi(x, y)$, to be:

$$C(X, Y) = \int \int \bar{\psi}\psi \log \left[\frac{\bar{\psi}\psi}{\int \bar{\psi}\psi \, dx \int \bar{\psi}\psi \, dy} \right] dx \, dy.$$

---

[k] Everett took his information theoretic measure of correlation to have several advantages over standard statistical measures of correlation. See (chapter 8, pg. 84) and the following for a discussion of its virtues. Most of Everett's discussion of information theory was dropped between the long and short versions of the thesis.

CHAPTER 6

Minipaper: Probability in Wave Mechanics (1955)

This minipaper develops the basic premise of Everett's thesis stripped of its formalism but laced with explanatory metaphors, such as splitting amoebas that share overlapping memories until diverging into separate futures recorded as consistent histories. There are several handwritten drafts of this minipaper and two typed versions, one turned into Wheeler. Wheeler made handwritten notes on his copy of this minipaper. These notes are presented here in the lettered footnotes.[1]

Prof. Wheeler   H. Everett

Probability in Wave Mechanics[m]

In present formulations of quantum mechanics there are two essentially different ways in which the state of a system changes, one continuous and causal, and the other discontinuous and probabilistic. Let ψ be the state of a system with energy operator H, then the two processes are:

1. The discontinuous change brought about by the measurement of a quantity with eigenstates $\{\phi_i\}$, in which case the state ψ will be changed to the state ϕ_j with probability $|(\psi, \phi_j)|^2$.

2. The continuous, causal change of the state of the system with time generated by the energy operator:[n]

$$\psi_t = e^{-\frac{iHt}{\hbar}} \psi_0$$

The question arises as to whether these two rules are compatible; in particular what occurs in the event that (2) is applied to the measurement process itself (i.e., to the state of the combined system of the original system

---

[1] See also the discussion of the minipapers in the biographical introduction (chapter 2, pg. 17). Digital scans of the earlier versions of this minipaper can be found at UCIspace (pg. xi).

[m] Everett's copy of this document has handwritten marginal notes from his advisor John Wheeler. These are included here as notes with minimal editing in their approximate location.

[n] Wheeler writes in the margin:

$$\frac{\partial}{\partial t}$$

plus apparatus and observer). In this case nothing like the discontinuity of (1) can occur, and one has to decide whether to abandon (1), and the statistical interpretation of quantum mechanics in favor of the purely causal description (2), or to limit the applicability of (2) to systems within which "measurements" are not taking place. If we were to deny the applicability of (2) to the measurement process, however, we are faced with the difficulty of how to distinguish a measurement process from other natural processes. For what n might a group of n particles be construed as forming a measuring apparatus or observer, and cease to be governed by (2)? We would be faced with the task of dividing all processes into two categories, the usual ones which are governed by (2), and the mysterious type called measurements which are immune to (2) and follow (1) instead. Still another alternative would be to attempt some sort of deterministic "hidden parameter" theory in which probabilities would arise as a result of the ignorance of the observer concerning the values of the hidden parameters.[o]

It is our purpose here to indicate that a purely causal theory which postulates the existence of some sort of wave function for the entire universe, and for which (2) alone holds, forms an adequate theory. That is, that by assuming the general validity of pure wave mechanics, without any initial statistical interpretation, we obtain a theory which is in principle applicable to all natural processes, and furthermore one which even leads to probabilistic aspects on a subjective level in a rather novel way (i.e., that we are able to deduce that (1) will appear to hold to observers).

We turn now to the formalism of quantum mechanics. We shall assume a particle model, in which we envisage the universe to be composed of a large number of elementary particles, possessing a single total wave function, which we assume to obey the Schrödinger equation. (No results will depend upon this, however, they will hold for field theories as well, and any wave equation, that is, any system of "quantum mechanics".)

The first question that arises is "What actually does happen in the process of measurement?" Several authors (Von Neumann, Bohm, etc.) treat this question to some degree.[p] Assume that we have a system, S, and an apparatus, A, and that the system variable of interest is x, while the apparatus variable of interest is y (position of a meter needle, spot on photographic film, etc.) and that prior to making a measurement the system is in a definite state ψ_S^0 and the apparatus in a state ψ_A^0, and furthermore that they are initially independent, so that the wave function of the whole system before the measurement begins is simply the product wave function. The measurement is then brought about by allowing the two systems to interact, by "turning on" a suitable interaction hamiltonian $H_I(x, y)$, which

---

[o] This would be something like Bohmian mechanics (Bohm, 1952), which Everett understood well and discussed in the long thesis (chapter 8, pgs. 153–54).

[p] Everett took von Neumann (1932, 1955) and Bohm (1952) as canonical texts on this topic.

is chosen so as to introduce a *correlation* between the system variable x and the apparatus variable y. However, in order that the measurement be "good", H_I must be chosen so that the system state will not be disturbed (except in phase) if it is an eigenstate of the measurement.

Now, the measurement is arranged so that corresponding to each system eigenstate ϕ_i, with value x_i, there will be a definite apparatus state with value y_i *after* the measurement. However, if the system is originally *not* in an eigenstate of x, but in a state of the form $\psi_S^0 = \sum_i a_i \phi_i$, then after the measurement the apparatus state will be *indefinite* to the same extent. This follows from the linearity of the wave equation and the superposition principle. In short, nothing discontinuous has happened, the system has not been forced to jump into an eigenstate, and, indeed, the relative amplitudes for the various eigenstates of the system remain unaltered. Nothing remotely resembling process (1) has taken place.

What has happened, however, is that the apparatus has become *correlated* to the system, even though neither is in a definite eigenstate of the variable under discussion. (Reminiscent of the example of Einstein, Rosen, and Podolsky.) This is possible because after the measurement the wave function for $S + A$ is in a higher dimensional space than that of S or A alone. If we look at a "cross section"[q] of the total wave function for which the variable x has the definite value x_i, we find that the apparatus has the definite value y_i, which corresponds, while if we choose to consider the "cross section" for x_j definite, we immediately find that y has the definite value y_j, etc. [r]

So we see that from the viewpoint of wave mechanics that when a measuring apparatus interacts with a system which is not in an eigenstate of the variable being measured that the apparatus itself "smears out" and is indefinite, no matter how large or "classical" it is.[s] Nevertheless, it is correlated with the system in the above sense, and it is this correlation which allows us to give an adequate interpretation of the theory.

How is it possible, this "smearing out" of even classical objects which is implied by wave mechanics, and which is seemingly so contrary to our experience? Does this mean that we must abandon our quantum mechanical description and say that it fails at a classical level? We shall see that all that is necessary is to carry the theory to its logical conclusion to see that it is consistent after all.

---

[q] Wheeler writes in the margin: "image unclear."

[r] This notion of the cross section of the total wave function is the basis for Everett's notion of relative states in the long and short theses. Here the object system has definite value x_i relative to the measuring apparatus recording the corresponding outcome y_i. Cross sections, relative states, elements, and branches provided Everett with a language for talking about correlations between subsystems of a larger composite system. That is, they provided alternative ways to talk about the correlation structure that models pure wave mechanics.

[s] Wheeler writes in the margin: "meaning of *smear*-out example showing how compatible with class. nature of meas & associated *amplification*."

Suppose that a human observer sets up his apparatus and makes a measurement on a system not in an eigenstate of the measurement, the result to appear as the position of a meter needle. According to what we have said, the meter needle itself will be "smeared out" after the measurement, but correlated to the system.[t] Why doesn't our observer see a smeared out needle?[u] The answer is quite simple. He behaves just like[v] the apparatus did. When he looks at the needle (interacts), he himself becomes smeared out,[w] but at the same time correlated to the apparatus, and hence to the system. If we reflect for a moment upon the total wave function of the situation system-apparatus-observer, and again consider "cross sections", we see that for the definite system value x_j the needle has definite position y_j, and there is a *definite observer* who perceives that the needle has the definite position y_j, and, of course, similarly for all the other values. In other words, the observer himself has split[x] into a number of observers, each of which sees a definite result of the measurement.[y] Furthermore, should our observer call over his lab assistant to look at the needle, the assistant would also split,[z] but be correlated in such a manner as always to agree with the first observer as to the position of the needle, so that no inconsistencies would ever arise.

In order to better illustrate the central role of correlations in quantum mechanics we consider the following example: In a box, say a one centimeter cube, we place a proton and an electron, each in momentum eigenstates, so that the position amplitude of each is uniform over the whole box. After a period of time we would expect a hydrogen atom to have formed.[aa] Nevertheless the position amplitude of the electron is still uniform over the whole box, just as that of the proton. All that has occurred is that the position densities have become correlated. All that is meant by the statement "There is a hydrogen atom in the box" is the existence of this correlation.[ab, ac]

[t] Wheeler writes in the margin: "X".
[u] Wheeler writes in the margin: "X".
[v] Wheeler changed "like" to "as."
[w] Wheeler writes in the margin: "X".
[x] Wheeler writes in the margin: "X".
[y] This passage suggests that the measurement process involves a physical splitting of the observer. Here Everett clearly describes the process as one where a single observer splits into multiple observers, each of which has a definite measurement outcome. One clear advantage of this line is that it makes the explanation for why an observer perceives a determinate measurement outcome entirely straightforward: each postmeasurement observer actually has a single determinate outcome.
[z] Wheeler writes in the margin: "X".
[aa] Wheeler writes in the margin:"X By radiation? What about the lack of correlation it introduces".
[ab] Wheeler writes in the margin: "X".
[ac] This example later plays a central role in the long thesis in explaining by analogy with the hydrogen atom what a composite physical object is in pure wave mechanics (pgs. 134–37).

In fact, it is clear from the circumstance that the wave equation is in $3N$ dimensional space, rather than 3 dimensional, that whenever several systems interact some degree of correlation is produced. Consider a large number of interacting particles. If we suppose them to be initially independent, then throughout the course of time the position amplitude of any single particle spreads out farther and farther, approaching uniformity over the whole universe, while at the same time, due to interactions, strong correlations will be built up, so that we might say that the particles have coalesced to form a solid object.[ad] That is, even though the position amplitude of any single particle would be "smeared out" over a vast region, if we consider a "cross section" of the total wave function for which one particle has a definite position, then we immediately find all the rest of the particles nearby, forming our solid object. It is this phenomenon which accounts for the classical appearance of the macroscopic world, the existence of definite solid objects, etc., since we ourselves are strongly correlated to our environment. Even though it is possible for a macroscopic object to "smear out", particularly if it is connected to an amplification device whose operation depends upon microscopic events, we would never be aware of it due to the fact that the interactions[ae] between the object and our senses are so strong that we become correlated almost instantly.[af]

We now see that the wave mechanical description is really compatible with our ideas about the definiteness on a classical level, due to the existence of strong correlations.

We must now turn around and try to see why process (1) has been so successful. Imagine an observer making a series of quantum mechanical measurements (such as the sequence of measuring the z component of spin of an electron, then its x component, then again its z component, etc.). From the point of view of wave mechanics he is splitting[ag] each time a measurement is made, so that after a number of measurements we could speak of his "life tree". If we focus our attention on any single "branch" of this tree we see an observer who always perceives definite (and unpredictable)[ah] results of his measurements,   and to whom the

---

[ad] Wheeler writes in the margin: "rad'n assumed? Analog to H_2? If so, caution on radiation uncertainties. A complication in the discussion. But without it, particles escape those [?] position correlations".

[ae] Wheeler writes in the margin: "spell out".

[af] Perhaps the best sense of what Everett had in mind is given by how he reworked and filled in the arguments here in the long thesis. See the discussion beginning on (pg. 134).

[ag] Wheeler writes in the margin: "Split? Better words needed. Do first an unconscious object to show ideas more objectively." From the earliest manuscripts both Wheeler and Everett were in search of appropriate words to describe pure wave mechanics. While he spoke most often in terms of elements, branches, and relative states, Everett changed his language regularly. For a few examples of language problems see pgs. 121, 206, 209–10, and 222.

[ah] Wheeler writes in the margin: "Elucidate details to show doubter how expt'lly to convince himself."

system has, with each measurement, apparently popped discontinuously into an eigenstate of the measurement. (Whereas from our point of view the observer himself has simply split into a number of observers, one for each eigenstate of the system, a process which is quite continuous and causal.)[ai] Furthermore, for almost all of the "branches" of his "life tree" which we might consider, the frequencies with which the observer sees the various results of his measurements will follow the probabilistic law of (1).[aj] Therefore, for practical considerations, an observer is justified in using (1) for calculations; not because the system undergoes any such probabilistic jumps, but simply because he himself will split into a number of observers, to each of which it *appears* that the system underwent a probabilistic jump.

We have, then, a theory which is objectively causal and continuous, while at the same time subjectively probabilistic and discontinuous. It can lay claim to a certain completeness, since it applies to all systems, of whatever size, and is still capable of explaining the appearance of the macroscopic world. The price, however, is the abandonment of the concept of the uniqueness of the observer, with its somewhat disconcerting philosophical implications.[ak]

As an analogy one can imagine an intelligent amoeba with a good memory. As time progresses the amoeba is constantly splitting, each time the resulting amoebas having the same memories as the parent.[al] Our amoeba hence does not have a life line, but a life tree. The question of the identity or non identity of two amoebas at a later time is somewhat vague. At any time we can consider two of them, and they will possess common memories up to a point (common parent) after which they will diverge according to their separate lives thereafter.

We can get a closer analogy if we were to take one of these intelligent amoebas, erase his past memories, and render him unconscious while he underwent fission, placing the two resulting amoebas in separate tanks, and repeating this process for all succeeding generations, so that none of them would be aware of their splitting. After a while we would have a large number of individuals, sharing some memories with one another, differing in others, each of which is completely unaware of his "other selves" and under the impression that he is a unique individual. It would be difficult

---

[ai] Wheeler writes in the margin: "Amplifier necessary for validity of what is said here, it seems to me."

[aj] Everett provides no argument for this point here nor is there an explanation of the specific sense of "almost all" or *typicality* for which this is true. It is likely, however, that Everett knew how he would argue for this later. See, for example, the discussions following pgs. 189 and 123.

[ak] Wheeler writes in the margin: "Careful examination needed of all the important apparent paradoxes".

[al] Wheeler writes in the margin: "This analogy seems to me quite capable of misleading readers in what is a very subtle point. Suggest omission".

indeed to convince such an amoeba of the true situation short of actually confronting him with his "other selves". The same is true if one accepts the hypothesis of the universal wave function. Each time an individual splits he is unaware of it, and any single individual is at all times unaware of his "other selves" with which he has no interaction from the time of splitting.

We have indicated that it is possible to have a complete, causal theory of quantum mechanics, which simultaneously displays probabilistic aspects on a subjective level, and that this theory does not involve any new postulates, but in fact results simply by taking seriously wave mechanics and assuming its general validity. The physical "reality" is assumed to be the wave function of the whole universe itself. By properly interpreting the internal correlations in this wave function it is possible to explain the appearance of the macroscopic world to us, as well as the apparent probabilistic aspects.[am]

[am] Wheeler writes at the bottom of the page: "Have to discuss questions of knowability of the universal ψ f'n.—And latitude with which we can ever determine it. Question of pooling of data by diff. observers. Question whether new view has any practical consequences. Also its implications for machinery of the world. Any special simplicity to be expected for *the* wave f'n? If not, why not? If so, what kind of simplicity? Any explanation then why world doesn't look simple?" The questions of how observers pool data in pure wave mechanics and the complex appearance of the world became central questions for Everett in the long thesis. See for example the discussions starting on pp. 194, 130, and 134.

CHAPTER 7

Correspondence: Wheeler to Everett (1955)

Wheeler thought that the correlation minipaper was close to being ready to be published. But he had serious reservations about "Probability and Wave Mechanics," especially the splitting metaphors like the amoeba story that Everett had used to describe the branching structure exhibited by the linear superpositions of states represented by the universal wave function (see Wheeler's notes on the paper itself for his cautions). More specifically, Wheeler said that the third minipaper was not ready for Bohr's inspection "because of parts subject to mystical misinterpretations by too many unskilled readers." Everpolitic, Wheeler was warning Everett that the metaphorical language he was using to frame his mathematically consistent theory was going to cause him serious professional problems unless he toned down the exposition.[an]

Hugh Everett—

I would very much like to discuss these two important papers with you. The correlation one seems to me practically ready to publish—(1) where would you publish it? (2) Can one generalize your definition of correlation, which is inv't. so to speak in the schema of spec. rel. (against linear transf) so it will be inv't in the sense of general relativity? Probably not except in a very artificial way—but what does this circumstance tell about the meaning of correlation?

As for the 2nd one, I am frankly bashful about showing it to Bohr in its present form, valuable & important as I consider it to be, because of parts subject to mystical misinterpretations by too many unskilled readers. I would welcome the chance to discuss this with you Mon.—1:30 if you have lunch engagement then, 12:30 otherwise—if you are free. Let me know if this is convenient.

John Wheeler.
21 Sept '55

The pencil notes will give some guidance as to my worries.

[an] See the discussion of the minipapers in the biographical introduction (pg. 17).

CHAPTER 8

Long Thesis: Theory of the Universal Wave Function (1956)

Everett's long thesis was submitted to John Archibald Wheeler, his doctoral thesis advisor, in January 1956 under the title "Quantum Mechanics by the Method of the Universal Wave Function." It was retitled "Wave Mechanics Without Probability" and circulated in April of that year to several prominent physicists, including Niels Bohr. The disapproval of Bohr and his colleagues, largely due to Everett's direct criticism of the Copenhagen interpretation in the long thesis and their sense that Everett did not properly understand the current orthodoxy, led to the later short thesis. The short thesis, which Everett defended for his Ph.D., was a redacted and re-focused version of the long thesis that was reworked under the direction of Wheeler to soften the force of Everett's attack on the orthodox Copenhagen interpretation. Everett's extended discussion of information theory, which has a prominent place in the long thesis, was cut. The long thesis itself was not published until 1973 in the DeWitt and Graham Princeton University Press anthology. It was published under the title "The Theory of the Universal Wave Function," and it includes a number of changes that Everett made on his own typescript copy of the version circulated in April 1956. This is the version of the thesis reproduced here.[ao]

The Theory of the Universal Wave Function
Hugh Everett, III

I. INTRODUCTION

We begin, as a way of entering our subject, by characterizing a particular interpretation of quantum theory which, although not representative of

---

[ao] For further historical details, see the discussions of the long thesis in chapters 1 and 2 of this volume.

the more careful formulations of some writers, is the most common form encountered in textbooks and university lectures on the subject.[ap]

A physical system is described completely by a state function ψ, which is an element of a Hilbert space, and which furthermore gives information only concerning the probabilities of the results of various observations which can be made on the system. The state function ψ is thought of as objectively characterizing the physical system, i.e., at all times an isolated system is thought of as possessing a state function, independently of our state of knowledge of it. On the other hand, ψ changes in a causal manner so long as the system remains isolated, obeying a differential equation. Thus there are two fundamentally different ways in which the state function can change:[1]

Process 1: The discontinuous change brought about by the observation of a quantity with eigenstates $\phi_1, \phi_2, \ldots$, in which the state ψ will be changed to the state ϕ_j with probability $|(\psi, \phi_j)|^2$.

Process 2: The continuous, deterministic change of state of the (isolated) system with time according to a wave equation $\frac{\partial \psi}{\partial t} = U\psi$, where U is a linear operator.

The question of the consistency of the scheme arises if one contemplates regarding the observer and his object-system as a single (composite) physical system. Indeed, the situation becomes quite paradoxical if we allow for the existence of more than one observer. Let us consider the case of one observer A, who is performing measurements upon a system S, the totality $(A+S)$ in turn forming the object-system for another observer, B.[aq]

If we are to deny the possibility of B's use of a quantum mechanical description (wave function obeying wave equation) for $A+S$, then we must be supplied with some alternative description for systems which contain observers (or measuring apparatus). Furthermore, we would have to have a criterion for telling precisely what type of systems would have the preferred

---

[1] We use here the terminology of von Neumann [J. von Neumann, *Mathematical Foundations of Quantum Mechanics*. (Translated by R. T. Beyer) Princeton University Press: 1955.].

[ap] This is the von Neumann–Dirac formation of quantum mechanics (von Neumann, 1955). See the discussion of these postulates in the conceptual introduction (chapter 3, pgs. 28–29). Everett later contrasted this with the Copenhagen interpretation. See for example pgs. 238–40. See also the discussion of Everett's understanding of the Copenhagen interpretation in the conceptual introduction (chapter 3, pg. 32).

[aq] What follows are two versions of what is now known as the Wigner's Friend story (Wigner, 1961). Everett uses these idealized thought experiments to argue for the inconsistency of the standard collapse formulation of quantum mechanics. See also Everett's presentation in the short thesis (chapter 6, pg. 176) and the discussion of Everett's understanding of the measurement problem in the conceptual introduction (chapter 3, pg. 30).

positions of "measuring apparatus" or "observer" and be subject to the alternate description. Such a criterion is probably not capable of rigorous formulation.

On the other hand, if we do allow B to give a quantum description to $A + S$, by assigning a state function ψ^{A+S}, then, so long as B does not interact with $A + S$, its state changes causally according to Process 2, *even though A may be performing measurements upon S*. From B's point of view, nothing resembling Process 1 can occur (there are no discontinuities), and the question of the validity of A's use of Process 1 is raised. That is, *apparently* either A is incorrect in assuming Process 1, with its probabilistic implications, to apply to his measurements, or else B's state function, with its purely causal character, is an inadequate description of what is happening to $A + S$.

To better illustrate the paradoxes which can arise from strict adherence to this interpretation we consider the following amusing, but *extremely hypothetical* drama.[ar]

Isolated somewhere out in space is a room containing an observer, A, who is about to perform a measurement upon a system S. After performing his measurement he will record the result in his notebook. We assume that he knows the state function of S (perhaps as a result of previous measurement), and that it is not an eigenstate of the measurement he is about to perform. A, being an orthodox quantum theorist, then believes that the outcome of his measurement is undetermined and that the process is correctly described by Process 1.

In the meantime, however, there is another observer, B, outside the room, who is in possession of the state function of the entire room, including S, the measuring apparatus, and A, just prior to the measurement. B is only interested in what will be found in the notebook one week hence, so he computes the state function of the room for one week in the future according to Process 2. One week passes, and we find B still in possession of the state function of the room, which this equally orthodox quantum theorist believes to be a complete description of the room and its contents. If B's state function calculation tells beforehand exactly what is going to be in the notebook, then A is incorrect in his belief about the indeterminacy of the outcome of his measurement. We therefore assume that B's state function contains non-zero amplitudes over several of the notebook entries.

At this point, B opens the door to the room and looks at the notebook (performs his observation). Having observed the notebook entry, he turns to A and informs him in a patronizing manner that since his (B's) wave function just prior to his entry into the room, which he knows to have been a complete

---

[ar] It is extremely hypothetical because such an experiment would be virtually impossible in practice due to environmental decoherence effects. Everett is concerned with a conceptual problem that has nothing to do with what measurements one might in fact be able to perform.

description of the room and its contents, had non-zero amplitude over other than the present result of the measurement, the result must have been decided only when B entered the room, so that A, his notebook entry, and his memory about what occurred one week ago had no independent objective existence until the intervention by B. In short, B implies that A owes his present objective existence to B's generous nature which compelled him to intervene on his behalf. However, to B's consternation, A does not react with anything like the respect and gratitude he should exhibit towards B, and at the end of a somewhat heated reply, in which A conveys in a colorful manner his opinion of B and his beliefs, he rudely punctures B's ego by observing that if B's view is correct, then he has no reason to feel complacent, since the whole present situation may have no objective existence, but may depend upon the future actions of yet another observer.

It is now clear that the interpretation of quantum mechanics with which we began is untenable if we are to consider a universe containing more than one observer. We must therefore seek a suitable modification of this scheme, or an entirely different system of interpretation. Several alternatives which avoid the paradox are:

Alternative 1: To postulate the existence of only one observer in the universe. This is the solipsist position, in which each of us must hold the view that he alone is the only valid observer, with the rest of the universe and its inhabitants obeying at all times Process 2 except when under his observation.

This view is quite consistent, but one must feel uneasy when, for example, writing textbooks on quantum mechanics, describing Process 1, for the consumption of other persons to whom it does not apply.

Alternative 2: To limit the applicability of quantum mechanics by asserting that the quantum mechanical description fails when applied to observers, or to measuring apparatus, or more generally to systems approaching macroscopic size.

If we try to limit the applicability so as to exclude measuring apparatus, or in general systems of macroscopic size, we are faced with the difficulty of sharply defining the region of validity. For what n might a group of n particles be construed as forming a measuring device so that the quantum description fails? And to draw the line at human or animal observers, i.e., to assume that all mechanical apparata obey the usual laws, but that they are somehow not valid for living observers, does violence to the so-called principle of psycho-physical parallelism,[2] and constitutes

---

[2] In the words of von Neumann ([J. von Neumann, *Mathematical Foundations of Quantum Mechanics*. (Translated by R. T. Beyer) Princeton University Press: 1955.], p. 418) : "...it is

a view to be avoided, if possible. To do justice to this principle we must insist that we be able to conceive of mechanical devices (such as servomechanisms), obeying natural laws, which we would be willing to call observers.

Alternative 3: To admit the validity of the state function description, but to deny the possibility that B could ever be in possession of the state function of $A + S$. Thus one might argue that a determination of the state of A would constitute such a drastic intervention that A would cease to function as an observer.

The first objection to this view is that no matter what the state of $A + S$ is, there is in principle a complete set of commuting operators for which it is an eigenstate, so that, at least, the determination of *these* quantities will not affect the state nor in any way disrupt the operation of A. There are no fundamental restrictions in the usual theory about the knowability of *any* state functions, and the introduction of any such restrictions to avoid the paradox must therefore require extra postulates.

The second objection is that it is not particularly relevant whether or not B actually *knows* the precise state function of $A + S$. If he merely *believes* that the system is described by a state function, which he does not presume to know, then the difficulty still exists. He must then believe that this state function changed deterministically, and hence that there was nothing probabilistic in A's determination.

Alternative 4: To abandon the position that the state function is a *complete* description of a system. The state function is to be regarded not as a description of a single system, but of an ensemble of systems, so that the probabilistic assertions arise naturally from the incompleteness of the description.[as]

It is assumed that the correct complete description, which would presumably involve further (hidden) parameters beyond the state function alone, would lead to a deterministic theory, from which the probabilistic aspects arise as a result of our ignorance of these extra parameters in the same manner as in classical statistical mechanics.

---

a fundamental requirement of the scientific viewpoint—the so-called principle of the psycho-physical parallelism—that it must be possible so to describe the extra-physical process of the subjective perception as if it were in reality in the physical world—*i.e.*, to assign to its parts equivalent physical processes in the objective environment, in ordinary space."

[as] This would be a hidden variable proposal like Bohmian mechanics. Everett has no quick argument against this proposal here. It is discussed later in this thesis (pg. 153). Everett's main objection is that, since pure wave mechanics works fine without such an assumption, postulating hidden variables is unnecessary.

Alternative 5: To assume the universal validity of the quantum description, by the complete abandonment of Process 1. The general validity of pure wave mechanics, *without any statistical assertions*, is assumed for *all* physical systems, including observers and measuring apparata. Observation processes are to be described completely by the state function of the composite system which includes the observer and his object-system, and which at all times obeys the wave equation (Process 2).

This brief list of alternatives is not meant to be exhaustive, but has been presented in the spirit of a preliminary orientation. We have, in fact, omitted one of the foremost interpretations of quantum theory, namely the position of Niels Bohr. The discussion will be resumed in the final chapter, when we shall be in a position to give a more adequate appraisal of the various alternate interpretations. For the present, however, we shall concern ourselves only with the development of Alternative 5.

It is evident that Alternative 5 is a theory of many advantages. It has the virtue of logical simplicity, and it is complete in the sense that it is applicable to the entire universe. All processes are considered equally (there are no "measurement processes" which play any preferred role), and the principle of psycho-physical parallelism is fully maintained. Since the universal validity of the state function description is asserted, one can regard the state functions themselves as the fundamental entities, and one can even consider the state function of the whole universe. In this sense this theory can be called the theory of the "universal wave function," since all of physics is presumed to follow from this function alone. There remains, however, the question whether or not such a theory can be put into correspondence with our experience.

The present thesis is devoted to showing that this concept of a universal wave mechanics, together with the necessary correlation machinery for its interpretation, forms a logically self consistent description of a universe in which several observers are at work.

We shall be able to introduce into the theory systems which represent observers. Such systems can be conceived as automatically functioning machines (servomechanisms) possessing recording devices (memory) and which are capable of responding to their environment. The behavior of these observers shall always be treated within the framework of wave mechanics. Furthermore, we shall deduce the probabilistic assertions of Process 1 as *subjective* appearances to such observers, thus placing the theory in correspondence with experience. We are then led to the novel situation in which the formal theory is objectively continuous and causal, while subjectively discontinuous and probabilistic. While this point of view thus shall ultimately justify our use of the statistical assertions of the orthodox view, it enables us to do so in a logically consistent manner, allowing for the existence of other observers. At the same time it gives a deeper insight into the meaning of quantized systems, and the role played by quantum mechanical correlations.

In order to bring about this correspondence with experience for the pure wave mechanical theory, we shall exploit the correlation between subsystems of a composite system which is described by a state function. A subsystem of such a composite system does not, in general, possess an independent state function. That is, in general a composite system cannot be represented by a single pair of subsystem states, but can be represented only by a *superposition* of such pairs of subsystem states. For example, the Schrödinger wave function for a pair of particles, $\psi(x_1, x_2)$, cannot always be written in the form $\psi = \phi(x_1)\eta(x_2)$, but only in the form $\psi = \sum_{i,j} a_{ij}\phi^i(x_1)\eta^j(x_2)$. In the latter case, there is no single state for Particle 1 alone or Particle 2 alone, but only the superposition of such cases.

In fact, to any arbitrary choice of state for one subsystem there will correspond a *relative state* for the other subsystem, which will generally be dependent upon the choice of state for the first subsystem, so that the state of one subsystem is not independent, but correlated to the state of the remaining subsystem. Such correlations between systems arise from interaction of the systems, and from our point of view all measurement and observation processes are to be regarded simply as interactions between observer and object-system which produce strong correlations.

Let one regard an observer as a subsystem of the composite system: observer + object-system. It is then an inescapable consequence that after the interaction has taken place there will not, generally, exist a single observer state. There will, however, be a superposition of the composite system states, each element of which contains a definite observer state and a definite relative object-system state. Furthermore, as we shall see, *each* of these relative object-system states will be, approximately, the eigenstates of the observation corresponding to the value obtained by the observer which is described by the same element of the superposition. Thus, each element of the resulting superposition describes an observer who perceived a definite and generally different result, and to whom it appears that the object-system state has been transformed into the corresponding eigenstate. In this sense the usual assertions of Process 1 appear to hold on a subjective level to each observer described by an element of the superposition. We shall also see that correlation plays an important role in preserving consistency when several observers are present and allowed to interact with one another (to "consult" one another) as well as with other object-systems.

In order to develop a language for interpreting our pure wave mechanics for composite systems we shall find it useful to develop quantitative definitions for such notions as the "sharpness" or "definiteness" of an operator A for a state ψ, and the "degree of correlation" between the subsystems of a composite system or between a pair of operators in the subsystems, so that we can use these concepts in an unambiguous manner. The mathematical development of these notions will be carried out in the next chapter (II) using some concepts borrowed from Information Theory.[3] We shall develop there the general definitions of information and correlation, as well as some of their more important properties. Throughout Chapter II we shall use the language of probability theory to facilitate the exposition, and because it enables us to introduce in a unified manner a number of concepts that will be of later use. We shall nevertheless subsequently apply the mathematical definitions directly to state functions, by replacing probabilities by square amplitudes, *without*, however, *making any reference to probability models.*

Having set the stage, so to speak, with Chapter II, we turn to quantum mechanics in Chapter III. There we first investigate the quantum formalism of composite systems, particularly the concept of relative state functions, and the meaning of the representation of subsystems by non-interfering mixtures of states characterized by density matrices. The notions of information and correlation are then applied to quantum mechanics. The final section of this chapter discusses the measurement process, which is regarded simply as a correlation-inducing interaction between subsystems of a single isolated system. A simple example of such a measurement is given and discussed, and some general consequences of the superposition principle are considered.

This will be followed by an abstract treatment of the problem of Observation (Chapter IV). In this chapter we make use only of the superposition principle, and general rules by which composite system states are formed of subsystem states, in order that our results shall have the greatest generality and be applicable to any form of quantum theory for which these principles hold. (Elsewhere, when giving examples, we restrict ourselves to the non-relativistic Schrödinger Theory for simplicity.) The validity of Process 1 as a subjective phenomenon is deduced, as well as the consistency of allowing several observers to interact with one another.

Chapter V supplements the abstract treatment of Chapter IV by discussing a number of diverse topics from the point of view of the theory of pure wave mechanics, including the existence and meaning of macroscopic objects in the light of their atomic constitution, amplification processes in measurement, questions of reversibility and irreversibility, and approximate measurement.

The final chapter summarizes the situation, and continues the discussion of alternate interpretations of quantum mechanics.

---

[3] The theory originated by Claude E. Shannon [C. E. Shannon, W. Weaver, *The Mathematical Theory of Communication*. University of Illinois Press: 1949].

II. PROBABILITY, INFORMATION, AND CORRELATION

The present chapter is devoted to the mathematical development of the concepts of information and correlation. As mentioned in the introduction we shall use the language of probability theory throughout this chapter to facilitate the exposition, although we shall apply the mathematical definitions and formulas in later chapters without reference to probability models. We shall develop our definitions and theorems in full generality, for probability distributions over arbitrary sets, rather than merely for distributions over real numbers, with which we are mainly interested at present. We take this course because it is as easy as the restricted development, and because it gives a better insight into the subject.[at]

The first three sections develop definitions and properties of information and correlation for probability distributions over *finite* sets only. In section four the definition of correlation is extended to distributions over arbitrary sets, and the general invariance of the correlation is proved. Section five then generalizes the definition of information to distributions over arbitrary sets. Finally, as illustrative examples, sections six and seven give brief applications to stochastic processes and classical mechanics, respectively.

§1. Finite joint distributions

We assume that we have a collection of finite sets, $\mathcal{X}, \mathcal{Y}, \ldots, \mathcal{Z}$, whose elements are denoted by $x_i \in \mathcal{X}$, $y_j \in \mathcal{Y}, \ldots, z_k \in \mathcal{Z}$, etc., and that we have a *joint probability distribution*, $P = P(x_i, y_j, \ldots, z_k)$, defined on the cartesian product of the sets, which represents the probability of the combined event $x_i, y_j, \ldots$, and z_k. We then denote by $X, Y, \ldots, Z$ the random variables whose values are the elements of the sets $\mathcal{X}, \mathcal{Y}, \ldots, \mathcal{Z}$, with probabilities given by P.

For any subset $Y, \ldots, Z$, of a set of random variables $W, \ldots, X, Y, \ldots, Z$, with joint probability distribution $P(w_i, \ldots, x_j, y_k, \ldots, z_l)$, the *marginal distribution*, $P(y_k, \ldots, z_l)$, is defined to be:

$$P(y_k, \ldots, z_l) = \sum_{i,\ldots,j} P(w_i, \ldots, x_j, y_k, \ldots, z_l), \qquad (1.1)$$

which represents the probability of the joint occurrence of $y_k, \ldots, z_l$, with no restrictions upon the remaining variables.

For any subset $Y, \ldots, Z$ of a set of random variables the *conditional distribution*, conditioned upon the values $W = w_i, \ldots, X = x_j$ for any remaining subset $W, \ldots, X$, and denoted by $P^{w_i, \ldots, x_j}(y_k, \ldots, z_l)$, is defined

---

[at] Although Everett took his discussion of information theory to be an important part of his project, most of the material from this chapter was cut in writing the short thesis.

to be:[1]

$$P^{w_i,\ldots,x_j}(y_k,\ldots,z_l) = \frac{P(w_i,\ldots,x_j,y_k,\ldots,z_l)}{P(w_i,\ldots,x_j)}, \tag{1.2}$$

which represents the probability of the joint event $Y = y_k,\ldots, Z = z_l$, conditioned by the fact that $W,\ldots, X$ are known to have taken the values $w_i,\ldots, x_j$, respectively.

For any numerical valued function $F(y_k,\ldots,z_l)$, defined on the elements of the cartesian product of $\mathcal{Y},\ldots,\mathcal{Z}$, the *expectation*, denoted by Exp[F], is defined to be:

$$\text{Exp}[F] = \sum_{k,\ldots,l} P(y_k,\ldots,z_l)F(y_k,\ldots,z_l). \tag{1.3}$$

We note that if $P(y_k,\ldots,z_l)$ is a marginal distribution of some larger distribution $P(w_i,\ldots,x_j,y_k,\ldots,z_l)$ then

$$\text{Exp}[F] = \sum_{k,\ldots,l}\left(\sum_{l,\ldots,j} P(w_i,\ldots,x_j,y_k,\ldots,z_l)\right)F(y_k,\ldots,z_l)$$

$$= \sum_{i,\ldots,j,k,\ldots,l} P(w_i,\ldots,x_j,y_k,\ldots,z_l)F(y_k,\ldots,z_l), \tag{1.4}$$

so that if we wish to compute Exp[F] with respect to some joint distribution it suffices to use *any* marginal distribution of the original distribution which contains at least those variables which occur in F.

We shall also occasionally be interested in *conditional expectations*, which we define as:

$$\text{Exp}^{w_i,\ldots,x_j}[F] = \sum_{k,\ldots,l} P^{w_i,\ldots,x_j}(y_k,\ldots,z_l)F(y_k,\ldots,z_l), \tag{1.5}$$

and we note the following easily verified rules for expectations:

$$\text{Exp}[\text{Exp}[F]] = \text{Exp}[F], \tag{1.6}$$

$$\text{Exp}^{u_i,\ldots,v_j}[\text{Exp}^{u_i,\ldots,v_j,w_k,\ldots,x_l}[F]] = \text{Exp}^{u_i,\ldots,v_j}[F], \tag{1.7}$$

$$\text{Exp}[F + G] = \text{Exp}[F] + \text{Exp}[G]. \tag{1.8}$$

---

[1] We regard it as undefined if $P(w_i,\ldots,x_j) = 0$. In this case $P(w_i,\ldots,x_j,y_k,\ldots,z_l)$ is necessarily zero also.

We should like finally to comment upon the notion of *independence*. Two random variables X and Y with joint distribution $P(x_i, y_j)$ will be said to be independent if and only if $P(x_i, y_j)$ is equal to $P(x_i)P(y_j)$ for all i, j. Similarly, the groups of random variables $(U \ldots V)$, $(W \ldots X)$, $\ldots, (Y \ldots Z)$ will be called *mutually independent groups* if and only if $P(u_i, \ldots, v_j, w_k, \ldots, x_l, \ldots, y_m, \ldots, z_n)$ is always equal to $P(u_i, \ldots, v_j)$ $P(w_k, \ldots, x_l) \ldots P(y_m, \ldots, z_n)$.

Independence means that the random variables take on values which are not influenced by the values of other variables with respect to which they are independent. That is, the conditional distribution of one of two independent variables, Y, conditioned upon the value x_i for the other, is independent of x_i, so that knowledge about one variable tells nothing of the other.

§2. Information for finite distributions

Suppose that we have a single random variable X, with distribution $P(x_i)$. We then define[2] a number, I_X, called the *information* of X, to be:

$$I_X = \sum_i P(x_i) \ln P(x_i) = \text{Exp}[\ln P(x_i)], \tag{2.1}$$

which is a function of the probabilities alone and not of any possible numerical values of the x_i's themselves.[3]

The information is essentially a measure of the sharpness of a probability distribution, that is, an inverse measure of its "spread." In this respect information plays a role similar to that of variance. However, it has a number of properties which make it a superior measure of the "sharpness" than the variance, not the least of which is the fact that it can be defined for distributions over arbitrary sets, while variance is defined only for distributions over real numbers.

Any change in the distribution $P(x_i)$ which "levels out" the probabilities decreases the information. It has the value zero for "perfectly sharp" distributions, in which probability is one for one of the x_i and zero for all others, and ranges downward to $-\ln n$ for distributions over n elements which are equal over all of the x_i. The fact that the information is nonpositive is no liability, since we are seldom interested in absolute information of a distribution, but only in differences.

---

[2] This definition corresponds to the negative of the *entropy* of a probability distribution as defined by Shannon [C. E. Shannon, W. Weaver, *The Mathematical Theory of Communication*. University of Illinois Press: 1949].

[3] A good discussion of information is to be found in Shannon [C. E. Shannon, W. Weaver, *The Mathematical Theory of Communication*. University of Illinois Press: 1949], or Woodward [P. M. Woodward, *Probability and Information Theory, with Applications to Radar*. McGraw-Hill, New York: 1953]. Note, however, that in the theory of communication one defines the information of a *state* x_i, which has *a priori* probability P_i, to be $-\ln P_i$. We prefer, however, to regard information as a property of the distribution itself.

We can generalize (2.1) to obtain the formula for the information of a *group* of random variables $X, Y, \ldots, Z$, with joint distribution $P(x_i, y_j, \ldots, z_k)$, which we denote by $I_{XY\ldots Z}$:

$$I_{XY\ldots Z} = \sum_{i,j,\ldots,k} P(x_i, y_j, \ldots, z_k) \ln P(x_i, y_j, \ldots, z_k) \qquad (2.2)$$
$$= \text{Exp}[\ln P(x_i, y_j, \ldots, z_k)],$$

which follows immediately from our previous definition, since the group of random variables $X, Y, \ldots, Z$ may be regarded as a single random variable W, which takes its values in the cartesian product $\mathcal{X} \times \mathcal{Y} \times \cdots \times \mathcal{Z}$.

Finally, we define a *conditional information*, $I_{XY\ldots Z}^{v_m,\ldots,w_n}$, to be:

$$I_{XY\ldots Z}^{v_m,\ldots,w_n} = \sum_{i,j,\ldots,k} P^{v_m,\ldots,w_n}(x_i, y_j, \ldots, z_k) \ln P^{v_m,\ldots,w_n}(x_i, y_j, \ldots, z_k)$$
$$= \text{Exp}^{v_m,\ldots,w_n}[\ln P^{v_m,\ldots,w_n}(x_i, y_j, \ldots, z_k)], \qquad (2.3)$$

a quantity which measures our information about $X, Y, \ldots, Z$ given that we know that $V \ldots W$ have taken the particular values $v_m, \ldots, w_n$.

For independent random variables $X, Y, \ldots, Z$, the following relationship is easily proved:

$$I_{XY\ldots Z} = I_X + I_Y + \ldots + I_Z \quad (X, Y, \ldots, Z \text{ independent}), \qquad (2.4)$$

so that the information of $XY \ldots Z$ is the sum of the individual quantities of information, which is in accord with our intuitive feeling that if we are given information about unrelated events, our total knowledge is the sum of the separate amounts of information. We shall generalize this definition later, in §5.

§3. Correlation for finite distributions

Suppose that we have a pair of random variables, X and Y, with joint distribution $P(x_i, y_j)$. If we say that X and Y are *correlated*, what we intuitively mean is that *one learns something about one variable when he is told the value of the other.*[au] Let us focus our attention upon the variable X. If we are not informed of the value of Y, then our information concerning X, I_X, is calculated from the marginal distribution $P(x_i)$. However, if we are now told that Y has the value y_j, then our information about X changes

---

[au] This intuition gives Everett a way to tie the notion of a correlation to the formal notion of information since correlations in this sense provide information one might use to infer one state from another. But Everett's use of standard epistemic language is misleading since the formal apparatus here refers to uninterpreted distributions and not to values. More specifically, information here is just a measure of the shape of a distribution. Similarly, correlation is just a measure of how the properties of two distributions covary.

to the information of the conditional distribution $P^{y_j}(x_i)$, $I_X^{y_j}$. According to what we have said, we wish the degree of correlation to measure how much we learn about X by being informed of Y's value. However, since the change of information, $I_X^{y_j} - I_X$, may depend upon the particular value, y_j, of Y which we are told, the natural thing to do to arrive at a single number to measure the strength of correlation is to consider the *expected* change in information about X, given that we are to be told the value of Y. This quantity we call the *correlation information*, or for brevity, the *correlation*, of X and Y, and denote it by $\{X, Y\}$. Thus:

$$\{X, Y\} = \text{Exp}[I_X^{y_j} - I_X] = \text{Exp}[I_X^{y_j}] - I_X. \tag{3.1}$$

Expanding the quantity $\text{Exp}[I_X^{y_j}]$ using (2.3) and the rules for expectations (1.6)–(1.8) we find:

$$\text{Exp}[I_X^{y_j}] = \text{Exp}[\text{Exp}^{y_j}[\ln P^{y_j}(x_i)]]$$
$$= \text{Exp}\left[\ln \frac{P(x_i, y_j)}{P(y_j)}\right] = \text{Exp}[\ln P(x_i, y_j)] - \text{Exp}[\ln P(y_j)]$$
$$= I_{XY} - I_Y, \tag{3.2}$$

and combining with (3.1) we have:

$$\{X, Y\} = I_{XY} - I_X - I_Y. \tag{3.3}$$

Thus the correlation is symmetric between X and Y, and hence also equal to the expected change of information about Y given that we will be told the value of X. Furthermore, according to (3.3) the correlation corresponds precisely to the amount of "missing information" if we possess only the marginal distributions, i.e., the loss of information if we choose to regard the variables as independent.

THEOREM 1. $\{X, Y\} = 0$ *if and only if X and Y are independent, and is otherwise strictly positive.* (Proof in Appendix I.)

In this respect the correlation so defined is superior to the usual correlation coefficients of statistics, such as covariance, etc., which can be zero even when the variables are not independent, and which can assume both positive and negative values. An inverse correlation is, after all, quite as useful as a direct correlation. Furthermore, it has the great advantage of depending upon the probabilities alone, and not upon any numerical values of x_i and y_j, so that it is defined for distributions over sets whose elements are of an arbitrary nature, and not only for distributions over numerical properties. For example, we might have a joint probability distribution for the political party and religious affiliation for individuals. Correlation and information are defined for such distributions, although they possess nothing like covariance or variance.

We can generalize (3.3) to define a *group correlation* for the groups of random variables $(U \dots V), (W \dots X), \dots, (Y \dots Z)$, denoted by $\{U \dots V, W \dots X, \dots, Y \dots Z\}$ (where the groups are separated by commas), to be:

$$\{U \dots V, W \dots X, \dots, Y \dots Z\} = I_{U \dots V W \dots X \dots Y \dots Z} \quad (3.4)$$
$$- I_{U \dots V} - I_{W \dots X} - \dots - I_{Y \dots Z},$$

again measuring the information deficiency for the group marginals. Theorem 1 is also satisfied by the group correlation, so that it is zero if and only if the groups are mutually independent. We can, of course, also define conditional correlations in the obvious manner, denoting these quantities by appending the conditional values as superscripts, as before.

We conclude this section by listing some useful formulas and inequalities which are easily proved:

$$\{U, V, \dots, W\} = \text{Exp}\left[\ln \frac{P(u_i, v_j, \dots, w_k)}{P(u_i)P(v_j) \dots P(w_k)}\right], \quad (3.5)$$

$$\{U, V, \dots, W\}^{x_i \dots y_j} = \text{Exp}^{x_i \dots y_j}\left[\ln \frac{P^{x_i \dots y_j}(u_k, v_l, \dots, w_m)}{P^{x_i \dots y_j}(u_k)P^{x_i \dots y_j}(v_l) \dots P^{x_i \dots y_j}(w_m)}\right]$$
$$\text{(conditional correlation)}, \quad (3.6)$$

$$\{\dots, U, V, \dots\} = \{\dots, UV, \dots\} + \{U, V\}, \quad (3.7)$$

$$\{\dots, U, V, \dots, W, \dots\} = \{\dots, UV \dots W, \dots\} + \{U, V, \dots, W\}$$

$$\text{(comma removal)} \quad (3.8)$$

$$\{\dots, U, VW, \dots\} - \{\dots, UV, W, \dots\} = \{U, V\} - \{V, W\}$$
$$\text{(commutator)},$$

$$\{X\} = 0 \quad \text{(definition of bracket with no commas)}, \quad (3.9)$$

$$\{\dots, XXV, \dots\} = \{\dots, XV, \dots\} \quad (3.10)$$
$$\text{(removal of repeated variable within a group)},$$

$$\{\dots, UV, VW, \dots\} = \{\dots, UV, W, \dots\} - \{V, W\} - I_V$$
$$\text{(removal of repeated variable in separate groups)}, \quad (3.11)$$

$$\{X, X\} = -I_X \quad \text{(self correlation)}, \tag{3.12}$$

$$\{U, VW, X\}^{\cdots w_j \cdots} = \{U, V, X\}^{\cdots w_j \cdots}, \tag{3.13}$$

$$\{U, W, X\}^{\cdots w_j \cdots} = \{U, X\}^{\cdots w_j \cdots} \quad \text{(removal of conditioned variables)}$$
$$\{XY, Z\} \geqq \{X, Z\}, \tag{3.14}$$

$$\{XY, Z\} \geqq \{X, Z\} + \{Y, Z\} - \{X, Y\}, \tag{3.15}$$

$$\{X, Y, Z\} \geqq \{X, Y\} + \{X, Z\}. \tag{3.16}$$

Note that in the above formulas any random variable W may be replaced by any group $XY \ldots Z$ and the relation holds true, since the set $XY \ldots Z$ may be regarded as the single random variable W, which takes its values in the cartesian product $\mathcal{X} \times \mathcal{Y} \times \ldots \times \mathcal{Z}$.

§4. Generalizations and further properties of correlation

Until now we have been concerned only with finite probability distributions, for which we have defined information and correlation. We shall now generalize the definition of correlation so as to be applicable to joint probability distributions over arbitrary sets of unrestricted cardinality.

We first consider the effects of refinement of a finite distribution. For example, we may discover that the event x_i is actually the disjunction of several exclusive events $\tilde{x}_i^1, \ldots, \tilde{x}_i^n$, so that x_i occurs if any one of the $\tilde{x}_i^\mu$ occurs, i.e., the single event x_i results from failing to distinguish between the $\tilde{x}_i^\mu$. The probability distribution which distinguishes between the $\tilde{x}_i^\mu$ will be called a *refinement* of the distribution which does not. In general, we shall say that a distribution $P' = P'(\tilde{x}_i^\mu, \ldots, \tilde{y}_j^\nu)$ is a refinement of $P = P(x_i, \ldots, y_j)$ if

$$P(x_i, \ldots, y_j) = \sum_{\mu \ldots \nu} P'(\tilde{x}_i^\mu, \ldots, \tilde{y}_j^\nu) \quad \text{(all } i, \ldots, j). \tag{4.1}$$

We now state an important theorem concerning the behavior of correlation under a refinement of a joint probability distribution:

THEOREM 2. *P' is a refinement of $P \Rightarrow \{X, \ldots, Y\}' \geqq \{X, \ldots, Y\}$ so that correlations never decrease upon refinement of a distribution.* (Proof in Appendix I, §3.)[av]

[av] The correlation measure is robust under refinement in the sense that it never decreases when one considers a distribution over a more fine-grained set of physical distinctions. Strong

As an example, suppose that we have a continuous probability density $P(x, y)$. By division of the axes into a finite number of intervals, $\bar{x}_i$, $\bar{y}_j$, we arrive at a finite joint distribution P_{ij}, by integration of $P(x, y)$ over the rectangle whose sides are the intervals $\bar{x}_i$ and $\bar{y}_j$, and which represents the probability that $X \in \bar{x}_i$ and $Y \in \bar{y}_j$. If we now subdivide the intervals, the new distribution P' will be a refinement of P, and by Theorem 2 the correlation $\{X, Y\}$ computed from P' will never be less than that computed from P. Theorem 2 is seen to be simply the mathematical verification of the intuitive notion that closer analysis of a situation in which quantities X and Y are dependent can never lessen the knowledge about Y which can be obtained from X.

This theorem allows us to give a general definition of correlation which will apply to joint distributions over completely arbitrary sets, i.e., for any probability measure[4] on an arbitrary product space, in the following manner:

Assume that we have a collection of arbitrary sets $\mathcal{X}, \mathcal{Y}, \ldots, \mathcal{Z}$, and a probability measure, $M_P(\mathcal{X} \times \mathcal{Y} \times \ldots \times \mathcal{Z})$, on their cartesian product. Let $\mathcal{P}^\mu$ be any finite partition of $\mathcal{X}$ into subsets $\mathcal{X}_i^\mu$, $\mathcal{Y}$ into subsets $\mathcal{Y}_j^\mu$, $\ldots$, and $\mathcal{Z}$ into subsets $\mathcal{Z}_k^\mu$, such that the sets $\mathcal{X}_i^\mu \times \mathcal{Y}_j^\mu \times \ldots \times \mathcal{Z}_k^\mu$ of the cartesian product are measurable in the probability measure M_P. Another partition $\mathcal{P}^\nu$ is a *refinement* of $\mathcal{P}^\mu$, $\mathcal{P}^\nu \subseteqq \mathcal{P}^\mu$, if $\mathcal{P}^\nu$ results from $\mathcal{P}^\mu$ by further subdivision of the subsets $\mathcal{X}_i^\mu, \mathcal{Y}_j^\mu, \ldots, \mathcal{Z}_k^\mu$. Each partition $\mathcal{P}^\mu$ results in a finite probability distribution, for which the correlation, $\{X, Y, \ldots, Z\}^{\mathcal{P}^\mu}$, is always defined through (3.3). Furthermore a refinement of a partition leads to a refinement of the probability distribution, so that by Theorem 2:

$$\mathcal{P}^\nu \subseteqq \mathcal{P}^\mu \Rightarrow \{X, Y, \ldots, Z\}^{\mathcal{P}^\nu} \geqq \{X, Y, \ldots, Z\}^{\mathcal{P}^\mu} \qquad (4.2)$$

Now the set of all partitions is partially ordered under the refinement relation. Moreover, because for any pair of partitions $\mathcal{P}, \mathcal{P}'$ there is always a third partition $\mathcal{P}''$ which is a refinement of both (common lower bound), the set of all partitions forms a *directed set*.[5] For a function, f, on a directed set, $\mathcal{S}$, one defines a directed set limit, lim f,: [*sic*]

DEFINITION. lim f exists and is equal to $a \Leftrightarrow$ for every $\varepsilon > 0$ there exists an $\alpha \in \mathcal{S}$ such that $|f(\beta) - a| < \varepsilon$ for every $\beta \in \mathcal{S}$ for which $\beta \leqq \alpha$.

---

[4] A measure is a non-negative, countably additive set function, defined on some subsets of a given set. It is a probability measure if the measure of the entire set is unity. See Halmos [P. R. Halmos, *Measure Theory*. Van Nostrand, New York: 1950].

[5] See Kelley [J. Kelley, *General Topology*. Van Nostrand, New York: 1955], p. 65.

---

correlations then remain strong under more precise physical specification. The monotonic behavior of the correlations under increasingly fine-grained partitions also supports Everett's ultimate definition of the general correlation between variables as the limit of the correlations between the variables under increasingly fine-grained partitions.

It is easily seen from the directed set property of common lower bounds that if this limit exists it is necessarily unique.

By (4.2) the correlation $\{X, Y, \ldots, Z\}^{\mathcal{P}}$ is a *monotone* function on the directed set of all partitions. Consequently the directed set limit, which we shall take as the basic definition of the correlation $\{X, Y, \ldots, Z\}$, *always exists*. (It may be infinite, but it is in every case well defined.) Thus:

DEFINITION. $\{X, Y, \ldots, Z\} = \lim\{X, Y, \ldots, Z\}^{\mathcal{P}}$,

and we have succeeded in our endeavor to give a completely general definition of correlation, applicable to all types of distributions.

It is an immediate consequence of (4.2) that this directed set limit is the supremum of $\{X, Y, \ldots, Z\}^{\mathcal{P}}$, so that:

$$\{X, Y, \ldots, Z\} = \sup_{\mathcal{P}}\{X, Y, \ldots, Z\}^{\mathcal{P}}, \qquad (4.3)$$

which we could equally well have taken as the definition.

Due to the fact that the correlation is defined as a limit for discrete distributions, Theorem 1 and all the relations (3.7) to (3.15), which contain only correlation brackets, remain true for arbitrary distributions. Only (3.11) and (3.12), which contain information terms, cannot be extended.

We can now prove an important theorem about correlation which concerns its invariant nature. Let $\mathcal{X}, \mathcal{Y}, \ldots, \mathcal{Z}$ be arbitrary sets with probability measure M_P on their cartesian product. Let f be any one-one mapping of $\mathcal{X}$ onto a set $\mathcal{U}$, g a one-one map of $\mathcal{Y}$ onto $\mathcal{V}, \ldots$, and h a map of $\mathcal{Z}$ onto $\mathcal{W}$. Then a joint probability distribution over $\mathcal{X} \times \mathcal{Y} \times \ldots \times \mathcal{Z}$ leads also to a one over $\mathcal{U} \times \mathcal{V} \times \ldots \times \mathcal{W}$ where the probability M'_P induced on the product $\mathcal{U} \times \mathcal{V} \times \ldots \times \mathcal{W}$ is simply the measure which assigns to each subset of $\mathcal{U} \times \mathcal{V} \times \ldots \times \mathcal{W}$ the measure which is the measure of its image set in $\mathcal{X} \times \mathcal{Y} \times \ldots \times \mathcal{Z}$ for the original measure M_P. (We have simply transformed to a new set of random variables: $U = f(X), V = g(Y), \ldots, W = h(Z)$.) Consider any partition $\mathcal{P}$ of $\mathcal{X}, \mathcal{Y}, \ldots, \mathcal{Z}$ into the subsets $\{\mathcal{X}_i\}, \{\mathcal{Y}_j\}, \ldots, \{\mathcal{Z}_k\}$ with probability distribution $P_{ij\ldots k} = M_P(\mathcal{X}_i \times \mathcal{Y}_j \times, \ldots, \times \mathcal{Z}_k)$. Then there is a corresponding partition $\mathcal{P}'$ of $\mathcal{U}, \mathcal{V}, \ldots, \mathcal{W}$ into the image sets of the sets of $\mathcal{P}$, $\{\mathcal{U}_i\}, \{\mathcal{V}_j\}, \ldots, \{\mathcal{W}_k\}$, where $\mathcal{U}_i = f(\mathcal{X}_i), \mathcal{V}_j = g(\mathcal{Y}_j), \ldots, \mathcal{W}_k = h(\mathcal{Z}_k)$. But the probability distribution for $\mathcal{P}'$ is the same as that for $\mathcal{P}$, since $P'_{ij\ldots k} = M'_P(\mathcal{U}_i \times \mathcal{V}_j \times \ldots \times \mathcal{W}_k) = M_P(\mathcal{X}_i \times \mathcal{Y}_j \times \ldots \times \mathcal{Z}_k) = P_{ij\ldots k}$, so that:

$$\{X, Y, \ldots, Z\}^{\mathcal{P}} = \{U, V, \ldots, W\}^{\mathcal{P}'} \qquad (4.4)$$

Due to the correspondence between the $\mathcal{P}$'s and $\mathcal{P}'$'s we have that:

$$\sup_{\mathcal{P}}\{X, Y, \ldots, Z\}^{\mathcal{P}} = \sup_{\mathcal{P}'}\{U, V, \ldots, W\}^{\mathcal{P}'}, \qquad (4.5)$$

and by virtue of (4.3) we have proved the following theorem:

THEOREM 3. $\{X, Y, \ldots, Z\} = \{U, V, \ldots, W\}$, *where* $U, V, \ldots, W$ *are any one-one images of* $X, Y, \ldots, Z$, *respectively. In other notation:* $\{X, Y, \ldots, Z\} = \{f(X), g(Y), \ldots, h(Z)\}$ *for all one-one functions* $f, g, \ldots, h$.

This means that changing variables to functionally related variables preserves the correlation. Again this is plausible on intuitive grounds, since a knowledge of $f(x)$ is just as good as knowledge of x, provided that f is one-one.

A special consequence of Theorem 3 is that for any continuous probability density $P(x, y)$ over real numbers the correlation between $f(x)$ and $g(y)$ is the same as between x and y, where f and g are any real valued one-one functions. As an example consider a probability distribution for the position of two particles, so that the random variables are the position coordinates. Theorem 3 then assures us that the position correlation is *independent of the coordinate system*, even if different coordinate systems are used for each particle! Also for a joint distribution for a pair of events in space-time the correlation is invariant to arbitrary space-time coordinate transformations, again even allowing different transformations for the coordinates of each event.

These examples illustrate clearly the *intrinsic nature* of the correlation of various groups for joint probability distributions, which is implied by its invariance against arbitrary (one-one) transformations of the random variables. These correlation quantities are thus fundamental properties of probability distributions. A correlation is an *absolute* rather than *relative* quantity, in the sense that the correlation between (numerical valued) random variables is completely independent of the scale of measurement chosen for the variables.

§5. Information for general distributions

Although we now have a definition of correlation applicable to all probability distributions, we have not yet extended the definition of information past finite distributions. In order to make this extension we first generalize the definition that we gave for discrete distributions to a definition of *relative* information for a random variable, relative to a given underlying measure, called the *information measure*, on the values of the random variable.

If we assign a measure to the set of values of a random variable, X, which is simply the assignment of a positive number a_i to each value x_i in the finite case, we define the information of a probability distribution $P(x_i)$ *relative* to this *information measure* to be:

$$I_X = \sum_i P(x_i) \ln \frac{P(x_i)}{a_i} = \mathrm{Exp}\left[\ln \frac{P(x_i)}{a_i}\right] \tag{5.1}$$

If we have a joint distribution of random variables $X, Y, \ldots, Z$, with information measures $\{a_i\}, \{b_j\}, \ldots, \{c_k\}$ on their values, then we define the total information relative to these measures to be:

$$
\begin{aligned}
I_{XY\ldots Z} &= \sum_{ij\ldots k} P(x_i, y_j, \ldots, z_k) \ln \frac{P(x_i, y_j, \ldots, z_k)}{a_i b_j \ldots c_k} \\
&= \mathrm{Exp}\left[\ln \frac{P(x_i, y_j, \ldots, z_k)}{a_i b_j \ldots c_k}\right],
\end{aligned}
\tag{5.2}
$$

so that the information measure on the cartesian product set is *always* taken to be the product measure of the individual information measures.

We shall now alter our previous position slightly and consider information as always being defined relative to some information measure, so that our previous definition of information is to be regarded as the information relative to the measure for which all the a_i's, b_j's, ... and c_k's are taken to be unity, which we shall henceforth call the *uniform* measure.[aw]

Let us now compute the correlation $\{X, Y, \ldots, Z\}'$ by (3.5) using the relative information:

$$
\begin{aligned}
\{X, Y, \ldots, Z\}' &= I'_{XY\ldots Z} - I'_X - I'_Y - \ldots - I'_Z \\
&= \mathrm{Exp}\left[\ln \frac{P(x_i, y_j, \ldots, z_k)}{a_i b_j \ldots c_k}\right] - \mathrm{Exp}\left[\ln \frac{P(x_i)}{a_i}\right] - \ldots \\
&\quad - \mathrm{Exp}\left[\ln \frac{P(s_k)}{c_k}\right] \\
&= \mathrm{Exp}\left[\ln \frac{P(x_i, y_j, \ldots, z_k)}{P(x_i) P(y_j) \ldots P(z_k)}\right] = \{X, Y, \ldots, Z\}, \quad (5.3)
\end{aligned}
$$

so that the correlation for discrete distributions, as defined by (3.5), is independent of the choice of information measure, and the correlation remains an absolute, not relative quantity. It can, however, be computed from the information relative to any information measure through (3.5).[ax]

If we consider refinements of our distributions, as before, and realize that such a refinement is also a refinement of the information measure, then we can prove a relation analogous to Theorem 2:

THEOREM 4. *The information of a distribution relative to a given information measure never decreases under refinement.* (Proof in Appendix I.)

[aw] This is a more general notion of *relative* information. Whereas the more general notion has the standard information measure as a special case (for finite distributions when the basic measure is uniform), it introduces a new parameter: *the basic measure*. Everett needs the more general notion to get an information measure for distributions over infinite sets that does not diverge on increasingly fine-grained partitions.

[ax] The claim here is that the more general notion of information introduced does not change the salient objective properties of the correlation measure.

Therefore, just as for correlation, we can define the information of a probability measure M_P on the cartesian product of arbitrary sets $\mathcal{X}, \mathcal{Y}, \ldots, \mathcal{Z}$, relative to the information measures $\mu_X, \mu_Y, \ldots, \mu_Z$, on the individual sets, by considering finite partitions $\mathcal{P}$ into subsets $\{\mathcal{X}_i\}, \{\mathcal{Y}_j\}, \ldots, \{\mathcal{Z}_k\}$, for which we take as the definition of the information:

$$I^{\mathcal{P}}_{XY\ldots Z} = \sum_{ij\ldots k} M_P(\mathcal{X}_i, \mathcal{Y}_j, \ldots, \mathcal{Z}_k) \ln \frac{M_P(\mathcal{X}_i, \mathcal{Y}_j, \ldots, \mathcal{Z}_k)}{\mu_X(\mathcal{X}_i)\mu_Y(\mathcal{Y}_j)\ldots\mu_Z(\mathcal{Z}_k)} \qquad (5.4)$$

Then $I^{\mathcal{P}}_{XY\ldots Z}$ is, as was $\{X, Y, \ldots, Z\}^P$, a monotone function upon the directed set of partitions (by Theorem 4), and as before we take the directed set limit for our definition:

$$I_{XY\ldots Z} = \lim I^{\mathcal{P}}_{XY\ldots Z} = \sup_{\mathcal{P}} I^{\mathcal{P}}_{XY\ldots Z}, \text{ }^{\text{ay}} \qquad (5.5)$$

which is then the information relative to the information measures $\mu_X, \mu_Y, \ldots, \mu_Z$.

Now, for functions f, g on a directed set the existence of $\lim f$ and $\lim g$ is a sufficient condition for the existence of $\lim(f + g)$, which is then $\lim f + \lim g$, provided that this is not indeterminate. Therefore:

THEOREM 5. $\{X, \ldots, Y\} = \lim\{X, \ldots, Y\}^P = \lim[I^{\mathcal{P}}_{X\ldots Y} - I^{\mathcal{P}}_X - \ldots - I^{\mathcal{P}}_Y] = I_{X\ldots Y} - I_X - \ldots - I_Y$, *where the information is taken relative to any information measure for which the expression is not indeterminate. It is sufficient for the validity of the above expression that the basic measures $\mu_X, \ldots, \mu_Y$ be such that none of the marginal informations $I_X \ldots I_Y$ shall be positively infinite.*

The latter statement holds since, because of the general relation $I_{X\ldots Y} \geqq I_X + \ldots + I_Y$, the determinateness of the expression is guaranteed so long as all of the $I_X, \ldots, I_Y$ are $< +\infty$.[az]

Henceforth, unless otherwise noted, we shall understand that information is to be computed with respect to the uniform measure for discrete distributions, and Lebesgue measure for continuous distributions over real numbers. In the case of a mixed distribution, with a continuous density $P(x, y, \ldots z)$ plus discrete "lumps" $P'(x_i, y_j, \ldots, z_k)$, we shall understand the information measure to be the uniform measure over the discrete range,

---

[ay] This is Everett's notion for distributions over infinite sets: The information of a distribution is the limit of information of distributions under increasingly fine-grained partitions.

[az] The information of a general distribution then is always relative to a set of basic measures on each of the variables. The idea is that there are choices of the basic measures for which the information for the distributions defined relative to the basic measures will be finite. In particular, Everett proposes using the uniform measure for distributions over finite sets and Lebesgue measure for distributions over infinite sets.

and the Lebesgue measure over the continuous range. These conventions then lead us to the expressions:

$$
I_{XY...Z} = \begin{cases}
\sum_{ij...k} P(x_i, y_j, \ldots, z_k) \ln P(x_i, y_j, \ldots, z_k) \} \text{ (discrete)} \\
\int P(x, y, \ldots, z) \ln P(x, y, \ldots, z) \, dxdy \ldots dz \} \text{ (cont.)} \\
\sum_{i...k} P(x_i, \ldots, z_k) \ln P(x_i, \ldots, z_k) \\
\quad + \int P(x, \ldots, z) \ln P(x, \ldots, z) \, dx \ldots dz
\end{cases} \Bigg\} \text{ (mixed)}
$$

(unless otherwise noted) \hfill (5.6)

The mixed case occurs often in quantum mechanics, for quantities which have both a discrete and continuous spectrum.

§6. Example: Information decay in stochastic processes

As an example illustrating the usefulness of the concept of relative information we shall consider briefly stochastic processes.[6] Suppose that we have a stationary Markov[7] process with a finite number of states S_i, and that the process occurs at discrete (integral) times $1, 2, \ldots, n, \ldots$, at which times the transition probability from the state S_i to the state S_j is T_{ij}. The probabilities T_{ij} then form what is called a *stochastic matrix*, i.e., the elements are between 0 and 1, and $\sum_i T_{ij}$ for all i. If at any time k the probability distribution over the states is $\{P_i^k\}$ then at the next time the probabilities will be $P_j^{k+1} = \sum_i P_i^k T_{ij}$.

In the special case where the matrix is *doubly-stochastic*, which means that $\sum_i T_{ij}$, as well as $\sum_j T_{ij}$, equals unity, and which amounts to a principle of detailed balancing holding, it is known that the entropy of a probability distribution over the states, defined as $H = -\sum_i P_i \ln P_i$, is a monotone increasing function of the time. This entropy is, however, simply the negative of the information relative to the uniform measure.

One can extend this result to more general stochastic processes only if one uses the more general definition of relative information. For an arbitrary stationary process the choice of an information measure which is stationary, i.e., for which

$$
a_j = \sum_i a_i T_{ij} \quad \text{(all } j\text{)} \tag{6.1}
$$

---

[6] See Feller [W. Feller, *An Introduction to Probability Theory and its Applications*. Wiley, New York: 1950], or Doob [J. L. Doob, *Stochastic Processes*. Wiley, New York: 1953].

[7] A Markov process is a stochastic process whose future development depends only upon its present state and not on its past history.

leads to the desired result. In this case the *relative* information,

$$I = \sum_i P_i \ln \frac{P_i}{a_i}, \tag{6.2}$$

is a monotone decreasing function of time and constitutes a suitable basis for the definition of the entropy $H = -I$. Note that this definition leads to the previous result for doubly-stochastic processes, since the uniform measure, $a_i = 1$ (all i), is obviously stationary in this case.

One can furthermore drop the requirement that the stochastic process be stationary, and even allow that there are completely different sets of states, $\{S_i^n\}$, at each time n, so that the process is now given by a sequence of matrices T_{ij}^n representing the transition probability at time n from state S_i^n to state S_j^{n+1}. In this case probability distributions change according to:

$$P_j^{n+1} = \sum_i P_i^n T_{ij}^n. \tag{6.3}$$

If we then choose *any* time-dependent information measure which satisfies the relations:

$$a_j^{n+1} = \sum a_i^n T_{ij}^n \quad (\text{all } j, n), \tag{6.4}$$

then the information of a probability distribution is again monotone decreasing with time. (Proof in Appendix I.)

All of these results are easily extended to the continuous case, and we see that the concept of relative information allows us to define entropy for quite general stochastic processes.

§7. Example: Conservation of information in classical mechanics

As a second illustrative example we consider briefly the classical mechanics of a group of particles. The system at any instant is represented by a point, $(x^1, y^1, z^1, p_x^1, p_y^1, p_z^1, \ldots, x^n, y^n, z^n, p_x^n, p_y^n, p_z^n)$ in the phase space of all position and momentum coordinates. The natural motion of the system then carries each point into another, defining a continuous transformation of the phase space into itself. According to Liouville's theorem the measure of a set of points of the phase space is invariant under this transformation.[8] This invariance of measure implies that if we begin with a probability distribution over the phase space, rather than a single point, the total information

$$I_{\text{total}} = I_{X^1 Y^1 Z^1 P_x^1 P_y^1 P_z^1 \ldots X^n Y^n Z^n P_x^n P_y^n P_z^n}, \tag{7.1}$$

[8] See Khinchin [A. I. Khinchin, *Mathematical Foundations of Statistical Mechanics*. (Translated by George Gamow) Dover, New York: 1949], p. 15.

which is the information of the *joint* distribution for all positions and momenta, remains *constant in time*.[ba]

In order to see that the total information is conserved, consider any partition $\mathcal{P}$ of the phase space at one time, t_0, with its information relative to the phase space measure, $I^{\mathcal{P}}(t_0)$. At a later time t_1 a partition $\mathcal{P}'$, into the image sets of $\mathcal{P}$ under the mapping of the space into itself, is induced, for which the probabilities for the sets of $\mathcal{P}'$ are the same as those of the corresponding sets of $\mathcal{P}$, and furthermore for which the measures are the same, by Liouville's theorem. Thus corresponding to each partition $\mathcal{P}$ at time t_0 with information $I^{\mathcal{P}}(t_0)$, there is a partition $\mathcal{P}'$ at time t_1 with information $I^{\mathcal{P}'}(t_1)$, which is the same:

$$I^{\mathcal{P}'}(t_1) = I^{\mathcal{P}}(t_0). \tag{7.2}$$

Due to the correspondence of the $\mathcal{P}$'s and $\mathcal{P}'$'s the supremums of each over all partitions must be equal, and by (5.5) we have proved that

$$I_{\text{total}}(t_1) = I_{\text{total}}(t_0), \tag{7.3}$$

and the total information is conserved.

Now it is known that the individual (marginal) position and momentum distributions tend to decay, except for rare fluctuations, into the uniform and Maxwellian distributions, respectively, for which the classical entropy is a maximum.[bb] This entropy is, however, except for the factor of Boltzmann's constant, simply the negative of the marginal information

$$I_{\text{marginal}} = I_{X^1} + I_{Y^1} + I_{Z^1} + \ldots + I_{P_x^n} + I_{P_y^n} + I_{P_z^n}, \tag{7.4}$$

which thus tends towards a minimum. But this decay of marginal information is exactly compensated by an increase of the total correlation information

$$\{\text{total}\} = I_{\text{total}} - I_{\text{marginal}}, \tag{7.5}$$

since the total information remains constant. Therefore, if one were to define the *total entropy* to be the negative of the total information, one could replace the usual second law of thermodynamics by a law of *conservation of total entropy*, where the increase in the standard (marginal) entropy is

---

[ba] That the total information is constant means that the information-theoretic entropy is also constant *for all classical systems*. Since one might expect the thermodynamic entropy typically to increase over time for macroscopic systems, this example illustrates why the information-theoretic entropy cannot be taken to determine the thermodynamic entropy.

[bb] Everett starts with this statement of the second law of thermodynamics as an empirically supported postulate. He then seeks to translate this statement of the second law into a form that is compatible with the fact that the total information-theoretic entropy is constant.

exactly compensated by a (negative) *correlation entropy*. The usual second law then results simply from our renunciation of all correlation knowledge (*Stosszahlansatz*) and not from any intrinsic behavior of classical systems. The situation for classical mechanics is thus in sharp contrast to that of stochastic processes, which are intrinsically irreversible.

III. Quantum Mechanics

Having mathematically formulated the ideas of information and correlation for probability distributions, we turn to the field of quantum mechanics. In this chapter we assume that the states of physical systems are represented by points in a Hilbert space, and that the time dependence of the state of an isolated system is governed by a linear wave equation.

It is well known that state functions lead to distributions over eigenvalues of Hermitian operators (square amplitudes of the expansion coefficients of the state in terms of the basis consisting of eigenfunctions of the operator) which have the mathematical properties of probability distributions (non-negative and normalized). The standard interpretation of quantum mechanics regards these distributions as actually giving the probabilities that the various eigenvalues of the operator will be observed, when a measurement represented by the operator is performed.

A feature of great importance to our interpretation is the fact that a state function of a *composite* system leads to *joint* distributions over subsystem quantities, rather than independent subsystem distributions, i.e., the quantities in different subsystems may be correlated with one another. The first section of this chapter is accordingly devoted to the development of the formalism of composite systems, and the connection of composite system states and their derived joint distributions with the various possible subsystem conditional and marginal distributions. We shall see that there exist *relative state functions* which correctly give the conditional distributions for all subsystem operators, while marginal distributions can *not* generally be represented by state functions, but only by *density matrices*.[bc]

In Section 2 the concepts of information and correlation, developed in the preceding chapter, are applied to quantum mechanics, by defining information and correlation for operators on systems with prescribed states. It is also shown that for composite systems there exists a quantity which can be thought of as the fundamental correlation between subsystems, and a closely related *canonical representation* of the composite system

---

[bc] The marginal distribution is the distribution of a proper subset of the total set of variables where one simply ignores any correlations outside the subset. Such distributions are typically not fully descriptive since one has thrown away correlation information.

state. In addition, a stronger form of the uncertainty principle, phrased in information language, is indicated.

The third section takes up the question of measurement in quantum mechanics, viewed as a correlation producing interaction between physical systems. A simple example of such a measurement is given and discussed. Finally some general consequences of the superposition principle are considered.

It is convenient at this point to introduce some notational conventions.[bd] We shall be concerned with points ψ in a Hilbert space $\mathcal{H}$, with scalar product (ψ_1, ψ_2). A *state* is a point ψ for which $(\psi, \psi) = 1$. For any linear operator A we define a functional, $\langle A \rangle \psi$, called the *expectation of A for ψ*, to be:

$$\langle A \rangle \psi = (\psi, A\psi).$$

A class of operators of particular interest is the class of *projection operators*. The operator $[\phi]$, called the projection on ϕ, is defined through:

$$[\phi]\psi = (\phi, \psi)\phi.$$

For a complete orthonormal set $\{\phi_i\}$ and a state ψ we define a *square-amplitude distribution*, P_i, called the distribution of ψ over $\{\phi_i\}$ through:

$$P_i = |(\phi_i, \psi)|^2 = \langle [\phi_i] \rangle \psi.$$

In the probabilistic interpretation this distribution represents the probability distribution over the results of a measurement with eigenstates ϕ_i, performed upon a system in the state ψ. (Hereafter when referring to the probabilistic interpretation we shall say briefly "the probability that the system will be found in ϕ_i", rather than the more cumbersome phrase "the probability that the measurement of a quantity B, with eigenfunctions $\{\phi_i\}$, shall yield the eigenvalue corresponding to ϕ_i," which is meant.)

For two Hilbert spaces $\mathcal{H}_1$ and $\mathcal{H}_2$, we form the *direct product* Hilbert space $\mathcal{H}_3 = \mathcal{H}_1 \otimes \mathcal{H}_2$ (tensor product) which is taken to be the space of all possible[1] sums of formal products of points of $\mathcal{H}_1$ and $\mathcal{H}_2$, i.e., the elements of $\mathcal{H}_3$ are those of the form $\sum_i a_i \xi_i \eta_i$ where $\xi_i \in \mathcal{H}_1$ and

---

[1] More rigorously, one considers only *finite* sums, then completes the resulting space to arrive at $\mathcal{H}_1 \otimes \mathcal{H}_2$.

[bd] What follows in the next few pages is an introduction of the basic notation and concepts of quantum mechanics. Everett favors the Schrödinger over the Heisenberg representation. Most of the orthodox quantum formalism is simply carried over into pure wave mechanics, but often with a different significance.

$\eta_i \in \mathcal{H}_2$. The scalar product in $\mathcal{H}_3$ is taken to be $(\sum_i a_i \xi_i \eta_i, \sum_j b_j \xi_j \eta_j) = \sum_{ij} a_i^* b_j (\xi_i, \xi_j)(\eta_i, \eta_j)$. It is then easily seen that if $\{\xi_i\}$ and $\{\eta_i\}$ form complete orthonormal sets in $\mathcal{H}_1$ and $\mathcal{H}_2$, respectively, then the set of all formal products $\{\xi_i \eta_j\}$ is a complete orthonormal set in $\mathcal{H}_3$. For any pair of operators A, B, in $\mathcal{H}_1$ and $\mathcal{H}_2$ there corresponds an operator $C = A \otimes B$, the direct product of A and B, in $\mathcal{H}_3$, which can be defined by its effect on the elements $\xi_i \eta_j$ of $\mathcal{H}_3$:

$$C\xi_i \eta_j = A \otimes B\xi_i \eta_j = (A\xi_i)(B\eta_j).$$

§1. Composite systems

It is well known that if the states of a pair of systems S_1 and S_2, are represented by points in Hilbert spaces $\mathcal{H}_1$ and $\mathcal{H}_2$, respectively, then the states of the *composite system* $S = S_1 + S_2$ (the two systems S_1 and S_2 regarded as a single system S) are represented correctly by points of the direct product $\mathcal{H}_1 \otimes \mathcal{H}_2$. This fact has far reaching consequences which we wish to investigate in some detail. Thus if $\{\xi_i\}$ is a complete orthonormal set for $\mathcal{H}_1$, and $\{\eta_j\}$ for $\mathcal{H}_2$, the general state of $S = S_1 + S_2$ has the form:

$$\psi^S = \sum_{ij} a_{ij} \xi_i \eta_j \qquad \left(\sum_{ij} a_{ij}^* a_{ij} = 1\right). \tag{1.1}$$

In this case we shall call $P_{ij} = a_{ij}^* a_{ij}$ the *joint square-amplitude distribution* of ψ^S over $\{\xi_i\}$ and $\{\eta_j\}$. In the standard probabilistic interpretation $a_{ij}^* a_{ij}$ represents the joint probability that S_1 will be found in the state ξ_i *and* S_2 will be found in the state η_j. Following the probabilistic model we now derive some distributions from the state ψ^S. Let A be a Hermitian operator in S_1 with eigenfunctions ϕ_i and eigenvalues λ_i, and B an operator in S_2 with eigenfunctions θ_j and eigenvalues μ_j. Then the joint distribution of ψ^S over $\{\phi_i\}$ and $\{\phi_j\}$,[be] P_{ij}, is:

$$P_{ij} = P(\phi_i \text{ and } \theta_j) = |(\phi_i \theta_j, \psi^S)|^2. \tag{1.2}$$

The *marginal* distributions of ψ^S over $\{\phi_i\}$ and of ψ^S over $\{\phi_j\}$ are:

$$P_i = P(\phi_i) = \sum_j P_{ij} = \sum_j |(\phi_i \theta_j, \psi^S)|^2,$$

$$P_j = P(\theta_j) = \sum_i P_{ij} = \sum_i |(\phi_i \theta_j, \psi^S)|^2, \tag{1.3}$$

[be] Everett alternates in the text between calling this eigenfunction ϕ_j and θ_j.

and the *conditional distributions* P_i^j and P_j^i are:

$$P_i^j = P(\phi_i \text{ conditioned on } \phi_j) = \frac{P_{ij}}{P_j},$$

$$P_j^i = P(\phi_j \text{ conditioned on } \phi_i) = \frac{P_{ij}}{P_i}. \tag{1.4}$$

We now define the *conditional expectation* of an operator A on S_1, conditioned on θ_j in S_2, denoted by $\text{Exp}^{\theta_j}[A]$, to be:

$$\begin{aligned}
\text{Exp}^{\theta_j}[A] &= \sum_i \lambda_i P_i^j = (1/P_j) \sum_i P_{ij} \lambda_i \\
&= (1/P_j) \sum_i \lambda_i |(\phi_i \theta_j, \psi^S)|^2 \\
&= (1/P_j) \sum_i |(\phi_i \theta_j, \psi^S)|^2 (\phi_i, A\phi_i),
\end{aligned} \tag{1.5}$$

and we define the *marginal expectation* of A on S_1 to be:

$$\text{Exp}[A] = \sum_i P_i \lambda_i = \sum_{ij} \lambda_i P_{ij} = \sum_{ij} |(\phi_i \theta_j, \psi^S)|^2 (\phi_i, A\phi_i) \tag{1.6}$$

We shall now introduce projection operators to get more convenient forms of the conditional and marginal expectations, which will also exhibit more clearly the degree of dependence of these quantities upon the chosen basis $\{\phi_i \theta_j\}$. Let the operators $[\phi_i]$ and $[\phi_j]$ be the projections on ϕ_i in S_1 and ϕ_j in S_2, respectively, and let I^1 and I^2 be the identity operators in S_1 and S_2. Then, making use of the identity $\psi^S = \sum_{ij}(\phi_i \theta_j, \psi^S)\phi_i \theta_j$ for any complete orthonormal set $\{\phi_i \theta_j\}$, we have:

$$\begin{aligned}
\langle [\phi_i][\theta_j] \rangle \psi^S &= (\psi^S, [\phi_i][\theta_j] \psi^S) \\
&= \left(\sum_{kl}(\phi_k \theta_l, \psi^S)\phi_k \theta_l, [\phi_i][\theta_j] \sum_{mn}(\phi_m \theta_n, \psi^S)\phi_m \theta_n \right) \\
&= \sum_{klmn}(\phi_k \theta_l, \psi^S)^* (\phi_m \theta_n, \psi^S)\delta_{km}\delta_{ln}\delta_{im}\delta_{jn} \\
&= (\phi_i \theta_j, \psi^S)^* (\phi_i \theta_j, \psi^S) = P_{ij},
\end{aligned} \tag{1.7}$$

so that the joint distribution is given simply by $\langle [\phi_i][\phi_j] \rangle \psi^S$.

For the marginal distributions we have:

$$\begin{aligned}
P_i &= \sum_j P_{ij} = \sum_j \langle [\phi_i][\theta_j] \rangle \psi^S \\
&= \left\langle [\phi_i]\left(\sum_j [\theta_i] \right) \right\rangle \psi^S = \langle [\phi_i]I^2 \rangle \psi^S,
\end{aligned} \tag{1.8}$$

and we see that the marginal distribution over the ϕ_i is *independent* of the set $\{\theta_j\}$ chosen in S_2. This result has the consequence in the ordinary interpretation that the expected outcome of measurement in one subsystem of a composite system is not influenced by the choice of quantity to be measured in the other subsystem. This expectation is, in fact, the expectation for the case in which no measurement at all (identity operator) is performed in the other subsystem. Thus no measurement in S_2 can affect the expected outcome of a measurement in S_1, *so long as the result of any S_2 measurement remains unknown.* The case is quite different, however, if this result *is* known, and we must turn to the conditional distributions and expectations in such a case.

We now introduce the concept of a *relative state-function*, which will play a central role in our interpretation of pure wave mechanics. Consider a composite system $S = S_1 + S_2$ in the state ψ^S. To every state η of S_2 we associate a state of S_1, ψ^η_{rel}, called the relative state in S_1 for η in S_2, through:

$$\text{DEFINITION. } \psi^\eta_{\mathrm{rel}} = N \sum_i (\phi_i \eta, \psi^S)\phi_i, \tag{1.9}$$

where $\{\phi_i\}$ is *any* complete orthonormal set in S_1 and N is a normalization constant.[2,bf]

The first property of ψ^η_{rel} is its uniqueness,[3] i.e., its dependence upon the choice of the basis $\{\phi_i\}$ is only apparent. To prove this, choose another basis $\{\xi_k\}$, with $\phi_i = \sum_k b_{ik}\xi_k$. Then $\sum_i b^*_{ij} b_{ik} = \delta_{jk}$, and:

$$\sum_i (\phi_i \eta, \psi^S)\phi_i = \sum_i \left(\sum_j b_{ij}\xi_j \eta, \psi^S \right) \left(\sum_k b_{ik}\xi_k \right)$$

$$= \sum_{jk} \left(\sum_i b^*_{ij} b_{ik} \right) (\xi_j \eta, \psi^S)\xi_k = \sum_{jk} \delta_{jk}(\xi_j \eta, \psi^S)\xi_k$$

$$= \sum_k (\xi_k \eta, \psi^S)\xi_k.$$

---

[2] In case $\sum_i (\phi_i \eta, \psi^S)\phi_i = 0$ (unnormalizable) then choose any function for the relative function. This ambiguity has no consequences of any importance to us. See in this connection the remarks on p. 165.

[3] Except if $\sum_i (\phi_i \eta, \psi^S)\phi_i = 0$. There is still, of course, no dependence upon the basis.

[bf] Relative states track the precise nature of the correlations between the subsystems of a composite system. For a countable dimensional space, the state ψ^η_{rel} of S_1 relative to state η of S_2 is the state one gets by expanding the total state of the composite system in a basis such that the state η occurs in precisely one term in the expansion and taking the rest of the renormalized term to be the relative state of S_1. See the conceptual introduction (pg. 34) for a discussion of how Everett understood relative states and the role they played in pure wave mechanics.

The second property of the relative state, which justifies its name, is that $\psi_{\text{rel}}^{\theta_j}$ correctly gives the *conditional expectations* of all operators in S_1, conditioned by the state θ_j in S_2. As before let A be an operator in S_1 with eigenstates ϕ_i and eigenvalues λ_i. Then:

$$\langle A \rangle \psi_{\text{rel}}^{\theta_j} = (\psi_{\text{rel}}^{\theta_j}, A\psi_{\text{rel}}^{\theta_j})$$

$$= \left(N \sum_i (\phi_i \theta_j, \psi^S) \phi_i, \, AN \sum_{im} (\phi_m \theta_j, \psi^S) \phi_m \right)$$

$$= N^2 \sum_{im} (\phi_i \theta_j, \psi^S)^* (\phi_m \theta_j, \psi^S) \lambda_m \delta_{im}$$

$$= N^2 \sum_i \lambda_i P_{ij}. \tag{1.10}$$

At this point the normalizer N^2 can be conveniently evaluated by using (1.10) to compute: $\langle I^1 \rangle \psi_{\text{rel}}^{\theta_j} = N^2 \sum_i 1 P_{ij} = N^2 P_j = 1$, so that

$$N^2 = 1/P_j. \tag{1.11}$$

Substitution of (1.11) in (1.10) yields:

$$\langle A \rangle \psi_{\text{rel}}^{\theta_j} = (1/P_j) \sum_i \lambda_i P_{ij} = \sum_i \lambda_i P_i^j = \text{Exp}^{\theta_j}[A], \tag{1.12}$$

and we see that the conditional expectations of operators are given by the relative states. (This includes, of course, the conditional distributions themselves, since they may be obtained as expectations of projection operators.)

An important representation of a composite system state ψ^S, in terms of an orthonormal set $\{\theta_j\}$ in one subsystem S_2 and the set of relative states $\{\psi_{\text{rel}}^{\theta_j}\}$ in S_1 is:

$$\psi^S = \sum_{ij} (\phi_i \theta_j, \psi^S) \phi_i \theta_j = \sum_j \left(\sum_i (\phi_i \theta_j, \psi^S) \phi_i \right) \theta_j$$

$$= \sum_j \frac{1}{N_j} \left[N_j \sum_i (\phi_i \theta_j, \psi^S) \phi_i \right] \theta_j$$

$$= \sum_j \frac{1}{N_j} \psi_{\text{rel}}^{\theta_j} \theta_j, \quad \text{where } 1/N_j^2 = P_j = \langle I^1[\theta_j] \rangle \psi^S. \tag{1.13}$$

Thus, for *any* orthonormal set in one subsystem, the state of the composite system is a single superposition of elements consisting of a state of the given set and its relative state in the other subsystem. (The relative states, however, are not necessarily orthogonal.) We notice further that a particular element, $\psi_{rel}^{\theta_j}\theta_j$, is quite independent of the choice of basis $\{\theta_k\}$, $k \neq j$, for the orthogonal space of θ_j, since $\psi_{rel}^{\theta_j}$ depends *only* on θ_j and not on the other θ_k for $k \neq j$. We remark at this point that the ambiguity in the relative state which arises when $\sum_i (\phi_i \theta_j, \psi^S)\phi_i = 0$ (see p. 163) is unimportant for this representation, since although *any* state $\psi_{rel}^{\theta_j}$ can be regarded as the relative state in this case, the term $\psi_{rel}^{\theta_j}\theta_j$ will occur in (1.13) with coefficient zero.

Now that we have found subsystem states which correctly give conditional expectations, we might inquire whether there exist subsystem states which give marginal expectations. The answer is, unfortunately, no. Let us compute the marginal expectation of A in S_1 using the representation (1.13):

$$\text{Exp}[A] = \langle AI^2 \rangle \psi^S = \left(\sum_j \frac{1}{N_j} \psi_{rel}^{\theta_j}\theta_j, \, AI^2 \sum_k \frac{1}{N_k} \psi_{rel}^{\theta_k}\theta_k \right)$$

$$= \sum_{jk} \frac{1}{N_j N_k} (\psi_{rel}^{\theta_j}, \, A\psi_{rel}^{\theta_j})\delta_{jk}$$

$$= \sum_j \frac{1}{N_j^2} (\psi_{rel}^{\theta_j}, \, A\psi_{rel}^{\theta_j}) = \sum_j P_j \langle A \rangle \psi_{rel}^{\theta_j}. \tag{1.14}$$

Now suppose that there exists a state in S_1, ψ', which correctly gives the marginal expectation (1.14) for *all* operators A (i.e., such that $\text{Exp}[A] = \langle A \rangle \psi'$ for all A). One such operator is $[\psi']$, the projection on ψ', for which $\langle [\psi'] \rangle \psi' = 1$. But, from (1.14) we have that $\text{Exp}[\psi'] = \sum_j P_j \langle \psi' \rangle \psi_{rel}^{\theta_j}$, which is < 1 unless, for all j, $P_j = 0$ or $\psi_{rel}^{\theta_j} = \psi'$, a condition which is not generally true. Therefore *there exists in general no state for S_1 which correctly gives the marginal expectations for all operators in S_1.*

However, even though there is generally no single state describing marginal expectations, we see that there is always a *mixture* of states, namely the states $\psi_{rel}^{\theta_j}$ *weighted* with P_j, which does yield the correct expectations. The distinction between a mixture, M, of states ϕ_i, weighted by P_i, and a *pure state* ψ which is a superposition, $\psi = \sum a_i \phi_i$, is that there are *no interference phenomena* between the various states of a mixture. The expectation of an operator A for the mixture is $\text{Exp}^M[A] = \sum_i P_i \langle A \rangle \phi_i = \sum_i P_i(\phi_i, A\phi_i)$, while the expectation for the pure state ψ is $\langle A \rangle \psi = (\sum_i a_i \phi_i, A\sum_j a_j \phi_j) = \sum_{ij} a_i^* a_j (\phi_i, A\phi_j)$, which is *not* the same as that of the mixture with weights $P_i = a_i^* a_i$, due to the presence of the interference terms $(\phi_i, A\phi_j)$ for $j \neq i$.

It is convenient to represent such a mixture by a *density matrix*,[4] ρ. If the mixture consists of the states ψ_j weighted by P_j, and if we are working in a

---

[4] Also called a *statistical operator* (von Neumann [J. von Neumann, *Mathematical Foundations of Quantum Mechanics*. (Translated by R. T. Beyer) Princeton University Press: 1955]).

basis consisting of the complete orthonormal set $\{\phi_i\}$, where $\psi_j = \sum_i a_i^j \phi_i$, then we define the elements of the density matrix for the mixture to be:

$$\rho_{kl} = \sum_j P_j a_l^{j*} a_k^j \qquad (a_i^j = (\phi_i, \psi_j)). \qquad (1.15)$$

Then if A is any operator, with matrix representation $A_{il} = (\phi_i, A\phi_l)$ in the chosen basis, its expectation for the mixture is:

$$\text{Exp}^M[A] = \sum_j P_j(\psi_j, A\psi_j) = \sum_j P_j \left[\sum_{il} a_i^{j*} a_l^j (\phi_i, A\phi_l) \right]$$

$$= \sum_{il} \left(\sum_j P_j a_i^{j*} a_l^j \right) (\phi_i, A\phi_l) = \sum_{i,l} \rho_{li} A_{il}$$

$$= \text{Trace } (\rho A). \qquad (1.16)$$

Therefore any mixture is adequately represented by a density matrix.[5] Note also that $\rho_{kl}^* = \rho_{lk}$, so that ρ is Hermitian.

Let us now find the density matrices ρ^1 and ρ^2 for the subsystems S_1 and S_2 of a system $S = S_1 + S_2$ in the state ψ^S. Furthermore, let us choose the orthonormal bases $\{\xi_i\}$ and $\{\eta_j\}$ in S_1 and S_2, respectively, and let A be an operator in S_1, B an operator in S_2. Then:

$$\text{Exp}[A] = \langle AI^2 \rangle \psi^S = \left(\sum_{ij} (\xi_i \eta_j, \psi^S) \xi_i \eta_j, \; AI \sum_{lm} (\xi_l \eta_m, \psi^S) \xi_l \eta_m \right)$$

$$= \sum_{ijlm} (\xi_i \eta_j, \psi^S)^* (\xi_l \eta_m, \psi^S)(\xi_i, A\xi_l)(\eta_j, \eta_m)$$

$$= \sum_{il} \left[\sum_j (\xi_i \eta_j, \psi^S)^* (\xi_l \eta_j, \psi^S) \right] (\xi_i, A\xi_l)$$

$$= \text{Trace } (\rho^1 A), \qquad (1.17)$$

where we have defined ρ^1 in the $\{\xi_i\}$ basis to be:

$$\rho_{li}^1 = \sum_j (\xi_i \eta_j, \psi^S)^* (\xi_l \eta_j, \psi^S). \qquad (1.18)$$

In a similar fashion we find that ρ^2 is given, in the $\{\eta_j\}$ basis, by:

$$\rho_{mn}^2 = \sum_i (\xi_i \eta_n, \psi^S)^* (\xi_i \eta_m, \psi^S). \qquad (1.19)$$

---

[5] A better, coordinate free representation of a mixture is in terms of the operator which the density matrix represents. For a mixture of states ψ_n (not necessarily orthogonal) with weights ρ_n, the density operator is $\rho = \sum_n \rho_n[\psi_n]$, where $[\psi_n]$ stands for the projection operator on ψ_n.

It can be easily shown that here again the dependence of ρ^1 upon the choice of basis $\{\eta_j\}$ in S_2, and of ρ^2 upon $\{\xi_i\}$, is only apparent.

In summary, we have seen in this section that a state of a composite system leads to *joint* distributions over subsystem quantities which are generally not independent. Conditional distributions and expectations for subsystems are obtained from *relative states*, and subsystem marginal distributions and expectations are given by *density matrices*.

There does not, in general, exist anything like a single state for one subsystem of a composite system. That is, subsystems do *not* possess states independent of the states of the remainder of the system, so that the subsystem states are generally *correlated*. One can arbitrarily choose a state for one subsystem and be led to the *relative state* for the other subsystem. Thus we are faced with a fundamental *relativity of states*, which is implied by the formalism of composite systems. It is meaningless to ask the absolute state of a subsystem—one can only ask the state relative to a given state of the remainder of the system.

§2. Information and correlation in quantum mechanics

We wish to be able to discuss information and correlation for Hermitian operators $A, B, \ldots$, with respect to a state function ψ. These quantities are to be computed, through the formulas of the preceding chapter, from the square amplitudes of the coefficients of the expansion of ψ in terms of the eigenstates of the operators.

We have already seen (p. 159) that a state ψ and an orthonormal basis $\{\phi_i\}$ leads to a square amplitude distribution of ψ over the set $\{\phi_i\}$:

$$P_i = |(\phi_i, \psi)|^2 = \langle[\phi_i]\rangle\psi, \tag{2.1}$$

so that we can define the *information of the basis* $\{\phi_i\}$ *for the state* ψ, $I_{\{\phi_i\}}(\psi)$, to be simply the information of this distribution relative to the uniform measure:

$$I_{\{\phi_i\}}(\psi) = \sum_i P_i \ln P_i = \sum_i |(\phi_i, \psi)|^2 \ln |(\phi_i, \psi)|^2.^{\text{bg}} \tag{2.2}$$

We define the *information of an operator* A, for the state ψ, $I_A(\psi)$, to be the information in the square amplitude distribution over its *eigenvalues*, i.e., the information of the probability distribution over the results of a

---

bg The more spread out over the elements of the basis, the lower the information of the basis for the state; the more focused, the higher the information. The information of the basis for the state is then a measure of how close the state is to being an element of the basis. The information of an operator for the state is defined similarly, but with respect to the eigenvectors associated with the operator.

determination of A which is prescribed in the probabilistic interpretation. For a *non-degenerate* operator A this distribution is the same as the distribution (2.1) over the eigenstates. But because the information is dependent only on the distribution, and not on numerical values, the information of the distribution over eigenvalues of A is precisely the information of the eigenbasis of A, $\{\phi_i\}$. Therefore:

$$I_A(\psi) = I_{\{\phi_i\}}(\psi) = \sum_i \langle[\phi_i]\rangle \psi \ln\langle[\phi_i]\rangle \psi \quad (A \text{ non-degenerate}). \qquad (2.3)$$

We see that for fixed ψ, the information of all non-degenerate operators having the same set of eigenstates is the same.

In the case of *degenerate* operators it will be convenient to take, as the definition of information, the information of the square amplitude distribution over the eigenvalues *relative* to the information measure which consists of the *multiplicity* of the eigenvalues, rather than the uniform measure. This definition preserves the choice of uniform measure over the *eigenstates*, in distinction to the eigenvalues. If ϕ_{ij} (j from 1 to m_i) are a complete orthonormal set of eigenstates for A', with distinct eigenvalues λ_i (degenerate with respect to j), then the multiplicity of the ith eigenvalue is m_i and the information $I_A, (\psi)^{\text{bh}}$ is defined to be:

$$I_{A'}(\psi) = \sum_i \left(\sum_j \langle[\phi_{ij}]\rangle \psi\right) \ln \frac{\sum_j \langle[\phi_{ij}]\rangle \psi}{m_i}. \qquad (2.4)$$

The usefulness of this definition lies in the fact that any operator A'' which distinguishes further between any of the degenerate states of A' leads to a refinement of the relative density, in the sense of Theorem 4, and consequently has equal or greater information. A non-degenerate operator thus represents the maximal refinement and possesses maximal information.

It is convenient to introduce a new notation for the projection operators which are *relevant* for a specified operator. As before let A have eigenfunctions ϕ_{ij} and distinct eigenvalues λ_i. Then define the projections A_i, the projections on the *eigenspaces* of different eigenvalues of A, to be:

$$A_i = \sum_{j=1}^{m_i} [\phi_{ij}]. \qquad (2.5)$$

To each such projection there is associated a number m_i, the multiplicity of the degeneracy, which is the dimension of the ith eigenspace. In this notation

---

bh Everett likely means $I_{A'}, (\psi)$.

the distribution over the eigenvalues of A for the state ψ, P_i, becomes simply:

$$P_i = P(\lambda_i) = \langle A_i \rangle \psi, \tag{2.6}$$

and the information, given by (2.4), becomes:

$$I_A = \sum_i \langle A_i \rangle \psi \ln \frac{\langle A_i \rangle \psi}{m_i}. \tag{2.7}$$

Similarly, for a pair of operators, A in S_1 and B in S_2, for the composite system $S = S_1 + S_2$ with state ψ^S, the *joint* distribution over eigenvalues is:

$$P_{ij} = P(\lambda_i, \mu_j) = \langle A_i B_j \rangle \psi^S, \tag{2.8}$$

and the marginal distributions are:

$$P_i = \sum_j P_{ij} = \left\langle A_i \left(\sum_j B_j \right) \right\rangle \psi^S = \langle A_i I^2 \rangle \psi^S,$$

$$P_j = \sum_i P_{ij} = \left\langle \left(\sum_i A_i \right) B_j \right\rangle \psi^S = \langle I^1 B_j \rangle \psi^S. \tag{2.9}$$

The *joint* information, I_{AB}, is given by:

$$I_{AB} = \sum_{ij} P_{ij} \ln \frac{P_{ij}}{m_i n_j} = \sum_{ij} \langle A_i B_j \rangle \psi^S \ln \frac{\langle A_i B_j \rangle \psi^S}{m_i n_j}, \tag{2.10}$$

where m_i and n_j are the multiplicities of the eigenvalues λ_i and μ_j. The marginal information quantities are given by:

$$I_A = \sum_i \langle A_i I^2 \rangle \psi^S \ln \frac{\langle A_i I^2 \rangle \psi^S}{m_i},$$

$$I_B = \sum_j \langle I^1 B_j \rangle \psi^S \ln \frac{\langle I^1 B_j \rangle \psi^S}{n_j}, \tag{2.11}$$

and finally the correlation, $\{A, B\}\psi^S$ is given by:

$$\{A, B\}\psi^S = \sum_{ij} P_{ij} \ln \frac{P_{ij}}{P_i P_j} = \sum_{ij} \langle A_i B_j \rangle \psi^S \ln \frac{\langle A_i B_j \rangle \psi^S}{\langle A_i I \rangle \psi^S \langle I B_j \rangle \psi^S}, \tag{2.12}$$

where we note that the expression does not involve the multiplicities, as do the information expressions, a circumstance which simply reflects the independence of correlation on any information measure. These expressions of course generalize trivially to distributions over more than two variables (composite systems of more than two subsystems).

In addition to the correlation of pairs of subsystem operators, given by (2.12), there always exists a unique quantity $\{S_1, S_2\}$, the *canonical correlation*, which has some special properties and may be regarded as the fundamental correlation between the two subsystems S_1 and S_2 of the composite system S. As we remarked earlier a density matrix is Hermitian, so that there is a representation in which it is diagonal.[6] In particular, for the decomposition of S (with state ψ^S) into S_1 and S_2, we can choose a representation in which both ρ^{S_1} and ρ^{S_2} are diagonal. (This choice is always possible because ρ^{S_1} is independent of the basis in S_2 and vice versa.) Such a representation will be called a *canonical representation*. This means that it is always possible to represent the state ψ^S by a *single* superposition:

$$\psi^S = \sum_i a_i \xi_i \eta_i, \tag{2.13}$$

where *both* the $\{\xi_i\}$ and the $\{\eta_i\}$ constitute orthonormal sets of states for S_1 and S_2, respectively.

To construct such a representation choose the basis $\{\eta_i\}$ for S_2 so that ρ^{S_2} is diagonal:

$$\rho_{ij}^{S_2} = \lambda_i \delta_{ij}, \tag{2.14}$$

and let the ξ_i be the *relative* states in S_1 for the η_i in S_2:

$$\xi_i = N_i \sum_j (\phi_j \eta_i, \psi^S) \phi_j \quad \text{(any basis } \{\phi_j\}). \tag{2.15}$$

Then, according to (1.13), ψ^S is represented in the form (2.13) where the $\{\eta_i\}$ are orthonormal by choice, and the $\{\xi_i\}$ are normal since they are relative states. We therefore need only show that the states $\{\xi_i\}$ are

---

[6] The density matrix of a subsystem always has a pure discrete spectrum, if the composite system is in a state. To see this we note that the choice of any orthonormal basis in S_2 leads to a discrete (i.e., denumerable) set of relative states in S_1. The density matrix in S_1 then represents *this* discrete mixture, $\psi_{rel}^{\theta_j}$ weighted by P_j. This means that the expectation of the identity, $\text{Exp}[I] = \sum_j P_j (\psi_{rel}^{\theta_j}, I\psi_{rel}^{\theta_j}) = \sum_j P_j = 1 = \text{Trace} (\rho I) = \text{Trace} (\rho)$. Therefore ρ has a finite trace and is a completely continuous operator, having necessarily a pure discrete spectrum. (See von Neumann [J. von Neumann, *Mathematical Foundations of Quantum Mechanics.* (Translated by R. T. Beyer) Princeton University Press: 1955], p. 89, footnote 115.)

orthogonal:

$$(\xi_j, \xi_k) = \left(N_j \sum_l (\phi_l \eta_j, \psi^S) \phi_l, N_k \sum_m (\phi_m \eta_k, \psi^S) \phi_m \right)$$

$$= \sum_{lm} N_j^* N_k (\phi_l \eta_j, \psi^S)^* (\phi_m \eta_k, \psi^S) \delta_{lm}$$

$$= N_j^* N_k \sum_l (\phi_l \eta_j, \psi^S)^* (\phi_l \eta_k, \psi^S)$$

$$= N_j^* N_k \rho_{kj}^{S_2} = N_j^* N_k \lambda_k \delta_{kj} = 0, \text{ for } j \neq k, \tag{2.16}$$

since we supposed ρ^{S_2} to be diagonal in this representation. We have therefore constructed a canonical representation (2.13).

The density matrix ρ^{S_1} is also automatically diagonal, by the choice of representation consisting of the basis in S_2 which makes ρ^{S_2} diagonal and the corresponding relative states in S_1. Since $\{\xi_i\}$ are orthonormal we have:

$$\rho^{S_1} = \sum_k (\xi_i \eta_k, \psi^S)^* (\xi_j \eta_k, \psi^S)$$

$$= \sum_k \left(\xi_i \eta_k, \sum_m a_m \xi_m \eta_m \right)^* \left(\xi_j \eta_k, \sum_l a_l \xi_l \eta_l \right)$$

$$= \sum_{klm} a_m^* a_l \delta_{im} \delta_{km} \delta_{jl} \delta_{kl} = \sum_k a_i^* a_j \delta_{ki} \delta_{kj}$$

$$= a_i^* a_i \delta_{ij} = P_i \delta_{ij}, \tag{2.17}$$

where $P_i = a_i^* a_i$ is the marginal distribution over the $\{\xi_i\}$. Similar computation shows that the elements of ρ^{S_2} are the *same*:

$$\rho_{kl}^{S_2} = a_k^* a_k \delta_{kl} = P_k \delta_{kl}. \tag{2.18}$$

Thus in the canonical representation both density matrices are diagonal and have the same elements, P_k, which give the marginal square amplitude distribution over both of the sets $\{\xi_i\}$ and $\{\eta_i\}$ forming the basis of the representation.

Now, any pair of operators, $\tilde{A}$ in S_1 and $\tilde{B}$ in S_2, which have as non-degenerate eigenfunctions the sets $\{\xi_i\}$ and $\{\eta_j\}$ (i.e., operators which define the canonical representation), are "perfectly" correlated in the sense that there is a one-one correspondence between their eigenvalues. The joint square amplitude distribution for eigenvalues λ_i of $\tilde{A}$ and μ_j of $\tilde{B}$ is:

$$P(\lambda_i \text{ and } \mu_j) = P(\xi_i \text{ and } \eta_j) = P_{ij} = a_i^* a_i \delta_{ij} = P_i \delta_{ij}. \tag{2.19}$$

Therefore, the correlation between these operators, $\{\tilde{A}, \tilde{B}\}\psi^S$ is:

$$\{\tilde{A}, \tilde{B}\}\psi^S = \sum_{ij} P(\lambda_i \text{ and } \mu_j) \ln \frac{P(\lambda_i \& \mu_j)}{P(\lambda_i)P(\mu_j)} = \sum_{ij} P_i \delta_{ij} \ln \frac{P_i \delta_{ij}}{P_i P_j}$$

$$= -\sum_i P_i \ln P_i. \qquad (2.20)$$

We shall denote this quantity by $\{S_1, S_2\}\psi^S$ and call it the *canonical correlation* of the subsystems S_1 and S_2 for the system state ψ^S. It is the correlation between any pair of non-degenerate subsystem operators which defines the canonical representation.

In the canonical representation, where the density matrices are diagonal ((2.17) and (2.18)), the canonical correlation is given by:

$$\{S_1, S_2\}\psi^S = -\sum_i P_i \ln P_i = -\text{Trace}\left(\rho^{S_1} \ln \rho^{S_1}\right)$$

$$= -\text{Trace}\left(\rho^{S_2} \ln \rho^{S_2}\right). \qquad (2.21)$$

But the trace is invariant for unitary transformations, so that (2.21) holds independently of the representation, and we have therefore established the *uniqueness* of $\{S_1, S_2\}\psi^S$.

It is also interesting to note that the quantity $-\text{Trace}\,(\rho \ln \rho)$ is (apart from a factor of Boltzmann's constant) just the *entropy* of a mixture of states characterized by the density matrix ρ.[7] Therefore the entropy of the mixture characteristic of a subsystem S_1 for the state $\psi^S = \psi^{S_1+S_2}$ is exactly matched by a correlation information $\{S_1, S_2\}$, which represents the correlation between any pair of operators $\tilde{A}, \tilde{B}$, which define the canonical representation. The situation is thus quite similar to that of classical mechanics.[8]

Another special property of the canonical representation is that any operators $\tilde{A}, \tilde{B}$ defining a canonical representation have *maximum marginal information*, in the sense that for any other discrete spectrum operators, A on S_1, B on S_2, $I_A \leq I_{\tilde{A}}$ and $I_B \leq I_{\tilde{B}}$. If the canonical representation is (2.13), with $\{\xi_i\}, \{\eta_i\}$ non-degenerate eigenfunctions of $\tilde{A}, \tilde{B}$, respectively, and A, B any pair of non-degenerate operators with eigenfunctions $\{\phi_k\}$ and $\{\theta_l\}$, where $\xi_i = \sum_k c_{ik}\phi_k$, $\eta_i = \sum_l d_{il}\theta_l$, then ψ^S in ϕ, θ representation is:

$$\psi^S = \sum_{ikl} a_i c_{ik} d_{il} \phi_k \theta_l = \sum_{kl} \left(\sum_i a_i c_{ik} d_{il}\right) \phi_k \theta_l, \qquad (2.22)$$

---

[7] See von Neumann [J. von Neumann, *Mathematical Foundations of Quantum Mechanics*. (Translated by R. T. Beyer) Princeton University Press: 1955], p. 296.

[8] Cf. Chapter II, §7.

and the joint square amplitude distribution for ϕ_k, θ_l is:

$$P_{kl} = \left| \left(\sum_i a_i c_{ik} d_{il} \right) \right|^2 = \sum_{im} a_i^* a_m c_{ik}^* c_{mk} d_{il}^* d_{ml}, \qquad (2.23)$$

while the marginals are:

$$P_k = \sum_l P_{kl} = \sum_{im} a_i^* a_m c_{ik}^* c_{mk} \sum_l d_{il}^* d_{ml}$$
$$= \sum_{im} a_i^* a_m c_{ik}^* c_{mk} \delta_{im} = \sum_i a_i^* a_i c_{ik}^* c_{ik}, \qquad (2.24)$$

and similarly

$$P_l = \sum_k P_{kl} = \sum_i a_i^* a_i d_{il}^* d_{il}. \qquad (2.25)$$

Then the marginal information I_A is:

$$I_A = \sum_k P_k \ln P_k = \sum_k \left(\sum_i a_i^* a_i c_{ik}^* c_{ik} \right) \ln \left(\sum_i a_i^* a_i c_{ik}^* c_{ik} \right)$$
$$= \sum_k \left(\sum_i a_i^* a_i T_{ik} \right) \ln \left(\sum_i a_i^* a_i T_{ik} \right), \qquad (2.26)$$

where $T_{ik} = c_{ik}^* c_{ik}$ is doubly-stochastic ($\sum_i T_{ik} = \sum_k T_{ik} = 1$ follows from the unitary nature of the c_{ik}). Therefore (by Corollary 2, §4, Appendix I):

$$I_A = \sum_k \left(\sum_i a_i^* a_i T_{ik} \right) \ln \left(\sum_i a_i^* a_i T_{ik} \right)$$
$$\leqq \sum_i a_i^* a_i \ln a_i^* a_i = I_{\tilde{A}}, \qquad (2.27)$$

and we have proved that $\tilde{A}$ has maximal marginal information among the discrete spectrum operators. Identical proof holds for $\tilde{B}$.

While this result was proved only for non-degenerate operators, it is immediately extended to the degenerate case, since as a consequence of our definition of information for a degenerate operator, (2.4), its information is still less that that of an operator which removes the degeneracy. We have thus proved:

THEOREM. $I_A \leqq I_{\tilde{A}}$, where $\tilde{A}$ is any non-degenerate operator defining the canonical representation, and A is any operator with discrete spectrum.

We conclude the discussion of the canonical representation by conjecturing that in addition to the maximum marginal information properties of

$\tilde{A}$, $\tilde{B}$, which define the representation, they are also *maximally correlated*, by which we mean that for any pair of operators C in S_1, D in S_2, $\{C, D\} \leqq \{\tilde{A}, \tilde{B}\}$, i.e.:

CONJECTURE.[9] $\{C, D\}\psi^S \leqq \{\tilde{A}, \tilde{B}\}\psi^S = \{S_1, S_2\}\psi^S$ (2.28)

for *all* C on S_1, D on S_2.

As a final topic for this section we point out that the uncertainty principle can probably be phrased in a stronger form in terms of information. The usual form of this principle is stated in terms of *variances*, namely:

$$\sigma_x^2 \sigma_k^2 \geqq \frac{1}{4} \quad \text{for all } \psi(x),$$ (2.29)

where $\sigma_x^2 = \langle x^2 \rangle \psi - [\langle x \rangle \psi]^2$ and

$$\sigma_k^2 = \left\langle \left(-i\frac{\partial}{\partial x}\right)^2 \right\rangle \psi - \left[\left\langle -i\frac{\partial}{\partial x} \right\rangle \psi\right]^2 = \left\langle \left(\frac{P}{\hbar}\right)^2 \right\rangle \psi - \left[\left\langle \frac{P}{\hbar} \right\rangle \psi\right]^2.$$

The conjectured information form of this principle is:

$$I_x + I_k \leqq \ln(1/\pi e) \quad \text{for all } \psi(x).$$ (2.30)

Although this inequality has not yet been proved with complete rigor, it is made highly probable by the circumstance that *equality* holds for $\psi(x)$ of the form $\psi(x) = (1/2\pi)^{\frac{1}{4}}$ exponent $[x^2/4\sigma_x^2]$ the so called "minimum uncertainty packets" which give normal distributions for both position and momentum, and that furthermore the first variation of $(I_x + I_k)$ vanishes for such $\psi(x)$. (See Appendix I, §6.) Thus, although $\ln(1/\pi e)$ has not been proved an absolute maximum of $I_x + I_k$, it is at least a stationary value.

The principle (2.30) is *stronger* than (2.29), since it implies (2.29) but is not implied by it. To see that it implies (2.29) we use the well-known fact (easily established by a variation calculation) that, for fixed variance σ^2, the distribution of minimum information is a normal distribution, which has information $I = \ln(1/\sigma\sqrt{2\pi e})$. This gives us the general inequality involving information and variance:

$$I \geq \ln(1/\sigma\sqrt{2\pi e}) \quad \text{(for all distributions).}$$ (2.31)

---

[9]The relations $\{C, \tilde{B}\} \leqq \{\tilde{A}, \tilde{B}\} = \{S_1, S_2\}$ and $\{\tilde{A}, D\} \leqq \{S_1, S_2\}$ for all C on S_1, D on S_2, can be proved easily in a manner analogous to (2.27). These do not, however, necessarily imply the general relation (2.29).

Substitution of (2.31) into (2.30) then yields:

$$\ln(1/\sigma_x \sqrt{2\pi e}) + \ln(1/\sigma_k \sqrt{2\pi e}) \leq I_x + I_k \leq \ln(1/\pi e)$$

$$\Rightarrow (1/\sigma_x \sigma_k 2\pi e) \leq (1/\pi e) \Rightarrow \sigma_x^2 \sigma_k^2 \geq \tfrac{1}{4}, \tag{2.32}$$

so that our principle implies the standard principle (2.29).

To show that (2.29) does *not imply* (2.30) it suffices to give a counterexample. The distributions $P(x) = \tfrac{1}{2}\delta(x) + \tfrac{1}{2}\delta(x - 10)$ and $P(k) = \tfrac{1}{2}\delta(k) + \tfrac{1}{2}\delta(k - 10)$, which consist simply of spikes at 0 and 10, clearly satisfy (2.29), while they both have infinite information and thus do *not* satisfy (2.30). Therefore it is possible to have arbitrarily high information about *both* x and k (or p) and still satisfy (2.13). We have, then, another illustration that information concepts are more powerful and more natural than the older measures based upon variance.

§3. Measurement

We now consider the question of measurement in quantum mechanics, which we desire to treat as a natural process within the theory of pure wave mechanics. From our point of view there is no fundamental distinction between "measuring apparata" and other physical systems. For us, therefore, a measurement is simply a special case of interaction between physical systems—an interaction which has the property of *correlating* a quantity in one subsystem with a quantity in another.

Nearly every interaction between systems produces *some* correlation however. Suppose that at some instant a pair of systems are independent, so that the composite system state function is a product of subsystem states ($\psi^S = \psi^{S_1} \psi^{S_2}$). Then this condition obviously holds only instantaneously if the systems are interacting[10]—the independence is immediately destroyed and the systems become correlated. We could, then, take the position that the two interacting systems are continually "measuring" one another, if we wished. At each instant t we could put the composite system into canonical representation, and choose a pair of operators $\tilde{A}(t)$ in S_1 and $\tilde{B}(t)$ in S_2 which define this representation. We might then reasonably assert that the quantity $\tilde{A}$ in S_1 is measured by $\tilde{B}$ in S_2 (or vice versa), since there is a one-one correspondence between their values.

Such a viewpoint, however, does not correspond closely with our intuitive idea of what constitutes "measurement," since the quantities $\tilde{A}$ and $\tilde{B}$ which turn out to be measured depend not only on the time, but also upon

---

[10] If U_t^S is the unitary operator generating the time dependence for the state function of the composite system $S = S_1 + S_2$, so that $\psi_t^S = U_t^S \psi_0^S$, then we shall say that S_1 and S_2 have not interacted during the time interval $[0, t]$ if and only if U_t^S is the direct product of two subsystem unitary operators, i.e., if $U_t^S = U_t^{S_1} \otimes U_t^{S_2}$.

the initial state of the composite system. A more reasonable position is to associate the term "measurement" with a fixed interaction H between systems,[11] and to define the "measured quantities" not as those quantities $\tilde{A}(t)$, $\tilde{B}(t)$ which are instantaneously canonically correlated, but as the limit of the instantaneous canonical operators as the time goes to infinity, $\tilde{A}_\infty$, $\tilde{B}_\infty$—provided that this limit exists and is independent of the initial state.[12] In such a case we are able to associate the "measured quantities," $\tilde{A}_\infty$, $\tilde{B}_\infty$, with the interaction H independently of the actual system states and the time. We can therefore say that H is an interaction which causes the quantity $\tilde{A}_\infty$ in S_1 to be measured by $\tilde{B}_\infty$ in S_2. For finite times of interaction the measurement is only approximate, approaching exactness as the time of interaction increases indefinitely.

There is still one more requirement that we must impose on an interaction before we shall call it a measurement. If H is to produce a measurement of A in S_1 by B in S_2, then we require that H shall never decrease the information in the marginal distribution of A. If H is to produce a measurement of A by correlating it with B, we expect that a knowledge of B shall give us more information about A than we had before the measurement took place, since otherwise the measurement would be useless. Now, H might produce a correlation between A and B by simply destroying the marginal information of A, without improving the expected conditional information of A given B, so that a knowledge of B would give us no more information about A than we possessed originally. Therefore in order to be sure that we will gain information about A by knowing B, when B has become correlated with A, it is necessary that the marginal information about A has not decreased. The expected information gain in this case is assured to be not less than the correlation $\{A, B\}$.

The restriction that H shall not decrease the marginal information of A has the interesting consequence that the eigenstates of A will not be disturbed, i.e., initial states of the form $\psi_0^S = \phi\eta_0$, where ϕ is an eigenfunction of A, must be transformed after any time interval into states of the form $\psi_t^S = \phi\eta_t$, since otherwise the marginal information of A, which was initially perfect, would be decreased. This condition, in turn, is connected with the *repeatability* of measurements, as we shall subsequently see, and could alternately have been chosen as the condition for measurement.

---

[11] Here H means the *total* Hamiltonian of S, not just an interaction part.

[12] Actually, rather than referring to canonical operators $\tilde{A}$, $\tilde{B}$, which are not unique, we should refer to the *bases* of the canonical representation, $\{\xi_i\}$ in S_1 and $\{\eta_j\}$ in S_2, since *any* operators $\tilde{A} = \sum_i \lambda_i[\xi_i]$, $\tilde{B} = \sum_j \mu_j[\eta_j]$, with the completely arbitrary eigenvalues λ_i, μ_j, are canonical. The limit then refers to the limit of the canonical bases, if it exists in some appropriate sense. However, we shall, for convenience, continue to represent the canonical bases by operators.

We shall therefore accept the following definition. An interaction H is a measurement of A in S_1 by B in S_2 if H does not destroy the marginal information of A (equivalently: if H does not disturb the eigenstates of A in the above sense) and if furthermore the correlation $\{A, B\}$ increases toward its maximum[13] with time.

We now illustrate the production of correlation with an example of a simplified measurement due to von Neumann.[14] Suppose that we have a system of only one coordinate, q (such as position of a particle), and an apparatus of one coordinate r (for example the position of a meter needle). Further suppose that they are initially independent, so that the combined wave function is $\psi_0^{S+A} = \phi(q)\eta(r)$, where $\phi(q)$ is the initial system wave function, and $\eta(r)$ is the initial apparatus function. Finally suppose that the masses are sufficiently large or the time of interaction sufficiently small that the kinetic portion of the energy may be neglected, so that during the time of measurement the Hamiltonian shall consist only of an interaction, which we shall take to be:

$$H_I = -i\hbar q \frac{\partial}{\partial t}. \tag{3.1}$$

Then it is easily verified that the state $\psi_t^{S+A}(q, r)$:

$$\psi_t^{S+A}(q, r) = \phi(q)\eta(r - qt) \tag{3.2}$$

is a solution of the Schrödinger equation

$$i\hbar \frac{\partial \psi_t^{S+A}}{\partial t} = H_I \psi_t^{S+A} \tag{3.3}$$

for the specified initial condition at time $t = 0$.

Translating (3.2) into square amplitudes we get:

$$P_t(q, r) = P_1(q) P_2(r - qt),$$

$$\text{where} \quad P_1(q) = \phi^*(q)\phi(q), \quad P_2(r) = \eta^*(t)\eta(r), \tag{3.4}$$

$$\text{and} \quad P_t(q, r) = \psi_t^{S+A*}(q, r)\psi_t^{S+A}(q, r),$$

and we note that for a fixed time, t, the conditional square amplitude distribution for r has been translated by an amount depending upon the value of q, while the marginal distribution for q has been unaltered. We see thus that a correlation has been introduced between q and r by this

[13] The maximum of $\{A, B\}$ is $-I_A$ if A has only a discrete spectrum, and ∞ if it has a continuous spectrum.

[14] von Neumann [J. von Neumann, *Mathematical Foundations of Quantum Mechanics.* (Translated by R. T. Beyer) Princeton University Press: 1955], p. 442.

interaction, which allows us to interpret it as a measurement. It is instructive to see quantitatively how fast this correlation takes place. We note that:

$$I_{QR}(t) = \int\int P_t(q,r) \ln P_t(q,r) \, dq \, dr \tag{3.5}$$

$$= \int\int P_1(q) P_2(r-qt) \ln P_1(q) P_2(r-qt) \, dq \, dr$$

$$= \int\int P_1(q) P_2(\omega) \ln P_1(q) P_2(\omega) \, dq \, d\omega$$

$$= I_{QR}(0),$$

so that the information of the joint distribution does not change. Furthermore, since the marginal distribution for q is unchanged:

$$I_Q(t) = I_Q(0), \tag{3.6}$$

and the only quantity which can change is the marginal information, I_R, of r, whose distribution is:

$$P_t(r) = \int P_t(r,q) \, dq = \int P_1(q) P_2(r-qt) \, dq. \tag{3.7}$$

Application of a special inequality (proved in §5, Appendix I) to (3.7) yields the relation:

$$I_R(t) \leqq I_Q(0) - \ln t, \tag{3.8}$$

so that, except for the additive constant $I_Q(0)$, the marginal information I_R tends to decrease at least as fast as $\ln t$ with time during the interaction. This implies the relation for the correlation:

$$\{Q,R\}_t = I_{QR}(t) - I_Q(t) - I_R(t) \geqq I_{RQ}(t) - I_Q(t) - I_Q(0) + \ln t. \tag{3.9}$$

But at $t = 0$ the distributions for R and Q were independent, so that $I_{RQ}(0) = I_R(0) + I_Q(0)$. Substitution of this relation, (3.5), and (3.6) into (3.9) then yields the final result:

$$\{Q,R\}_t \geqq I_R(0) - I_Q(0) + \ln t. \tag{3.10}$$

Therefore the correlation is built up at least as fast as $\ln t$, except for an additive constant representing the difference of the information of the initial distributions $P_2(r)$ and $P_1(q)$. Since the correlation goes to infinity with increasing time, and the marginal system distribution is not changed, the interaction (3.1) satisfies our definition of a measurement of q by r.

Even though the apparatus does not indicate any definite system value (since there are no independent system or apparatus states), one can nevertheless look upon the total wave function (3.2) as a *superposition* of pairs of subsystem states, each element of which has a definite q value and a correspondingly displaced apparatus state.[15,bi] Thus we can write (3.2) as:

$$\psi_t^{S+A} = \int \phi(q')\delta(q - q')\eta(r - q't)\,dq', \qquad (3.11)$$

which is a superposition of states $\psi_{q'} = \delta(q - q')\eta(r - q't)$. Each of these elements, $\psi_{q'}$, of the superposition describes a state in which the system has the definite value $q = q'$, and in which the apparatus has a state that is displaced from its original state by the amount $q't$. These elements $\psi_{q'}$ are then superposed with coefficients $\phi(q')$ to form the total state (3.11).

Conversely, if we transform to the representation where the *apparatus* is definite, we write (3.2) as:

$$\psi_t^{S+A} = \int (1/N_{r'})\xi^{r'}(q)\delta(r - r')\,dr', \qquad (3.12)$$

$$\text{where} \quad \xi^{r'}(q) = N_{r'}\phi(q)\eta(r' - qt)$$

$$\text{and} \quad (1/N_{r'})^2 = \int \phi^*(q)\phi(q)\eta^*(r' - qt)\eta(r - qt)\,dq.$$

Then the $\xi^{r'}(q)$ are the relative system state functions for the apparatus states $\delta(r - r')$ of definite value $r = r'$.

We notice that these relative system states, $\xi^{r'}(q)$, are nearly eigenstates for the values $q = r'/t$, if the degree of correlation between q and r is sufficiently high, i.e., if t is sufficiently large, or $\eta(t)$ sufficiently sharp, near $\delta(r)$, then $\xi^{r'}(q)$ is nearly $\delta(q - r'/t)$.

This property, that the relative system states become approximate eigenstates of the measurement, is in fact common to all measurements.[bi] If we adopt as a measure of the nearness of a state ψ to being an eigenfunction of an operator A the information $I_A(\psi)$, which is reasonable because $I_A(\psi)$ measures the sharpness of the distribution of A for ψ, then it is a consequence of our definition of a measurement that the relative system states tend to become eigenstates as the interaction proceeds. Since $\text{Exp}[I_Q^r] = I_Q + \{Q, R\}$, and I_Q remains constant while $\{Q, R\}$ tends toward

---

[15] See discussion of relative states, p. 99.

[bi] See the conceptual introduction, (chapter 3, pg. 34), for a discussion of Everett's notion of relative states.

[bj] For approximately correlating interactions, there are always relative system states that are approximate eigenstates of the measurement. Typically, in realistic interactions there also are relative states where neither the measurement pointer nor the observed quantity of the object system are approximately determinate and relative states where they are both approximately determinate but not well correlated. Consider, for example, the state of the pointer on the measuring device relative to the experimenter having tripped and fallen onto the measuring device while performing the measurement.

its maximum (or infinity) during the interaction, we have that $\text{Exp}[I_Q^r]$ tends to a maximum (or infinity). But I_Q^r is just the information in the relative system states, which we have adopted as a measure of the nearness to an eigenstate. Therefore, at least in expectation, the relative system states approach eigenstates.

We have seen that (3.13) is a superposition of states $\psi_{r'}$, *for each of which* the apparatus has recorded a definite value r', and the system is left in approximately the eigenstate of the measurement corresponding to $q = r'/t$. The discontinuous "jump" into an eigenstate is thus only a relative proposition, dependent upon our decomposition of the total wave function into the superposition, and relative to a particularly chosen apparatus value. So far as the complete theory is concerned all elements of the superposition exist simultaneously, and the entire process is quite continuous.

We have here only a special case of the following general principle which will hold for any situation which is treated entirely wave mechanically:

PRINCIPLE. *For any situation in which the existence of a property R_i for a subsystem S_1 of a composite system S will imply the later property Q_i for S, then it is also true that an initial state for S_1 of the form $\psi^{S_1} = \sum_i a_i \psi_{[R_i]}^{S_1}$ which is a superposition of states with the properties R_i, will result in a later state for S of the form $\psi^S = \sum_i a_i \psi_{[Q_i]}^S$, which is also a superposition, of states with the property Q_i. That is, for any arrangement of an interaction between two systems S_1 and S_2, which has the property that each initial state $\phi_i^{S_1} \psi^{S_2}$ will result in a final situation with total state $\psi_i^{S_1+S_2}$, an initial state of S_1 of the form $\sum_i a_i \phi_i^{S_1}$ will lead, after interaction, to the superposition $\sum_i a_i \psi_i^{S_1+S_2}$ for the whole system.*

This follows immediately from the superposition principle for solutions of a linear wave equation. It therefore holds for *any* system of quantum mechanics for which the superposition principle holds, both particle and field theories, relativistic or not, and is applicable to all physical systems, regardless of size.

This principle has the far-reaching implication that for any possible measurement for which the initial system state is not an eigenstate, the resulting state of the composite system leads to *no* definite system state nor any definite apparatus state. The system will not be put into one or another of its eigenstates with the apparatus indicating the corresponding value, and nothing resembling Process 1 can take place.

To see that this is indeed the case, suppose that we have a measuring arrangement with the following properties. The initial apparatus state is ψ_0^A. If the system is initially in an eigenstate of the measurement, ϕ_i^S, then after a specified time of interaction the total state $\phi_i^S \psi_0^A$ will be transformed into a state $\phi_i^S \psi_i^A$, i.e., the system eigenstate shall not be disturbed, and the apparatus state is changed to ψ_i^A, which is different for each ϕ_i^S. (ψ_i^A

may for example be a state describing the apparatus as indicating, by the position of a meter needle, the eigenvalue of ϕ_i^S.) However, if the initial system state is *not an eigenstate* but a superposition $\sum_i a_i \phi_i^S$, then the final composite system state is *also a superposition*, $\sum_i a_i \phi_i^S \psi_i^\lambda$. This follows from the superposition principle since all we need do is superpose our solutions for the eigenstates, $\phi_i^S \psi_0^A \rightarrow \phi_i^S \psi_i^A$, to arrive at the solution, $\sum_i a_i \phi_i^S \psi_0^A \rightarrow \sum_i a_i \phi_i^S \psi_i^A$, for the general case. Thus in general after a measurement has been performed there will be no definite system state nor any definite apparatus state, even though there is a correlation. It seems as though nothing can ever be settled by such a measurement. Furthermore this result is independent of the *size* of the apparatus and remains true for apparatus of quite macroscopic dimensions.

Suppose, for example, that we coupled a spin measuring device to a cannonball, so that if the spin is up the cannonball will be shifted one foot to the left, while if the spin is down it will be shifted an equal distance to the right. If we now perform a measurement with this arrangement upon a particle whose spin is a superposition of up and down, then the resulting total state will also be a superposition of two states, one in which the cannonball is to the left, and one in which it is to the right. There is no definite position for our macroscopic cannonball!

This behavior seems to be quite at variance with our observations, since macroscopic objects always appear to us to have definite positions. Can we reconcile this prediction of the purely wave mechanical theory with experience, or must we abandon it as untenable? In order to answer this question we must consider the problem of observation itself within the framework of the theory.[bk]

IV. OBSERVATION

We shall now give an abstract treatment of the problem of observation. In keeping with the spirit of our investigation of the consequences of pure wave mechanics we have no alternative but to introduce observers, considered as purely physical systems, into the theory.

We saw in the last chapter that in general a measurement (coupling of system and apparatus) had the outcome that neither the system nor the apparatus had any definite state after the interaction—a result seemingly at variance with our experience. However, we do not do justice to the theory

---

[bk] The prediction of pure wave mechanics is entirely straightforward. The question is whether one can reconcile the entangled state prediction with experience. On the face of it, this is simply a question of empirical adequacy. The goal is to show how to deduce the standard predictions of quantum mechanics for observers in pure wave mechanics or, more specifically, to show how to find the determinate measurement records we take ourselves to have in the structure described by pure wave mechanics. See the discussion of empirical adequacy in the second appendix (pg. 168).

of pure wave mechanics until we have investigated what the theory itself says about the *appearance* of phenomena to observers, rather than hastily concluding that the theory must be incorrect because the actual states of systems as given by the theory seem to contradict our observations.

We shall see that the introduction of observers can be accomplished in a reasonable manner and that the theory then predicts that the *appearance* of phenomena, as the subjective experience of these observers, is precisely in accordance with the predictions of the usual probabilistic interpretation of quantum mechanics.

§1. Formulation of the problem

We are faced with the task of making deductions about the appearance of phenomena on a subjective level, to observers which are considered as purely physical systems and are treated within the theory. In order to accomplish this it is necessary to identify some objective properties of such an observer (states) with the subjective knowledge (i.e., perceptions). Thus, in order to say that an observer O has observed the event α, it is necessary that the state of O has become changed from its former state to a new state which is dependent upon α.

It will suffice for our purposes to consider our observers to possess memories (i.e., parts of a relatively permanent nature whose states are in correspondence with the past experience of the observer). In order to make deductions about the subjective experience of an observer it is sufficient to examine the contents of the memory.

As models for observers we can, if we wish, consider automatically functioning machines, possessing sensory apparata and coupled to recording devices capable of registering past sensory data and machine configurations. We can further suppose that the machine is so constructed that its present actions shall be determined not only by its present sensory data, but by the contents of its memory as well. Such a machine will then be capable of performing a sequence of observations (measurements) and furthermore of deciding upon its future experiments on the basis of past results. We note that if we consider that current sensory data, as well as machine configuration, is immediately recorded in the memory, then the actions of the machine at a given instant can be regarded as a function of the memory contents only, and all relevant experience of the machine is contained in the memory.

For such machines we are justified in using such phrases as "the machine has perceived A" or "the machine is aware of A" if the occurrence of A is represented in the memory, since the future behavior of the machine will be based upon the occurrence of A. In fact, all of the customary language of subjective experience is quite applicable to such machines and forms the most natural and useful mode of expression when dealing with their behavior, as is well known to individuals who work with complex automata.

When dealing quantum mechanically with a system representing an observer we shall ascribe a state function, ψ^O, to it. When the state ψ^O describes an observer whose memory contains representations of the events $A, B, \ldots, C$ we shall denote this fact by appending the memory sequence in brackets as a subscript, writing:

$$\psi^O_{[A, B, \ldots, C]}.$$

The symbols $A, B, \ldots, C$, which we shall assume to be ordered time wise, shall therefore stand for memory configurations which are in correspondence with the past experience of the observer. These configurations can be thought of as punches in a paper tape, impressions on a magnetic reel, configurations of a relay switching circuit, or even configurations of brain cells. We only require that they be capable of the interpretation "The observer has experienced the succession of events $A, B, \ldots, C$." (We shall sometimes write dots in a memory sequence, $[\ldots A, B, \ldots, C]$, to indicate the possible presence of previous memories which are irrelevant to the case being considered.)

Our problem is, then, to treat the interaction of such observer-systems with other physical systems (observations), within the framework of wave mechanics, and to deduce the resulting memory configurations, which we can then interpret as the subjective experiences of the observers.

We begin by defining what shall constitute a "good" observation. A good observation of a quantity A, with eigenfunctions $\{\phi_i\}$ for a system S, by an observer whose initial state is $\psi^O_{[\ldots]}$, shall consist of an interaction which, in a specified period of time, transforms each (total) state

$$\psi^{S+O} = \phi_i \psi^O_{[\ldots]}$$

into a new state

$$\psi^{S+O'} = \phi_i \psi^O_{i[\ldots, \alpha_i]},$$

where α_i characterizes the state ϕ_i. (It might stand for a recording of the eigenvalue, for example.) That is, our requirement is that the system state, *if it is an eigenstate*, shall be unchanged, and that the observer state shall change so as to describe an observer that is "aware" of which eigenfunction it is, i.e., some property is recorded in the memory of the observer which characterizes ϕ_i, such as the eigenvalue. The requirement that the eigenstates for the system be unchanged is necessary if the observation is to be significant (repeatable), and the requirement that the observer state change in a manner which is different for each eigenfunction is necessary if we are to be able to call the interaction an observation at all.

§2. Deductions[bl]

From these requirements we shall first deduce the result of an observation upon a system which is *not* in an eigenstate of the observation. We know, by our previous remark upon what constitutes a good observation that the interaction transforms states $\phi_i \psi^O_{[\ldots]}$ into states $\phi_i \psi^O_{i[\ldots,\alpha_i]}$. Consequently we can simply superpose these solutions of the wave equation to arrive at the final state for the case of an arbitrary initial system state. Thus if the initial system state is not an eigenstate, but a general state $\sum_i a_i \phi_i$, we get for the final total state:

$$\psi^{S+O} = \sum_i a_i \phi_i \psi^O_{i[\ldots,\alpha_i]}. \tag{2.1}$$

This remains true also in the presence of further systems which do not interact for the time of measurement. Thus, if systems $S_1, S_2, \ldots, S_n$ are present as well as O, with original states $\psi^{S_1}, \psi^{S_2}, \ldots, \psi^{S_n}$, and the only interaction during the time of measurement is between S_1 and O, the result of the measurement will be the transformation of the initial total state:

$$\psi^{S_1+S_2+\ldots+S_n+O} = \psi^{S_1} \psi^{S_2} \ldots \psi^{S_n} \psi^O_{[\ldots]}$$

into the final state:

$$\psi'^{S_1+S_2+\ldots+S_n+O} = \sum_i a_i \phi_i^{S_1} \psi^{S_2} \ldots \psi^{S_n} \psi^O_{i[\ldots,\alpha_i]} \tag{2.2}$$

where $a_i = (\phi_i^{S_1}, \psi^{S_1})$ and $\phi_i^{S_1}$ are eigenfunctions of the observation.

Thus we arrive at the general rule for the transformation of total state functions which describe systems within which observation processes occur:

Rule 1. The observation of a quantity A, with eigenfunctions $\phi_i^{S_1}$, in a system S_1 by the observer O, transforms the total state according to:

$$\psi^{S_1} \psi^{S_2} \ldots \psi^{S_n} \psi^O_{[\ldots]} \rightarrow \sum_i a_i \phi_i^{S_1} \psi^{S_2} \ldots \psi^{S_n} \psi^O_{i[\ldots,\alpha_i]},$$

where $a_i = (\phi_i^{S_1}, \psi^{S_1})$.

If we next consider a *second* observation to be made, where our total state is now a superposition, we can apply *Rule* 1 separately to each element of

---

[bl] Each of the following deductions identifies a property of the correlation structure of pure wave mechanics. These properties follow from the definition of a good measurement and the linearity of the dynamics. The significance of the properties for Everett's project, however, depends on how one interprets Everett. On the bare theory, as a simple case, each of these properties describes the surefire dispositions of observers within pure wave mechanics. See the discussion of the bare theory in (Barrett, 1999).

the superposition, since each element separately obeys the wave equation and behaves independently of the remaining elements, and then superpose the results to obtain the final solution. We formulate this as:

Rule 2. Rule 1 may be applied separately to each element of a superposition of total system states, the results being superposed to obtain the final total state. Thus, a determination of B, with eigenfunctions $\eta_j^{S_2}$, on S_2 by the observer O transforms the total state

$$\sum_i a_i \phi_i^{S_1} \psi^{S_2} \dots \psi^{S_n} \psi_{i[\dots,\alpha_i]}^O$$

into the state

$$\sum_{i,j} a_i b_j \phi_i^{S_1} \eta_j^{S_2} \psi^{S_3} \dots \psi^{S_n} \psi_{ij[\dots,\alpha_i,\beta_j]}^O$$

where $b_j = (\eta_j^{S_2}, \psi^{S_2})$, which follows from the application of *Rule* 1 to each element $\phi_i^{S_1} \psi^{S_2} \dots \psi^{S_n} \psi_{i[\dots,\alpha_i]}^O$, and then superposing the results with the coefficients a_i.

These two rules, which follow directly from the superposition principle, give us a convenient method for determining final total states for any number of observation processes in any combinations. We must now seek the interpretation of such final total states.

Let us consider the simple case of a single observation of a quantity A, with eigenfunctions ϕ_i, in the system S with initial state ψ^S, by an observer O whose initial state is $\psi_{[\dots]}^O$. The final result is, as we have seen, the superposition:

$$\psi'^{S+O} = \sum_i a_i \phi_i \psi_{i[\dots,\alpha_i]}^O. \tag{2.3}$$

We note that there is no longer any independent system state or observer state, although the two have become correlated in a one-one manner. However, in each *element* of the superposition (2.3), $\phi_i \psi_{i[\dots,\alpha_i]}^O$, the object-system state is a particular eigenstate of the observer, and *furthermore the observer-system state describes the observer as definitely perceiving that particular system state.*[1] It is this correlation which allows one to maintain the interpretation that a measurement has been performed.

---

[1] At this point we encounter a language difficulty. Whereas before the observation we had a single observer state afterwards there were a number of different states for the observer, all occurring in a superposition. Each of these separate states is a state for an observer, so that we can speak of the different observers described by the different states. On the other hand, the same physical system is involved, and from this viewpoint it is the *same* observer, which is in different states for different elements of the superposition (i.e., has had different experiences in the separate elements of the superposition). In this situation we shall use the singular when we

We now carry the discussion a step further and allow the observer-system to repeat the observation. Then according to *Rule 2* we arrive at the total state after the second observation:

$$\psi''^{S+O} = \sum_i a_i \phi_i \psi^O_{ii[...,\alpha_i,\alpha_i]}. \tag{2.4}$$

Again, we see that each element of (2.4), $\phi_i \psi^O_{ii[...,\alpha_i,\alpha_i]}$, describes a system eigenstate, but this time also describes the observer as having obtained the *same result* for each of the two observations. Thus for every separate state of the observer in the final superposition, the result of the observation was repeatable, even though different for different states. This repeatability is, of course, a consequence of the fact that after an observation the *relative* system state for a particular observer state is the corresponding eigenstate.

Let us suppose now that an observer-system O, with initial state $\psi^O_{[...]}$, measures the *same* quantity A in a number of separate identical systems which are initially in the same state, $\psi^{S_1} = \psi^{S_2} = \ldots = \psi^{S_n} = \sum_i a_i \phi_i$ (where the ϕ_i are, as usual, eigenfunctions of A). The initial total state function is then

$$\psi_0^{S_1+S_2+\ldots+S_n+O} = \psi^{S_1} \psi^{S_2} \ldots \psi^{S_n} \psi^O_{[...]}.$$

We shall assume that the measurements are performed on the systems in the order $S_1, S_2, \ldots, S_n$. Then the total state after the first measurement will be, by *Rule 1*,

$$\psi_1^{S_1+S_2+\ldots+S_n+O} = \sum_i a_i \phi_i^{S_1} \psi^{S_2} \ldots \psi^{S_n} \psi^O_{i[...,\alpha_i^1]}$$

(where α_i^1 refers to the first system, S_i).

After the second measurement it will be, by *Rule 2*,

$$\psi_2^{S_1+S_2+\ldots S_n+O} = \sum_{i,j} a_i a_j \phi_i^{S_1} \phi_j^{S_2} \psi^{S_3} \ldots \psi^{S_n} \psi^O_{ij[...,\alpha_i^1,\alpha_j^2]} \tag{2.5}$$

---

wish to emphasize that a single physical system is involved, and the plural when we wish to emphasize the different experiences for the separate elements of the superposition. (e.g., "The observer performs an observation of the quantity A, after which each of the observers of the resulting superposition has perceived an eigenvalue.")[bm]

[bm] For Everett a measurement is fully constituted by the correlation that is produced in the absolute state between the observer and the object system. This correlation, as he explains in the footnote here, leads to a postmeasurement observer for whom multiple relative states obtain. Everett then associates the relative states of the observer with different experiences. See the discussion of relative states in the conceptual introduction, chapter 3, (pg. 34).

and in general, after r measurements have taken place ($r \leq n$) *Rule* 2 gives the result:

$$\psi_r = \sum_{i,j,\dots,k} a_i a_j \dots a_k \phi_i^{S_1} \phi_j^{S_2} \dots \phi_k^{S_r} \psi^{S_{r+1}} \dots \psi^{S_n} \psi_{ij\dots k[\dots,\alpha_i^1,\alpha_j^2,\dots,\alpha_k^r]}^{O}. \quad (2.6)$$

We can give this state, ψ_r, the following interpretation. It consists of a superposition of states:

$$\psi'_{ij\dots k} = \phi_i^{S_1} \phi_j^{S_2} \dots \phi_k^{S_r} \psi^{S_{r+1}} \dots \psi^{S_n} \psi_{ij\dots k[\dots,\alpha_i^1,\alpha_j^2,\dots,\alpha_k^r]}^{O}, \quad (2.7)$$

each of which describes the observer with a definite memory sequence $[\dots,\alpha_i^1,\alpha_j^2,\dots,\alpha_k^r]$, and relative to whom the observed system states are the corresponding eigenfunctions $\phi_i^{S_1}, \phi_j^{S_2}, \dots, \phi_k^{S_r}$, the remaining systems, $S_{r+1},\dots S_n$, being unaltered.

In the language of subjective experience, the observer which is described by a typical element, $\psi'_{ij\dots k}$, of the superposition has perceived an apparently random sequence of definite results for the observations. It is furthermore true, since in each element the system has been left in an eigenstate of the measurement, that if at this stage a redetermination of an earlier system observation (S_l) takes place, every element of the resulting final superposition will describe the observer with a memory configuration of the form $[\dots,\alpha_i^1,\dots,\underline{\alpha}_j^l,\dots,\alpha_k^r,\underline{\alpha}_j^l]$ in which the earlier memory coincides with the later—i.e., the memory states are *correlated*. It will thus *appear* to the observer which is described by a typical element of the superposition that each initial observation on a system caused the system to "jump" into an eigenstate in a random fashion and thereafter remain there for subsequent measurements on the same system. Therefore, qualitatively, at least, the probabilistic assertions of Process 1 *appear* to be valid to the observer described by a typical element of the final superposition.

In order to establish quantitative results, we must put some sort of measure (weighting) on the elements of a final superposition.[bn] This is

---

[bn] This is the beginning of Everett's extended argument for the special status of the norm-squared measure of typicality in pure wave mechanics. A version of this argument can also be found in the short thesis, chapter 9 (pg. 188). Everett considered his discussion of typicality to be of central importance. See pgs. 273–75 and 295 for descriptions of the role of these considerations in pure wave mechanics. See also the discussion of empirical faithfulness in the conceptual introduction (pg. 53). Establishing the uniqueness of the norm-squared measure required a special set of background assumptions. Everett was well aware that there were many ways one might introduce a typicality measure over branches. Although the norm-squared measure is perhaps particularly natural, that there are other measures one might adopt indicates that there can be no way to deduce this measure from pure wave mechanics alone; rather, to argue for the uniqueness of the norm-squared measure, one must add something to the theory that constrains one's selection. The question then becomes one of the naturalness of

necessary to be able to make assertions which will hold for almost all of the observers described by elements of a superposition. In order to make quantitative statements about the relative frequencies of the different possible results of observation which are recorded in the memory of a typical observer we must have a method of selecting a *typical* observer.

Let us therefore consider the search for a general scheme for assigning a measure to the elements of a superposition of orthogonal states $\sum a_i \phi_i$.[bo] We require then a positive function $\mathcal{M}$ of the complex coefficients of the elements of the superposition, so that $\mathcal{M}(a_i)$ shall be the measure assigned to the element ϕ_i. In order that this general scheme shall be unambiguous we must first require that the states themselves always be normalized, so that we can distinguish the coefficients from the states. However, we can still only determine the *coefficients*, in distinction to the states, up to an arbitrary phase factor, and hence the function $\mathcal{M}$ must be a function of the amplitudes of the coefficients alone, (i.e., $\mathcal{M}(a_i) = \mathcal{M}\left(\sqrt{a_i^* a_i}\right)$), in order to avoid ambiguities.

If we now impose the additivity requirement that if we regard a subset of the superposition, say $\sum_{i=1}^{n} a_i \phi_i$, as a single element $\alpha \phi'$:

$$\alpha \phi' = \sum_{i=1}^{n} a_i \phi_i , \qquad (2.8)$$

then the measure assigned to ϕ' shall be the sum of the measures assigned to the ϕ_i (i from 1 to n):

$$\mathcal{M}(\alpha) = \sum_i \mathcal{M}(a_i), \qquad (2.9)$$

then we have already restricted the choice of $\mathcal{M}$ to the square amplitude alone. ($\mathcal{M}(a_i) = a_i^* a_i$), apart from a multiplicative constant.)

To see this we note that the normality of ϕ' requires that $|\alpha| = \sqrt{\sum_{i=1}^{n} a_i^* a_i}$. From our remarks upon the dependence of $\mathcal{M}$ upon the amplitude alone, we replace the a_i by their amplitudes $\mu_i = |a_i|$.

the principles one adds. The methodological problem is that when one knows what measure of typicality one wants, the principles one adds to get that measure invariably seem natural.

[bo] This is a search for an appropriate measure of typicality on the set of elements of the superposition. The constraints Everett imposes are that (1) the measure be a positive function over the elements of the superposition for each possible expansion of the state, (2) each such measure must depend only on the magnitude, not the phase, of the coefficients associated with the terms describing the elements on the particular expansion, and (3) the measures associated with different expansions of the absolute state must be related by a nesting requirement so that the measure assigned to a term in a coarser-grained expansion that represents a linear combination of individual terms in a finer-grained expansion is equal to the sum of the finer-grained measures on the individual terms. This last condition represents a constraint on how the measures associated with different expansions of the absolute state are related. Everett then argues that the norm-squared coefficient measure is the unique measure up to a multiplicative constant satisfying these requirements.

(2.9) then requires that

$$M(\alpha) = M\left(\sqrt{\sum a_i^* a_i}\right) = M\left(\sqrt{\sum \mu_i^2}\right)$$
$$= \sum M(\mu_i) = \sum M\left(\sqrt{\mu_i^2}\right). \tag{2.10}$$

Defining a new function $g(x)$:

$$g(x) = M(\sqrt{x}), \tag{2.11}$$

we see that (2.10) requires that

$$g\left(\sum \mu_i^2\right) = \sum g(\mu_i^2), \tag{2.12}$$

so that g is restricted to be linear and necessarily has the form:

$$g(x) = cx \quad (c \text{ constant}). \tag{2.13}$$

Therefore $g(x^2) = cx^2 = M\sqrt{x^2} = M(x)$ and we have deduced that M is restricted to the form

$$M(a_i) = M(\mu_i) = c\mu_i^2 = ca_i^* a_i, \tag{2.14}$$

and we have shown that the only choice of measure consistent with our additivity requirement is the square amplitude measure, apart from an arbitrary multiplicative constant which may be fixed, if desired, by normalization requirements. (The requirement that the total measure be unity implies that this constant is 1.)

The situation here is fully analogous to that of classical statistical mechanics, where one puts a measure on trajectories of systems in the phase space by placing a measure on the phase space itself, and then making assertions which hold for "almost all" trajectories (such as ergodicity, quasi-ergodicity, etc).[2] This notion of "almost all" depends here also upon the choice of measure, which is in this case taken to be Lebesgue measure on the phase space. One could, of course, contradict the statements of classical statistical mechanics by choosing a measure for which only the exceptional trajectories had nonzero measure. Nevertheless the choice of Lebesgue measure on the phase space can be justified by the fact that it is the only choice for which the "conservation of probability" holds (Liouville's theorem), and hence the only choice which makes possible any reasonable statistical deductions at all.

[2] See Khinchin [A. I. Khinchin, *Mathematical Foundations of Statistical Mechanics.* (Translated by George Gamow) Dover, New York: 1949].

In our case, we wish to make statements about "trajectories" of observers. However, for us a trajectory is constantly branching (transforming from state to superposition) with each successive measurement. To have a requirement analogous to the "conservation of probability" in the classical case, we demand that the measure assigned to a trajectory at one time shall equal the sum of the measures of its separate branches at a later time. This is precisely the additivity requirement which we imposed and which leads uniquely to the choice of square-amplitude measure. Our procedure is therefore quite as justified as that of classical statistical mechanics.[bp]

Having deduced that there is a unique measure which will satisfy our requirements, the square-amplitude measure, we continue our deduction.[bq] This measure then assigns to the $i, j, \ldots, k^{\text{th}}$ element of the superposition (2.6),

$$\phi_i^{S_1} \phi_j^{S_2} \ldots \phi_k^{S_r} \psi^{S_{r+1}} \ldots \psi^{S_n} \psi_{ij\ldots k[\ldots,\alpha_i^1,\alpha_j^2,\ldots,\alpha_k^r]}^{O}, \tag{2.15}$$

the measure (weight)

$$M_{ij\ldots k} = (a_i a_j \ldots a_k)^* (a_i a_j \ldots a_k), \tag{2.16}$$

so that the observer state with memory configuration $[\ldots, \alpha_i^1, \alpha_j^2, \ldots, \alpha_k^r]$ is assigned the measure $a_i^* a_i a_j^* a_j \ldots a_k^* a_k = M_{ij\ldots k}$. We see immediately that this is a product measure, namely

$$M_{ij\ldots k} = M_i M_j \ldots M_k, \tag{2.17}$$

where

$$M_l = a_l^* a_l,$$

so that the measure assigned to a particular memory sequence $[\ldots, \alpha_i^1, \alpha_j^2, \ldots, \alpha_k^r]$ is simply the product of the measures for the individual components of the memory sequence.

[bp] See also the discussion of the analogy between probability in pure wave mechanics and probability in statistical mechanics in the short thesis (pgs. 191–92). Everett further explained the parallel with statistical mechanics in his comments at the Xavier conference in 1961 (pg. 275). A disanalogy, however, is that whereas the probabilities in statistical mechanics might be thought of as epistemic, resulting from one not knowing the microstate of a system, it is unclear what the corresponding epistemic consideration might be in pure wave mechanics.

[bq] The original version of the long thesis reads here: "We choose for this measure the square amplitude of the coefficients of the superposition, a choice which we shall subsequently see is not as arbitrary as it appears" (Everett 1956, pg. 98). Everett later changed this to the stronger statement for the version included in the DeWitt and Graham anthology (DeWitt and Graham, 1973).

We notice now a direct correspondence of our measure structure to the probability theory of random sequences. Namely, *if we were to regard* the $M_{ij...k}$ as probabilities for the sequences $[..., \alpha_i^1, \alpha_j^2, ..., \alpha_k^r]$, then the sequences are equivalent to the random sequences which are generated by ascribing to each term the *independent* probabilities $M_l = a_l^* a_l$.[br] Now the probability theory is equivalent to measure theory mathematically, so that we can make use of it, while keeping in mind that all results should be translated back to measure theoretic language.

Thus, in particular, if we consider the sequences to become longer and longer (more and more observations performed) *each* memory sequence of the final superposition will satisfy any given criterion for a randomly generated sequence, generated by the independent probabilities $a_i^* a_i$, except for a set of total measure which tends toward zero as the number of observations becomes unlimited. Hence all averages of functions over *any* memory sequence, including the special case of frequencies, can be computed from the probabilities $a_i^* a_i$, except for a set of memory sequences of measure zero. We have therefore shown that the statistical assertions of Process 1 will appear to be valid to *almost all* observers described by separate elements of the superposition (2.6), in the limit as the number of observations goes to infinity.[bs]

While we have so far considered only sequences of observations of the same quantity upon identical systems, the result is equally true for arbitrary sequences of observations. For example, the sequence of observations of the quantities $A^1, A^2, ..., A^n, ...$ with (generally different) eigenfunction sets $\{\phi_i^1\}, \{\phi_j^2\}, ..., \{\phi_k^n\}, ...$ applied successively to the systems $S_1, S_2, ..., S_n, ...$, with (arbitrary) initial states $\psi^{S_1}, \psi^{S_2}, ..., \psi^{S_n}, ...$ transforms the total initial state:

$$\psi^{S_1 + ... + S_n + O} = \psi^{S_1} \psi^{S_2} ... \psi^{S_n} \psi_{[...]}^O \tag{2.18}$$

by Rules 1 and 2, into the final state:

$$\psi'^{S_1 + S_2 + ... + S_n + O} = \sum_{i,j,...,k} (\phi_i^1, \psi^{S_1})(\phi_j^2, \psi^{S_2}) ... (\phi_k^n, \psi^{S_n}) ...$$
$$\times \phi_i^1 \phi_j^2 ... \phi_k^n ... \psi_{[..., \alpha_i^1, \alpha_j^2, ..., \alpha_k^n, ...]}^O, \tag{2.19}$$

where the memory sequence element α_l^r characterizes the lth eigenfunction, ϕ_l^r of the operator A^r. Again the square amplitude measure for each element of the superposition (2.19) reduces to the product measure of the individual

---

[br] See the discussion in fn. ea on pg. 193.

[bs] See also Everett's discussion of this result in the short thesis (chapter 6) and Barrett (1999, limiting properties of the bare theory) for more details.

memory element measures, $|(\phi_l^r, \psi^{S_r})|^2$, for the memory sequence element α_l^r. Therefore, the memory sequence of a *typical* element of (2.19) has all the characteristics of a random sequence, with individual, independent (and now different), probabilities $|(\phi_l^r, \psi^{S_r})|^2$ for the rth memory state.

Finally, we can generalize to the case where several observations are allowed to be performed upon the *same* system. For example, if we permit the observation of a new quantity B (eigenfunctions η_m, memory characterization β_i) upon the system S_r for which A^r has already been observed, then the state (2.19):

$$
\psi' = \sum_{i,l,\dots,k} (\phi_i^1, \psi^{S_1}) \dots (\phi_l^r, \psi^{S_r}) \dots (\phi_k^n, \psi^{S_n})
$$
$$
\times \phi_i^1 \dots \phi_l^r \dots \phi_k^n \dots \psi_{[\dots,\alpha_i^1,\dots,\alpha_l^r,\dots,\alpha_k^n,\dots]}^O \tag{2.20}
$$

is transformed by *Rule* 2 into the state:

$$
\psi' = \sum_{i,\dots,l,\dots,k,\underline{m}} (\phi_i^1, \psi^{S_1}) \dots (\phi_l^r, \psi^{S_r}) \dots (\phi_k^n, \psi^{S_n})(\eta_m^r, \phi_l^r)
$$
$$
\times \phi_i^1 \dots \phi_\mu^{r-1} \dots \underline{\eta_m^r} \dots \phi_\nu^{r+1} \dots \phi_k^n \dots \psi_{[\dots,\alpha_i^1,\dots,\alpha_l^r,\dots,\alpha_k^n,\dots,\underline{\beta_m^r}\dots]}^O. \tag{2.21}
$$

The *relative* system states for S have been changed from the eigenstates of A^r, $\{\phi_i^r\}$, to the eigenstates of B^r, $\{\eta_m^r\}$. We notice further that, with respect to our measure on the superposition, the memory sequences still have the character of random sequences, but of random sequences for which the individual terms are no longer independent. The memory states β_m^r now depend upon the memory states α_l^r which represent the result of the previous measurement upon the same system, S_r. The *joint* (normalized) measure for this pair of memory states, conditioned by fixed values for remaining memory states is:

$$
M^{\alpha_i^1 \dots \alpha_\mu^{r-1} \alpha_\nu^{r+1} \dots \alpha_k^n}(\alpha_l^r, \beta_m^r) = \frac{M(\alpha_i^1, \dots, \alpha_l^r, \dots, \alpha_k^n, \beta_m^r)}{\sum_{l,m} M(\alpha_i^2, \dots, \alpha_l^r, \dots, \alpha_k^n, \beta_m^r)}
$$
$$
= \frac{|(\phi_i^1, \psi^{S_1}) \dots (\phi_l^r, \psi^{S_r}) \dots (\phi_k^n, \psi^{S_n})(\eta_m^r, \phi_l^r)|^2}{\sum_{l,m} |(\phi_i^1, \psi^{S_1}) \dots (\phi_l^r, \psi^{S_r}) \dots (\phi_k^n, \psi^{S_n})(\eta_m^r, \phi_l^r)|^2}
$$
$$
= |(\phi_l^r, \psi^{S_r})|^2 |(\eta_m^r, \phi_l^r)|^2. \tag{2.22}
$$

The joint measure (2.15) is, first of all, independent of the memory states for the remaining systems ($S_1 \dots S_n$, excluding S_r). Second, the dependence of β_m^r on α_l^r is *equivalent*, measure theoretically, to that given by the *stochastic process*[3] which converts the states ϕ_l^r into the states η_m^r with transition probabilities:

$$T_{lm} = \text{Prob.}(\phi_l^r \rightarrow \eta_m^r) = |(\eta_m^r, \phi_l^r)|^2. \tag{2.23}$$

If we were to allow yet another quantity C to be measured in S_r, the new memory states α_p^r corresponding to the eigenfunctions of C would have a similar dependence upon the previous states β_m^r, but *no direct dependence* on the still earlier states α_l^r. This dependence upon only the previous result of observation is a consequence of the fact that the *relative* system states are completely determined by the last observation.

We can therefore summarize the situation for an arbitrary sequence of observations, upon the same or different systems in any order, and for which the number of observations of each quantity in each system is very large, with the following result:

Except for a set of memory sequences of measure nearly zero, the averages of any functions over a memory sequence can be calculated approximately by the use of the independent probabilities given by Process 1 for each initial observation, on a system, and by the use of the transition probabilities (2.23) for succeeding observations upon the same system. In the limit, as the number of all types of observations goes to infinity the calculation is exact and the exceptional set has measure zero.

This prescription for the calculation of averages over memory sequences by probabilities assigned to individual elements is precisely that of the orthodox theory (Process 1). Therefore all predictions of the usual theory will appear to be valid to the observer in almost all observer states, since these predictions hold for almost all memory sequences.

In particular, the uncertainty principle is never violated, since, as above, the latest measurement upon a system supplies all possible information about the relative system state, so that there is no direct correlation between any earlier results of observation on the system and the succeeding observation. Any observation of a quantity B between two successive observations of quantity A (all on the same system) will destroy the one-one correspondence between the earlier and later memory states for the result of A. Thus for alternating observations of different quantities there are fundamental limitations upon the correlations between memory states for the same observed quantity, these limitations expressing the content of the uncertainty principle.

In conclusion, we have described in this section processes involving an idealized observer, processes which are entirely deterministic and contin-uous from the over-all viewpoint (the total state function is presumed to satisfy a wave equation at all times) but whose result is a superposition,

---

[3] Cf. Chapter II, §6.

each element of which describes the observer with a different memory state. We have seen that in almost all of these observer states it *appears* to the observer that the probabilistic aspects of the usual form of quantum theory are valid. We have thus seen how pure wave mechanics, without any initial probability assertions, can lead to these notions on a subjective level, as appearances to observers.

§3. Several observers

We shall now consider the consequences of our scheme when several observers are allowed to interact with the same systems, as well as with one another (communication). In the following discussion observers shall be denoted by $O_1, O_2, \ldots$, other systems by $S_1, S_2, \ldots$, and observables by operators A, B, C, with eigenfunctions $\{\phi_i\}, \{\eta_j\}, \{\xi_k\}$, respectively. The symbols $\alpha_i, \beta_j, \gamma_k$ occurring in memory sequences shall refer to characteristics of the states ϕ_i, η_j, ξ_k, respectively. ($\psi^O_{i[\ldots,\alpha_i]}$ is interpreted as describing an observer, O_j, who has just observed the eigenvalue corresponding to ϕ_i, i.e., who is "aware" that the system is in state ϕ_i.)

We shall also wish to allow communication among the observers, which we view as an interaction by means of which the memory sequences of different observers become correlated. (For example, the transfer of impulses from the magnetic tape memory of one mechanical observer to that of another constitutes such a transfer of information.)[4] We shall regard these processes as observations made by one observer on another and shall use the notation that

$$\psi^{O_j}_{i[\ldots,\alpha_i^{O_k}]}$$

represents a state function describing an observer O_j who has obtained the information α_i from another observer, O_k. Thus the obtaining of information about A from O_1 by O_2 will transform the state

$$\psi^{O_1}_{i[\ldots,\alpha_i]}\psi^{O_2}_{[\ldots]}$$

into the state

$$\psi^{O_1}_{i[\ldots,\alpha_i]}\psi^{O_2}_{i[\ldots,\alpha_i^{O_1}]}. \tag{3.1}$$

Rules 1 and 2 are, of course, equally applicable to these interactions. We shall now illustrate the possibilities for several observers by considering several cases.

---

[4] We assume that such transfers merely duplicate, but do not destroy, the original information.

Case 1. We allow two observers to separately observe the same quantity in a system and then compare results.

We suppose that first observer O_1 observes the quantity A for the system S. Then by *Rule* 1 the original state

$$\psi^{S+O_1+O_2} = \psi^S \psi^{O_1}_{[\ldots]} \psi^{O_2}_{[\ldots]}$$

is transformed into the state

$$\psi' = \sum_i (\phi_i^S, \psi^S) \phi_i^S \psi^{O_1}_{i[\ldots,\alpha_i]} \psi^{O_2}_{[\ldots]}. \tag{3.2}$$

We now suppose that O_2 observes A, and by *Rule* 2 the state becomes:

$$\psi'' = \sum_i (\phi_i^S, \psi^S) \phi_i^S \psi^{O_1}_{i[\ldots,\alpha_i]} \psi^{O_2}_{i[\ldots,\alpha_i]}. \tag{3.3}$$

We now allow O_2 to "consult" O_1, which leads in the same fashion from (3.2) and *Rule* 2 to the final state

$$\psi''' = \sum_i (\phi_i^S, \psi^S) \phi_i^S \psi^{O_1}_{i[\ldots,\alpha_i]} \psi^{O_2}_{ii[\ldots,\alpha_i,\alpha_i^{O_1}]}. \tag{3.4}$$

Thus, for every element of the superposition the information obtained from O_1 agrees with that obtained directly from the system. This means that observers who have separately observed the same quantity will *always* agree with each other.

Furthermore, it is obvious at this point that the same result, (3.4), is obtained if O_2 *first* consults O_1, then performs the direct observation, except that the memory sequence for O_2 is reversed ($[\ldots, \alpha_i^{O_1}, \alpha_i]$ instead of $[\ldots, \alpha_i, \alpha_i^{O_1}]$). There is still perfect agreement in every element of the superposition. Therefore, information obtained from another observer is always reliable, since subsequent direct observation will always verify it. We thus see the central role played by correlations in wave functions for the preservation of consistency in situations where several observers are allowed to consult one another. It is the transitivity of correlation in these cases (that if S_1 is correlated to S_2, and S_2 to S_3, then so is S_1 to S_2)[bt] which is responsible for this consistency.

Case 2. We allow two observers to measure separately two different, non-commuting quantities in the same system.

Assume that first O_1 observes A for the system, so that, as before, the initial state $\psi^S \psi^{O_1} \psi^{O_2}$ is transformed to:

$$\psi' = \sum_i (\phi_i, \psi^S) \phi_i \psi^{O_1}_{i[\ldots,\alpha_i]} \psi^{O_2}_{[\ldots]}. \tag{3.5}$$

[bt] Everett likely means "$\ldots S_1$ to S_3."

Next let O_2 determine β for the system, where $\{\eta_j\}$ are the eigenfunctions of β. Then by application of *Rule 2* the result is

$$\psi'' = \sum_{i,j} (\phi_i, \psi^S)(\eta_j, \phi_i)(\eta_j \psi^{O_1}_{i[\ldots,\alpha_i]} \psi^{O_2}_{j[\ldots,\beta_j]}. \tag{3.6}$$

O_2 is now perfectly correlated with the system, since a redetermination by him will lead to agreeing results. This is no longer the case for O_1, however, since a redetermination of A by him will result in (by *Rule 2*)

$$\psi'' = \sum_{i,j,k} (\phi_i, \psi^S)(\eta_j, \phi_i)(\phi_k, \eta_j)\phi_k^S \psi^{O_2}_{j[\ldots,\beta_j]} \psi^{O_1}_{ik[\ldots,\alpha_j,\alpha_k]}. \tag{3.7}$$

Hence the second measurement of O_1 does not in all cases agree with the first, and has been upset by the intervention of O_2.

We can deduce the statistical relation between O_1's first and second results (α_i and α_k) by our previous method of assigning a measure to the elements of the superposition (3.7). The measure assigned to the $(i, j, k)^{\text{th}}$ element is then:

$$M_{ijk} = |(\phi_i, \psi^S)(\eta_j, \phi_i)(\phi_k, \eta_j)|^2. \tag{3.8}$$

This measure is equivalent, in this case, to the probabilities assigned by the orthodox theory (Process 1), where O_2's observation is regarded as having converted each state ϕ_i into a non-interfering mixture of states η_j, weighted with probabilities $|(\eta_j, \phi_i)|^2$, upon which O_1 makes his second observation.

Note, however, that this equivalence with the statistical results obtained by considering that O_2's observation changed the system state into a mixture, holding true *only so long as* O_1's *second observation is restricted to the system*. If he were to attempt to simultaneously determine a property of the system as well as of O_2, interference effects might become important. The description of the states relative to O_1, after O_2's observation, as non-interfering mixtures is therefore incomplete.

Case 3. We suppose that two systems S_1 and S_2 are correlated but no longer interacting, and that O_1 measures property A in S_1, and O_2 property β in S_2.[bu]

We wish to see whether O_2's intervention with S_2 can in any way affect O_1's results in S_1, so that perhaps signals might be sent by these means. We shall assume that the initial state for the system pair is

$$\psi^{S_1+S_2} = \sum_i a_i \phi_i^{S_1} \phi_i^{S_2}. \tag{3.9}$$

[bu] This is Everett's description of the EPR experiment (Einstein et al., 1935).

We now allow O_1 to observe A in S_1, so that after this observation the total state becomes:

$$\psi'^{S_1+S_2+O_1+O_2} = \sum_i a_i \phi_i^{S_1} \phi_i^{S_2} \psi_{i[...,\alpha_i]}^{O_1} \psi_{[...]}^{O_2}. \qquad (3.10)$$

O_1 can of course continue to repeat the determination, obtaining the same result each time.

We now suppose that O_2 determines β in S_2, which results in

$$\psi'' = \sum_{i,j} a_i(\eta_j^2, \phi_i^2) \phi_i^1 \eta_j^2 \psi_{i[...,\alpha_i]}^{O_1} \psi_{j[...,\beta_j]}^{O_2}. \qquad (3.11)$$

However, in this case, as distinct from *Case 2*, we see that the intervention of O_2 in no way affects O_1's determinations, since O_1 is still perfectly correlated to the states $\phi_i^{S_1}$ of S_1, and any further observations by O_1 will lead to the same results as the earlier observations. Thus each memory sequence for O_1 continues without change due to O_2's observation, and such a scheme could not be used to send any signals.

Furthermore, we see that the result (3.11) is arrived at even in the case that O_2 should make his determination before that of O_1. Therefore any expectations for the outcome of O_1's first observation are in no way affected by whether or not O_2 performs his observation before that of O_1. This is true because the expectation of the outcome for O_1 can be computed from (3.10), which is the same whether or not O_2 performs his measurement before or after O_1.

It is therefore seen that one observer's observation upon one system of a correlated, but non-interacting, pair of systems has no effect on the remote system, in the sense that the outcome or expected outcome of any experiments by another observer on the remote system are not affected. Paradoxes like that of Einstein-Rosen-Podolsky[5] which are concerned with such correlated, non-interacting, systems are thus easily understood in the present scheme.

Many further combinations of several observers and systems can be easily studied in the present framework, and all questions answered by first writing down the final state for the situation with the aid of *Rules* 1 and 2, and then noticing the relations between the elements of the memory sequences.

V. SUPPLEMENTARY TOPICS

We have now completed the abstract treatment of measurement and observation, with the deduction that the statistical predictions of the usual

---

[5] Einstein [A. Einstein, B. Podolsky, N. Rosen, *Phys. Rev.* 47, 777, 1935.].

form of quantum theory (Process 1) will appear to be valid to all observers. We have therefore succeeded in placing our theory in correspondence with experience, at least insofar as the ordinary theory correctly represents experience.

We should like to emphasize that this deduction was carried out by using only the principle of superposition, and the postulate that an observation has the property that *if* the observed variable has a definite value in the object-system then it will remain definite and the observer will perceive this value. This treatment is therefore valid for any possible quantum interpretation of observation processes, i.e., any way in which one can interpret wave functions as describing observers, as well as for any form of quantum mechanics for which the superposition principle for states is maintained. Our abstract discussion of observation is therefore logically complete, in the sense that our results for the subjective experiences of observers are correct, if there are any observers at all describable by wave mechanics.[1]

In this chapter we shall consider a number of diverse topics from the point of view of our pure wave mechanics, in order to supplement the abstract discussion and give a feeling for the new viewpoint. Since we are now mainly interested in elucidating the reasonableness of the theory, we shall often restrict ourselves to plausibility arguments, rather than detailed proofs.

§1. Macroscopic objects and classical mechanics

In the light of our knowledge about the atomic constitution of matter, any "object" of macroscopic size is composed of an enormous number of constituent particles. The wave function for such an object is then in a space of fantastically high dimension ($3N$, if N is the number of particles). Our present problem is to understand the existence of macroscopic objects and to relate their ordinary (classical) behavior in the three-dimensional world to the underlying wave mechanics in the higher dimensional space.

Let us begin by considering a relatively simple case. Suppose that we place in a box an electron and a proton, each in a definite momentum state, so that the position amplitude density of each is uniform over the whole box. After a time we would expect a hydrogen atom in the ground state to form, with ensuing radiation. We notice, however, that the position amplitude density of each particle is *still* uniform over the whole box. Nevertheless the amplitude distributions are now no longer independent, but correlated. In particular, the *conditional* amplitude density for the electron, conditioned by any definite proton (or centroid) position, is *not* uniform, but is given by the familiar ground state wave function for the hydrogen atom. What

---

[1] They are, of course, vacuously correct otherwise.

we mean by the statement, "a hydrogen atom has formed in the box," is just that this correlation has taken place—a correlation which insures that the *relative* configuration for the electron, for a definite proton position, conforms to the customary ground state configuration.

The wave function for the hydrogen atom can be represented as a product of a centroid wave function and a wave function over relative coordinates, where the centroid wave function obeys the wave equation for a particle with mass equal to the total mass of the proton-electron system. Therefore, if we now open our box, the centroid wave function will spread with time in the usual manner of wave packets, to eventually occupy a vast region of space. The *relative* configuration (described by the *relative coordinate* state function) has, however, a permanent nature, since it represents a bound state, and it is this relative configuration which we usually think of as the object called the hydrogen atom.[bv] Therefore, no matter how indefinite the positions of the individual particles become in the total state function (due to the spreading of the centroid), this state can be regarded as giving (through the centroid wave function) an amplitude distribution over a comparatively definite object, the tightly bound electron-proton system. The general state, then, does not describe any single such definite object, but a superposition of such cases with the object located at different positions.

In a similar fashion larger and more complex objects can be built up through strong correlations which bind together the constituent particles. It is still true that the general state function for such a system may lead to marginal position densities for any single particle (or centroid) which extend over large regions of space. Nevertheless we can speak of the existence of a relatively definite object, since the specification of a single position for a particle, or the centroid, leads to the case where the *relative* position densities of the remaining particles are distributed closely about the specified one, in a manner forming the comparatively definite object spoken of.

Suppose, for example, we begin with a cannonball located at the origin, described by a state function:

$$\psi_{[c_j(0,0,0)]},$$

where the subscript indicates that the total state function ψ describes a system of particles bound together so as to form an object of the size and shape of a cannonball, whose centroid is located (approximately) at the origin, say in the form of a real gaussian wave packet of small dimensions, with variance σ_0^2 for each dimension.

[bv] The stable relative configuration in the correlation structure represents the hydrogen atom as a classical object with a diachronic identity. For Everett, a complex object is fully determined by the internal correlations between its parts. This holds for microscopic objects like the hydrogen atom and by direct analogy for macroscopic systems like the cannonball below.

If we now allow a long lapse of time, the centroid of the system will spread in the usual manner to occupy a large region of space. (The spread in each dimension after time t will be given by $\sigma_t^2 = \sigma_0^2 + (\hbar^2 t^2 / 4\sigma_0^2 m^2)$, where m is the mass.) Nevertheless, for any *specified* centroid position, the particles, since they remain in bound states, have distributions which again correspond to the fairly well defined size and shape of the cannonball. Thus the total state can be regarded as a (continuous) superposition of states

$$\psi = \int a_{xyz} \psi_{[c_j(x,y,z)]} \, dx \, dy \, dz,$$

each of which ($\psi_{[c_j(x,y,z)]}$) describes a cannonball at the position (x, y, z). The coefficients a_{xyz} of the superposition then correspond to the centroid distribution.

It is *not* true that each individual particle spreads independently of the rest, in which case we would have a final state which is a grand superposition of states in which the particles are located independently everywhere. The fact that they are in bound states restricts our final state to a superposition of "cannonball" states. The wave function for the centroid can therefore be taken as a representative wave function for the whole object.

It is thus in this sense of correlations between constituent particles that definite macroscopic objects can exist within the framework of pure wave mechanics. The building up of correlations in a complex system supplies us with a mechanism which also allows us to understand how condensation phenomena (the formation of spatial boundaries which separate phases of different physical or chemical properties) can be controlled by the wave equation, answering a point raised by Schrödinger.

Classical mechanics, also, enters our scheme in the form of correlation laws. Let us consider a system of objects (in the previous sense), such that the centroid of each object has initially a fairly well defined position and momentum (e.g., let the wave function for the centroids consist of a product of gaussian wave packets). As time progresses, the centers of the square amplitude distributions for the objects will move in a manner approximately obeying the laws of motion of classical mechanics, with the degree of approximation depending upon the masses and the length of time considered, as is well known. (Note that we do not mean to imply that the wave packets of the individual objects remain independent if they are interacting. They do not. The motion that we refer to is that of the centers of the *marginal* distributions for the centroids of the bodies.)

The general state of a system of macroscopic objects does not, however, ascribe any nearly definite positions and momenta to the individual bodies. Nevertheless, any general state can at any instant be analyzed into a *superposition* of states each of which *does* represent the bodies with fairly well defined positions and momenta.[2] Each of these states then propagates approximately according to classical laws, so that the general state can be viewed as a superposition of quasi-classical states propagating according to nearly classical trajectories. In other words, if the masses are large or the time short, there will be strong correlations between the initial (approximate) positions and momenta and those at a later time, with the dependence being given approximately by classical mechanics.[bw]

Since large scale objects obeying classical laws have a place in our theory of pure wave mechanics, we have justified the introduction of models for observers consisting of classically describable, automatically functioning machinery, and the treatment of Chapter IV is non-vacuous.

Let us now consider the result of an observation (considered along the lines of Chapter IV) performed upon a system of macroscopic bodies in a general state. The observer will *not* become aware of the fact that the state does not correspond to definite positions and momenta (i.e., he will not see the objects as "smeared out" over large regions of space) but will himself simply become correlated with the system—after the observation the composite system of objects + observer will be in a superposition of states, each element of which describes an observer who has perceived that the objects have nearly definite positions and momenta, and for whom the relative system state is a quasi-classical state in the previous sense, and furthermore to whom the system will appear to behave according to classical mechanics if his observation is continued. We see, therefore, how the classical appearance of the macroscopic world to us can be explained in the wave theory.

---

[2] For any ε one can construct a complete orthonormal set of (one particle) states $\phi_{\mu,\nu}$, where the double index μ, ν refers to the approximate position and momentum, and for which the expected position and momentum values run independently through sets of approximately uniform density, such that the position and momentum uncertainties, σ_x and σ_p, satisfy $\sigma_x \leqq C\varepsilon$ and $\sigma_p \leqq C(\hbar/2\varepsilon)$ for each $\phi_{\mu,\nu}$, where C is a constant ~ 60. The uncertainty product then satisfies $\sigma_x\sigma_p \leqq C^2(\hbar/2)$, about 3,600 times the minimum allowable, but still sufficiently low for macroscopic objects. This set can then be used as a basis for our decomposition into states where every body has a roughly defined position and momentum. For a more complete discussion of this set see von Neumann [J. von Neumann, *Mathematical Foundations of Quantum Mechanics*. (Translated by R. T. Beyer) Princeton University Press: 1955.], pp. 406–407.

[bw] While correspondents complained that Everett did not address macrophysical phenomena, see for example pg. 228, this was in fact something he believed he had fully addressed. As explained here, Everett's account of the quasi-classicality of branch states depended on the persistence of approximately determinate relative positions and approximately determinate relative momenta on each branch, for systems with large masses over short times. Indeed, it is the relative quasi-classical behavior of macrosystems that often allows one to identify the *same branch* at different times. On this view, classical laws describe the regular behavior of relative quasi-classical properties of macroscopic systems on each branch. See also pg. 158 and the discussion of classicality in the conceptual introduction (pgs. 49–50).

§2. Amplification processes

In Chapters III and IV we discussed abstract measuring processes, which were considered to be simply a direct coupling between two systems, the object-system and the apparatus (or observer). There is, however, in actuality a whole chain of intervening systems linking a microscopic system to a macroscopic observer. Each link in the chain of intervening system becomes correlated to its predecessor, so that the result is an amplification of effects from the microscopic object-system to a macroscopic apparatus, and then to the observer.

The amplification process depends upon the ability of the state of one micro-system (particle, for example) to become correlated with the states of an enormous number of other microscopic systems, the totality of which we shall call a detection system. For example, the totality of gas atoms in a Geiger counter, or the water molecules in a cloud chamber, constitute such a detection system.

The amplification is accomplished by arranging the condition of the detection system so that the states of the individual micro-systems of the detector are *metastable*, in a way that if one micro-system should fall from its metastable state it would influence the reduction of others. This type of arrangement leaves the entire detection system metastable against chain reactions which involve a large number of its constituent systems. In a Geiger counter, for example, the presence of a strong electric field leaves the gas atoms metastable against ionization. Furthermore, the products of the ionization of one gas atom in a Geiger counter can cause further ionizations, in a cascading process. The operation of cloud chambers and photographic films is also due to metastability against such chain reactions.

The chain reactions cause large numbers of the micro-systems of the detector to behave as a unit, all remaining in the metastable state, or all discharging. In this manner the states of a sufficiently large number of micro-systems are correlated, so that one can speak of the whole ensemble being in a state of discharge, or not.

For example, there are essentially only two macroscopically distinguishable states for a Geiger counter; discharged or undischarged. The correlation of large numbers of gas atoms, due to the chain reaction effect, implies that either very few, or else very many of the gas atoms are ionized at a given time. Consider the complete state function ψ^G of a Geiger counter, which is a function of all the coordinates of all the constituent particles. Because of the correlation of the behavior of a large number of the constituent gas atoms, the total state ψ^G can always be written as a superposition of two states

$$\psi^G = a_1 \psi^1_{[U]} + a_2 \psi^2_{[D]}, \tag{2.1}$$

where $\psi^1_{[U]}$ signifies a state where only a small number of gas atoms are ionized, and $\psi^2_{[D]}$ a state for which a large number are ionized.

To see that the decomposition (2.1) is valid, expand ψ^G in terms of individual gas atom stationary states:

$$\psi^G = \sum_{i,j,\ldots,k} a_{ij\ldots k} \psi_i^{S_1} \psi_j^{S_2} \ldots \psi_k^{S_n}, \tag{2.2}$$

where $\psi_l^{S_r}$ is the lth state of atom r. Each element of the superposition (2.2)

$$\psi_i^{S_1} \psi_j^{S_2} \ldots \psi_k^{S_n} \tag{2.3}$$

must contain either a very large number of atoms in ionized states, or else a very small number, because of the chain reaction effect. By choosing some medium-sized number as a dividing line, each element of (2.2) can be placed in one of the two categories, high number or low number of ionized atoms. If we then carry out the sum (2.2) over only those elements of the first category, we get a state (and coefficient)

$$a_1 \psi^1_{[D]} = \sum_{ij\ldots k}{}' a_{ij\ldots k} \psi_i^{S_1} \psi_j^{S_2} \ldots \psi_k^{S_n}. \tag{2.4}$$

The state $\psi^1_{[D]}$ is then a state where a large number of particles are ionized. The subscript $[D]$ indicates that it describes a Geiger counter which has discharged. If we carry out the sum over the remaining terms of (2.2) we get in a similar fashion:

$$a_2 \psi^2_{[U]} = \sum_{ij\ldots k}{}'' a_{ij\ldots k} \psi_i^{S_1} \psi_j^{S_2} \ldots \psi_k^{S_n} \tag{2.5}$$

where $[U]$ indicates the undischarged condition. Combining (2.4) and (2.5) we arrive at the desired relation (2.1). So far, this method of decomposition can be applied to any system, whether or not it has the chain reaction property. However, in our case, more is implied, namely that the spread of the number of ionized atoms in both $\psi_{[D]}$ and $\psi_{[U]}$ will be small compared to the separation of their averages, due to the fact that the existence of the chain reactions means that either many or else few atoms will be ionized, with the middle ground virtually excluded.

This type of decomposition is also applicable to all other detection devices which are based upon this chain reaction principle (such as cloud chambers, photo plates, etc.).[bx]

We consider now the coupling of such a detection device to another micro-system (object-system) for the purpose of measurement. If it is true

---

[bx] It is, in Everett's words, the exclusion of the middle ground that does the work here. Such systems allow for *sharp* records.

that the initial object-system state ϕ_1 will at some time t trigger the chain reaction, so that the state of the counter becomes $\psi^1_{[D]}$, while the object-system state ϕ_2 will not, then it is still true that the initial object-system state $a_1\phi_1 + a_2\phi_2$ will result in the superposition

$$a_1\phi'_1\psi^1_{[D]} + a_2\phi'_2\psi^2_{[U]} \tag{2.6}$$

at time t.

For example, let us suppose that a particle whose state is a wave packet ϕ, of linear extension greater than that of our Geiger counter, approaches the counter. Just before it reaches the counter, it can be decomposed into a superposition $\phi = a_1\phi_1 + a_2\phi_2$ (ϕ_1, ϕ_2 orthogonal) where ϕ_1 has non-zero amplitude only in the region before the counter and ϕ_2 has non-zero amplitude elsewhere (so that ϕ_1 is a packet which will entirely pass through the counter while ϕ_2 will entirely miss the counter). The initial total state for the system particle + counter is then:

$$\phi\psi_{[U]} = (a_1\phi_1 + a_2\phi_2)\psi_{[U]},$$

where $\phi_{[U]}$ is the initial (assumed to be undischarged) state of the counter.

But at a slightly later time ϕ_1 is changed to ϕ'_1, after traversing the counter and causing it to go into a discharged state $\psi^1_{[D]}$, while ϕ_2 passes by into a state ϕ'_2 leaving the counter in an undischarged state $\psi^2_{[U]}$. Superposing these results, the total state at the later time is

$$a_1\phi'_1\psi^1_{[D]} + a_2\phi'_2\psi^2_{[U]} \tag{2.7}$$

in accordance with (2.6). Furthermore, the relative particle state for $\psi^1_{[D]}$, ϕ'_1, is a wave packet emanating from the counter, while the relative state for $\psi^2_{[U]}$ is a wave with a "shadow" cast by the counter. The counter therefore serves as an apparatus which performs an approximate position measurement on the particle.

No matter what the complexity or exact mechanism of a measuring process, the general superposition principle as stated in Chapter III, §3, remains valid, and our abstract discussion is unaffected. It is a vain hope that somewhere embedded in the intricacy of the amplification process is a mechanism which will somehow prevent the macroscopic apparatus state from reflecting the same indefiniteness as its object-system.

§3. Reversibility and irreversibility

Let us return, for the moment, to the probabilistic interpretation of quantum mechanics based on Process 1 as well as Process 2. Suppose that we have a large number of identical systems (ensemble), and that the jth system is in the state ψ^j. Then for purposes of calculating expectation values for operators over the ensemble, the ensemble is represented by the mixture of states ψ^j weighted with $1/N$, where N is the number of systems, for which the density operator[3] is:

$$\rho = \frac{1}{N} \sum_j [\psi^j], \tag{3.1}$$

where $[\psi^j]$ denotes the projection operator on ψ^j. This density operator, in turn, is equivalent to a density operator which is a sum of projections on orthogonal states (the eigenstates of ρ):[4]

$$\rho = \sum_i P_i[\eta_i], \quad (\eta_i, \eta_j) = \delta_{ij}, \quad \sum_i P_i = 1, \tag{3.2}$$

so that any ensemble is always equivalent to a mixture of orthogonal states, which representation we shall henceforth assume.

Suppose that a quantity A, with (non-degenerate) eigenstates $\{\phi_j\}$ is measured in each system of the ensemble. This measurement has the effect of transforming each state η_i into the state ϕ_j, with probability $|(\phi_j, \eta_i)|^2$; i.e., it will transform a large ensemble of systems in the state η_i into an ensemble represented by the mixture whose density operator is $\sum_j |(\phi_j, \eta_i)|^2[\phi_j]$. Extending this result to the case where the original ensemble is a mixture of the η_i weighted by P_i ((3.2)), we find that the density operator ρ is transformed by the measurement of A into the new density operator ρ':

$$\rho' = \sum_i P_i \sum_j |(\eta_i, \phi_j)|^2[\phi_j] = \sum_j \left(\sum_i P_i(\phi_j, (\eta_i, \phi_j)\eta_i) \right) [\phi_j]$$

$$= \sum_j \left(\phi_j, \sum_i P_i[\eta_i]\phi_j \right) [\phi_j] = \sum_j (\phi_j, \rho\phi_j)[\phi_j]. \tag{3.3}$$

This is the general law by which mixtures change through Process 1.

However, even when no measurements are taking place, the states of an ensemble are changing according to Process 2, so that after a time interval t each state ψ will be transformed into a state $\psi' = U_t\psi$, where U_t is a unitary operator. This natural motion has the consequence that each mixture $\rho = \sum_i P_i[\eta_i]$ is carried into the mixture $\rho' = \sum_i P_i[U_t\eta_i]$ after

---

[3] Cf. Chapter III, §1.
[4] See Chapter III, §2, particularly footnote 6, p. 106.

a time t. But for every state ξ,

$$\rho'\xi = \sum_i P_i[U_t\eta_i]\xi = \sum_i P_i(U_t\eta_i, \xi)U_t\eta_i \qquad (3.4)$$

$$= U_t \sum_i P_i(\eta_i, U_t^{-1}\xi)\eta_i = U_t \sum_i P_i[\eta_i](U_t^{-1}\xi)$$

$$= (U_t\rho U_t^{-1})\xi.$$

Therefore

$$\rho' = U_t\rho U_t^{-1}, \qquad (3.5)$$

which is the general law for the change of a mixture according to Process 2.

We are now interested in whether or not we get from any mixture to another by means of these two processes, i.e., if for any pair ρ, ρ', there exist quantities A which can be measured and unitary (time dependence) operators U such that ρ can be transformed into ρ' by suitable applications of Processes 1 and 2. We shall see that this is not always possible and that Process 1 can cause irreversible changes in mixtures.

For each mixture ρ we define a quantity I_ρ:

$$I_\rho = \text{Trace} \, (\rho \ln \rho). \qquad (3.6)$$

This number, I_ρ, has the character of information. If $\rho = \sum_i P_i[\eta_i]$, a mixture of orthogonal states η_i weighted with P_i, then I_ρ is simply the information of the distribution P_i over the eigenstates of ρ (relative to the uniform measure). (Trace $(\rho \ln \rho)$ is a unitary invariant and is proportional to the negative of the entropy of the mixture, as discussed in Chapter III, §2.)

Process 2 therefore has the property that it leaves I_ρ unchanged, because

$$I_{\rho'} = \text{Trace} \, (\rho' \ln \rho') = \text{Trace} \, (U_t\rho U_t^{-1} \ln U_t\rho U_t^{-1})$$

$$= \text{Trace} \, (U_t\rho \ln \rho U_t^{-1}) = \text{Trace} \, (\rho \ln \rho) = I_\rho. \qquad (3.7)$$

Process 1, on the other hand, can decrease I_ρ but never increase it. According to (3.3):

$$\rho' = \sum_j (\phi_j, \rho\phi_j)[\phi_j] = \sum_{i,j} P_i|(\eta_i, \phi_j)|^2[\phi_j] = \sum_j P'_j[\phi_j], \qquad (3.8)$$

where $\rho'_j \sum_i P_i T_{ij}$ and $T_{ij} = |(\eta_i, \phi_j)|^2$ is a doubly-stochastic matrix.[5] But $I_{\rho'} = \sum_j P'_j \ln P'_j$ and $I_\rho = \sum_i P_i \ln P_i$, with the P_i, P'_j connected by T_{ij},

---

[5] Since $\sum_i T_{ij} = \sum_i |(\eta_i, \phi_j)|^2 = \sum_i (\phi_j, [\eta_i]\phi_j) = (\phi_j, \sum_i [\eta_i]\phi_j) = (\phi_j, I\phi_j) = 1$, and similarly $\sum_j T_{ij} = 1$ because T_{ij} is symmetric.

implies, by the theorem of information decrease for stochastic processes (II-§6), that:

$$I_{\rho'} \leqq I_{\rho} . \tag{3.9}$$

Moreover, it can easily be shown by a slight strengthening of the theorems of Chapter II, §6 that *strict* inequality must hold unless (for each i such that $\rho_i > 0$) $T_{ij} = 1$ for one j and 0 for the rest ($T_{ij} = \delta_{ikj}$). This means that $|(\eta_i, \phi_j)|^2 = \delta_{ikj}$, which implies that the original mixture was already a mixture of eigenstates of the measurement.

We have answered our question, and it is *not* possible to get from any mixture to another by means of Processes 1 and 2. There is an essential irreversibility to Process 1, since it corresponds to a stochastic process, which cannot be compensated by Process 2, which is reversible, like classical mechanics.[6]

Our theory of pure wave mechanics, to which we now return, must give equivalent results on the subjective level, since it leads to Process 1 there. Therefore, measuring processes will appear to be irreversible to any observers (even though the composite system including the observer changes its state reversibly).

There is another way of looking at this apparent irreversibility within our theory which recognizes only Process 2. When an observer performs an observation the result is a superposition, each element of which describes an observer who has perceived a particular value. From this time forward there is no interaction between the separate elements of the superposition (which describe the observer as having perceived different results), since each element separately continues to obey the wave equation.[by] Each observer described by a particular element of the superposition behaves in the future completely independently of any events in the remaining elements, and he can no longer obtain any information whatsoever concerning these other elements (they are completely unobservable to him).

The irreversibility of the measuring process is therefore, within our framework, simply a subjective manifestation reflecting the fact that in observation processes the state of the observer is transformed into a

---

[6] For another, more complete, discussion of this topic in the probabilistic interpretation see von Neumann [J. von Neumann, *Mathematical Foundations of Quantum Mechanics.* (Translated by R. T. Beyer) Princeton University Press: 1955.], Chapter V, §4.

[by] Since the dynamics is linear, there is a precise formal sense in which each element of the superposition can be thought of as following the dynamics separately, but this cannot be understood to preclude the possibility of interference between elements when predicted by the linear dynamics. It was essential to Everett's understanding of pure wave mechanics that interference between branches always be possible, at least in principle. See for example pgs. 149–50.

superposition of observer states, each element of which describes an observer who is irrevocably cut off from the remaining elements.[bz] While it is conceivable that some outside agency could reverse the total wave function, such a change cannot be brought about by any observer which is represented by a single element of a superposition, since he is entirely powerless to have any influence on any other elements. There are, therefore, fundamental restrictions to the knowledge that an observer can obtain about the state of the universe.[ca] It is impossible for any observer to discover the total state function of any physical system, since the process of observation itself leaves no independent state for the system or the observer, but only a composite system state in which the object-system states are inextricably bound up with the observer states. As soon as the observation is performed, the composite state is split into a superposition for which each element describes a different object-system state and an observer with (different) knowledge of it. Only the totality of these observer states, with their diverse knowledge, contains complete information about the original object-system state—but there is no possible communication between the observers described by these separate states. Any single observer can therefore possess knowledge only of the relative state function (relative to his state) of any systems, which is in any case all that is of any importance to him.[cb]

We conclude this section by commenting on another question which might be raised concerning irreversible processes: Is it necessary for the existence of measuring apparata, which can be correlated to other systems, to have frictional processes which involve systems of a large number of degrees of freedom? Are such thermodynamically irreversible processes possible in the framework of pure wave mechanics with a reversible wave equation, and if so, does this circumstance pose any difficulties for our treatment of measuring processes?

In the first place, it is certainly not necessary for dissipative processes involving additional degrees of freedom to be present before an interaction which correlates an apparatus to an object-system can take place. The counter-example is supplied by the simplified measuring process of III-§3,

---

[bz] See Everett's footnote regarding the language difficulty (pg. 121). See Everett's other discussions of reversibility and his discussions of interference between branches (pgs. 224, 240, 287, and 150).

[ca] The question of what one can affect and what one can know under the linear dynamics is somewhat more subtle than suggested by what Everett says here. Although there is a clear sense in which a relative observer cannot influence another element of the absolute state, he might at least in principle know the relative states associated with other elements of the absolute state by knowing something concerning the absolute state itself. See the discussion following pg. 73 and pg. 176, Wigner (1961), and the following footnote cb.

[cb] See Albert (1992, Ch. 8) for a description of what more one might know and Monton (1998) for further discussion of this point.

which involves only a system of one coordinate and an apparatus of one coordinate and no further degrees of freedom.[cc]

To the question whether such processes are possible within reversible wave mechanics, we answer *yes*, in the same sense that they are present in classical mechanics, where the microscopic equations of motion are also reversible. This type of irreversibility, which might be called *macroscopic irreversibility*, arises from a failure to separate "macroscopically indistinguishable" states into "true" microscopic states.[7] It has a fundamentally different character from the irreversibility of Process 1, which applies to micro-states as well and is peculiar to quantum mechanics. Macroscopically irreversible phenomena are common to both classical and quantum mechanics, since they arise from our incomplete information concerning a system, not from any intrinsic behavior of the system.[8]

Finally, even when such frictional processes are involved, they present no new difficulties for the treatment of measuring and observation processes given here. We imposed no restrictions on the complexity or number of degrees of freedom of measuring apparatus or observers, and if any of these processes are present (such as heat reservoirs, etc.) then these systems are to be simply included as part of the apparatus or observer.

§4. Approximate measurement

A phenomenon which is difficult to understand within the framework of the probabilistic interpretation of quantum mechanics is the result of an approximate measurement. In the abstract formulation of the usual theory there are two fundamental processes: the discontinuous, probabilistic Process 1 corresponding to precise measurement, and the continuous, deterministic Process 2 corresponding to absence of any measurement. What mixture of probability and causality are we to apply to the case where only an approximate measurement is effected (i.e., where the apparatus or observer interacts only weakly and for a finite time with the object-system)?[cd]

---

[7] See any textbook on statistical mechanics, such as ter Haar [D. ter Haar, *Elements of Statistical Mechanics*. Rinehart, New York, 1954.], Appendix I.

[8] Cf. the discussion of Chapter II, §7. See also von Neumann [J. von Neumann, *Mathematical Foundations of Quantum Mechanics*. (Translated by R. T. Beyer) Princeton University Press: 1955.], Chapter V, §4.

[cc] Everett is clear here that irreversible processes are not required for an interaction to count as a measurement—only an appropriate correlation between the pointer variable and the system being measured.

[cd] This is one of the central problems Everett starts with in his short thesis but does not discuss in detail there, chapter 9 (pg. 196). The point of this section is that, unlike Bohr's interpretation, Everett's relative-state interpretation provides compelling models for all correlating, or measurement-like, interactions.

In the case of approximate measurement, we need to be supplied with rules which will tell us, for any initial object-system state, first, with what probability can we expect the various possible apparatus readings, and second, what new state to ascribe to the system after the value has been observed. We shall see that it is generally impossible to give these rules within a framework which considers the apparatus or observer as performing an (abstract) observation subject to Process 1, and that it is necessary, in order to give a full account of approximate measurements, to treat the entire system, including apparatus or observer, wave mechanically.

The position that an approximate measurement results in the situation that the object-system state is changed into an eigenstate of the exact measurement, but for which particular one the observer has only imprecise information, is manifestly false. It is a fact that we can make successive approximate position measurements of particles (in cloud chambers, for example) and use the results for somewhat reliable predictions of future positions. However, if either of these measurements left the particle in an "eigenstate" of position (δ function), even though the particular one remained unknown, the momentum would have such a variance that no such prediction would be possible. (The possibility of such predictions lies in the correlations between position and momentum at one time with position and momentum at a later time for wave packets[9]—correlations which are totally destroyed by precise measurements of either quantity.)

Instead of continuing the discussion of the inadequacy of the probabilistic formulation, let us first investigate what actually happens in approximate measurements, from the viewpoint of pure wave mechanics. An approximate measurement consists of an interaction, for a finite time, which only imperfectly correlates the apparatus (or observer) with the object-system. We can deduce the desired rules in any particular case by the following method: For fixed interaction and initial apparatus state and for any initial object-system state we solve the wave equation for the time of interaction in question. The result will be a superposition of apparatus (observer) states and relative object-system states. Then (according to the method of Chapter IV for assigning a measure to a superposition) we assign a probability to each observed result equal to the square-amplitude of the coefficient of the element which contains the apparatus (observer) state representing the registering of that result. Finally, the object-system is assigned the new state which is its *relative state* in that element.

For example, let us consider the measuring process described in Chapter III-§3, which is an excellent model for an approximate measurement. After the interaction, the total state was found to be (pg. 183):

$$\psi_t^{S+A} = \int \frac{1}{N_{r'}} \xi^{r'}(q)\delta(r - r')\, dr'. \tag{3.10}$$

[9] See Bohm [D. Bohm, *Quantum Theory*. Prentice-Hall, New York: 1951.], p. 202.

Then, according to our prescription, we assign the probability density $P(r')$ to the observation of the apparatus coordinate r'

$$P(r') = \left| \frac{1}{N_{r'}} \right|^2 = \int \phi^* \phi(q) \eta^* \eta(r' - qt) \, dq, \qquad (3.11)$$

which is the square amplitude of the coefficient $(1/N_{r'})$ of the element $\xi^{r'}(q)\delta(r - r')$ of the superposition (3.10) in which the apparatus coordinate has the value $r = r'$. Then, depending upon the observed apparatus coordinate r', we assign the object-system the new state

$$\xi^{r'}(q) = N_{r'}\phi(q)\eta(r' - qt) \qquad (3.12)$$

(where $\phi(q)$ is the old state, and $\eta(r)$ is the initial apparatus state) which is the relative object-system state in (3.10) for apparatus coordinate r'.

This example supplies the counter-example to another conceivable method of dealing with approximate measurement within the framework of Process 1. This is the position that when an approximate measurement of a quantity Q is performed, in actuality another quantity Q' is precisely measured, where the eigenstates of Q' correspond to fairly well-defined (i.e., sharply peaked distributions for) Q values.[10] However, any such scheme based on Process 1 always has the prescription that after the measurement, the (unnormalized) new state function results from the old by a projection (on an eigenstate or eigenspace), which depends upon the observed value. If this is true, then in the above example the new state $\xi^{r'}(q)$ must result from the old, $\phi(q)$, by a projection E:

$$\xi^{r'}(q) = NE\phi(q) = N_{r'}\phi(q)\eta(r' - qt) \qquad (3.13)$$

where N, $N_{r'}$ are normalization constants. But E is only a projection if $E^2 = E$. Applying the operation (3.13) twice, we get:

$$E(NE\phi(q)) = NE^2\phi(q) = N'\phi(q)\eta^2(r' - qt) \Rightarrow E^2\phi(q)$$
$$= \frac{N'}{N}\phi(q)\eta^2(r' - qt), \qquad (3.14)$$

and we see that E cannot be a projection unless $\eta(q) = \eta^2(q)$ for all q (i.e., $\eta(q) = 0$ or 1 for all q) and we have arrived at a contradiction to the assumption that in all cases the changes of state for approximate measurements are governed by projections. (In certain special cases, such as approximate position measurements with slits or Geiger counters,[11] the new functions arise from the old by multiplication by sharp cutoff

---

[10] Cf. von Neumann [J. von Neumann, *Mathematical Foundations of Quantum Mechanics.* (Translated by R. T. Beyer) Princeton University Press: 1955.], Chapter IV, §4.

[11] Cf. §2, this chapter.

functions which are 1 over the slit or counter and 0 elsewhere, so that these measurements *can* be handled by projections.)

One cannot, therefore, account for approximate measurements by any scheme based on Process 1, and it is necessary to investigate these processes entirely wave-mechanically. Our viewpoint constitutes a framework in which it is possible to make precise deductions about such measurements and observations, since we can follow in detail the interaction of an observer or apparatus with an object-system.

§5. Discussion of a spin measurement example

We shall conclude this chapter with a discussion of an instructive example of Bohm.[12] Bohm considers the measurement of the z component of the angular momentum of an atom, whose total angular momentum is $\frac{\hbar}{2}$, which is brought about by a Stern-Gerlach experiment. The measurement is accomplished by passing an atomic beam through an inhomogeneous magnetic field, which has the effect of giving the particle a momentum which is directed up or down depending upon whether the spin was up or down.

The measurement is treated as impulsive, so that during the time that the atom passes through the field the Hamiltonian is taken to be simply the interaction:

$$H_I = \mu(\vec{\delta} \cdot \vec{\mathcal{H}}), \quad \mu = -\frac{e\hbar}{2mc} \tag{3.15}$$

where $\mathcal{H}$ is the magnetic field and $\vec{\delta}$ is the spin operator for the atom. The particle is presumed to pass through a region of the field where the field is in the z direction, so that during the time of transit the field is approximately $\mathcal{H}_z \cong \mathcal{H}_0 + z\mathcal{H}'_0 \left(\mathcal{H}_0 = (\mathcal{H}_z)_{z=0} \text{ and } \mathcal{H}'_0 = \left(\frac{\partial \mathcal{H}_z}{\partial z} \right)_{z=0} \right)$, and hence the interaction is approximately:

$$H_I \cong \mu(\mathcal{H}_0 + z\mathcal{H}'_0)S_z, \tag{3.16}$$

where S_z denotes the operator for the z component of the spin.

It is assumed that the state of the atom, just prior to entry into the field, is a wave packet of the form:

$$\psi_0 = f_0(z)(c_+v_+ + c_-v_-) \tag{3.17}$$

where v_+ and v_- are the spin functions for $S_z = 1$ and -1, respectively. Solving the Schrödinger equation for the Hamiltonian (3.16) and initial

---

[12] Bohm [D. Bohm, *Quantum Theory*. Prentice-Hall, New York: 1951.], p. 593.

condition (3.17) yields the state for a later time t:

$$\psi = f_0(z)(c_+ e^{-i\mu(\mathcal{H}_0 + z\mathcal{H}'_0)t/\hbar} v_+ + c_- e^{+i\mu(\mathcal{H}_0 + z\mathcal{H}'_0)t/\hbar} v_-). \qquad (3.18)$$

Therefore, if Δt is the time that it takes the atom to traverse the field,[13] each component of the wave packet has been multiplied by a phase factor $e^{\pm i\mu(\mathcal{H}_0 + z\mathcal{H}'_0)\Delta t/\hbar}$, i.e., has had its mean momentum in the z direction changed by an amount $\pm\mathcal{H}'_0\mu\Delta t$, depending upon the spin direction. Thus the initial wave packet (with mean momentum zero) is split into a superposition of two packets, one with mean z-momentum $+\mathcal{H}'_0\mu\Delta t$ and spin up, and the other with spin down and mean z-momentum $-\mathcal{H}'_0\mu\Delta t$.

The interaction (3.16) has therefore served to correlate the spin with the momentum in the z-direction. These two packets of the resulting superposition now move in opposite z-directions, so that after a short time they become widely separated (provided that the momentum changes $\pm\mathcal{H}'_0\mu\Delta t$ are large compared to the momentum spread of the original packet), and the z-coordinate is itself then correlated with the spin— representing the "apparatus" coordinate in this case. The Stern-Gerlach apparatus therefore splits an incoming wave packet into a superposition of two diverging packets, corresponding to the two spin values.

We take this opportunity to caution against a certain viewpoint which can lead to difficulties. This is the idea that, after an apparatus has interacted with a system, in "actuality" one or another of the elements of the resultant superposition described by the composite state-function has been realized to the exclusion of the rest, the existing one simply being unknown to an external observer (i.e., that instead of the superposition there is a genuine mixture). This position must be erroneous since there is always the possibility for the external observer to make use of interference properties between the elements of the superposition.

In the present example, for instance, it is in principle possible to deflect the two beams back toward one another with magnetic fields and recombine them in another inhomogeneous field, which duplicates the first, in such a manner that the original spin state (before entering the apparatus) is restored.[14] This would not be possible if the original Stern-Gerlach apparatus performed the function of converting the original wave packet into a non-interfering mixture of packets for the two spin cases. Therefore the position that after the atom has passed through the inhomogeneous field it is "really" in one or the other beam with the corresponding spin, although we are ignorant of which one, is incorrect.

---

[13] This time is, strictly speaking, not well defined. The results, however, do not depend critically upon it.

[14] As pointed out by Bohm [D. Bohm, *Quantum Theory*. Prentice-Hall, New York: 1951.], p. 604.

After two systems have interacted and become correlated it is true that marginal expectations for *subsystem* operators can be calculated correctly when the composite system is represented by a certain non-interfering mixture of states. Thus if the composite system state is $\psi^{S_1+S_2} = \sum_i a_i \phi_i^{S_1} \eta_i^{S_2}$, where the $\{\eta_i\}$ are orthogonal, then for purposes of calculating the expectations of operators on S_1 the state $\psi^{S_1+S_2}$ is equivalent to the non-interfering mixture of states $\phi_i^{S_1} \eta_i^{S_2}$ weighted by $P_i = a_i^* a_i$, and one can take the picture that one or another of the cases $\phi_i^{S_1} \eta_i^{S_2}$ has been realized to the exclusion of the rest, with probabilities P_i.[15]

However, this representation by a mixture must be regarded as only a mathematical artifice which, although useful in many cases, is an *incomplete description* because it ignores phase relations between the separate elements which actually exist, and which become important in any interactions which involve more than just a subsystem.

In the present example, the "composite system" is made of the "sub-systems" spin value (object-system) and z-coordinate (apparatus), and the superposition of the two diverging wave packets is the state after interaction. It is only correct to regard this state as a mixture so long as any contemplated future interactions or measurements will involve only the spin value or only the z-coordinate, but not both simultaneously. As we saw, phase relations between the two packets are present and become important when they are deflected back and recombined in another inhomogeneous field—a process involving the spin values and z-coordinate simultaneously.

It is therefore improper to attribute any less validity or "reality" to any element of a superposition than any other element, due to this ever present possibility of obtaining interference effects between the elements. All elements of a superposition must be regarded as simultaneously existing.[ce]

At this time we should like to add a few remarks concerning the notion of *transition probabilities* in quantum mechanics. Often one considers a system, with Hamiltonian H and stationary states $\{\phi_i\}$, to be perturbed for a time by a time-dependent addition to the Hamiltonian, $H_I(t)$. Then under the action of the perturbed Hamiltonian $H' = H + H_I(t)$ the states $\{\phi_i\}$ are generally no longer stationary but change after time t into new states $\{\psi_i(t)\}$:

$$\phi_i \to \psi_i(t) = \sum_j (\phi_j, \psi_i(t)) \phi_j = \sum_j a_{ij}(t) \phi_j, \tag{3.19}$$

[15] See Chapter III, §1.

[ce] Everett's argument for the operational reality of all branches then was that the linear dynamics *requires* that it is always possible in principle that one might observe interference between branches.

which can be represented as a superposition of the old stationary states with time-dependent coefficients $a_{ij}(t)$.

If at time τ a measurement with eigenstates ϕ_j is performed, such as an energy measurement (whose operator is the original H), then according to the probabilistic interpretation the probability for finding the state ϕ_j, given that the state was originally ϕ_i, is $P_{ij}(\tau) = |a_{ij}(\tau)|^2$. The quantities $|a_{ij}(\tau)|^2$ are often referred to as *transition probabilities*. In this case, however, the name is a misnomer, since it carries the connotation that the original state ϕ_i is transformed into a *mixture* (of the ϕ_j weighted by $P_{ij}(\tau)$) and gives the erroneous impression that the quantum formalism itself implies the existence of quantum-jumps (stochastic processes) independent of acts of observation. This is incorrect since there is still a pure state $\sum_i a_{ij}(\tau)\phi_j$ with phase relations between the ϕ_j, and expectations of operators other than the energy *must* be calculated from the superposition and not the mixture.

There is another case, however, the one usually encountered in fact, where the transition probability concept is somewhat more justified. This is the case in which the perturbation is due to interaction of the system s_1 with another system s_2, and not simply a time dependence of s_1's Hamiltonian as in the case just considered. In this situation the interaction produces a *composite system state*, for which there are in general no independent subsystem states. However, as we have seen, for purposes of calculating expectations of operators on s_1 alone, we can regard s_1 as being represented by a certain mixture. According to this picture the states of subsystem s_1 are gradually converted into mixtures by the interaction with s_2, and the concept of transition probability makes some sense. Of course, it must be remembered that this picture is only justified so long as further measurements on s_1 alone are contemplated, and any attempt to make a simultaneous determination in s_1 and s_2 involves the composite state where interference properties may be important.

An example is a hydrogen atom interacting with the electromagnetic field. After a time of interaction we can picture the atom as being in a mixture of its states, so long as we consider future measurements on the atom only. But in actuality the state of the atom is dependent upon (correlated with) the state of the field, and some process involving both atom and field could conceivably depend on interference effects between the states of the alleged mixture. With these restrictions, however, the concept of transition probability is quite useful and justified.

VI. Discussion

We have shown that our theory based on pure wave mechanics, which takes as the basic description of physical systems the state function—supposed to be an *objective* description (i.e., in one-one, rather than statistical, correspondence to the behavior of the system)—can be put in satisfactory correspondence with experience. We saw that the probabilistic assertions of the usual interpretation of quantum mechanics can be *deduced* from this theory, in a manner analogous to the methods of classical statistical mechanics, as subjective appearances to observers—observers which were regarded simply as physical systems subject to the same type of description and laws as any other systems, and having no preferred position. The theory is therefore capable of supplying us with a complete conceptual model of the universe, consistent with the assumption that it contains more than one observer.

Because the theory gives us an objective description, it constitutes a framework in which a number of puzzling subjects (such as classical level phenomena, the measuring process itself, the interrelationship of several observers, questions of reversibility and irreversibility, etc.) can be investigated in detail in a logically consistent manner. It supplies a new way of viewing processes, which clarifies many apparent paradoxes of the usual interpretation[1]—indeed, it constitutes an objective framework in which it is possible to understand the general consistency of the ordinary view.

We shall now resume our discussion of alternative interpretations. There has been expressed lately a great deal of dissatisfaction with the present form of quantum theory by a number of authors, and a wide variety of new interpretations have sprung into existence. We shall now attempt to classify briefly a number of these interpretations and comment upon them.

    a. *The "popular" interpretation.* This is the scheme alluded to in the introduction, where ψ is regarded as objectively characterizing the single system, obeying a deterministic wave equation when the system is isolated but changing probabilistically and discontinuously under observation.

In its unrestricted form this view can lead to paradoxes like that mentioned in the introduction and is therefore untenable. However, this view *is* consistent so long as it is assumed that there is only one observer in the universe (the solipsist position—Alternative 1 of the Introduction). This consistency is most easily understood from the viewpoint of our own theory, where we were able to show that all phenomena will *seem* to follow the predictions of this scheme to any observer. Our theory therefore justifies the personal adoption of this probabilistic interpretation, for purposes of making practical predictions, from a more satisfactory framework.

    b. *The Copenhagen interpretation.* This is the interpretation developed by Bohr. The ψ function is not regarded as an objective description of a physical system (i.e., it is in no sense a conceptual model), but is

---

[1] Such as that of Einstein, Rosen, and Podolsky [A. Einstein, B. Podolsky, N. Rosen, *Phys. Rev.* 47, 777, 1935.], as well as the paradox of the introduction.

regarded as merely a mathematical artifice which enables one to make statistical predictions, albeit the best predictions which it is possible to make. This interpretation in fact denies the very possibility of a single conceptual model applicable to the quantum realm and asserts that the totality of phenomena can only be understood by the use of different, mutually exclusive (i.e., "complementary") models in different situations. All statements about microscopic phenomena are regarded as meaningless unless accompanied by a complete description (classical) of an experimental arrangement.

While undoubtedly safe from contradiction, due to its extreme conservatism, it is perhaps overcautious.[cf] We do not believe that the primary purpose of theoretical physics is to construct "safe" theories at severe cost in the applicability of their concepts, which is a sterile occupation, but to make useful models which serve for a time and are replaced as they are outworn.[2,cg]

Another objectionable feature of this position is its strong reliance upon the classical level from the outset, which precludes any possibility of explaining this level on the basis of an underlying quantum theory. (The deduction of classical phenomena from quantum theory is impossible simply because no meaningful statements can be made without pre-existing classical apparatus to serve as a reference frame). This interpretation suffers from the dualism of adhering to a "reality" concept (i.e., the possibility of objective description) on the classical level but renouncing the same in the quantum domain.

  c. *The "hidden variables" interpretation.* This is the position (Alternative 4 of the Introduction) that ψ is not a complete description of a single system. It is assumed that the correct complete description, which would involve further (hidden) parameters, would lead to a deterministic theory, from which the probabilistic aspects arise as a result of our ignorance of these extra parameters in the same manner as in classical statistical mechanics.

---

[2] Cf. Appendix II.

[cf] Everett's criticisms of the Copenhagen interpretation led to conflict with Bohr and the Copenhagen colleagues. Wheeler, as his adviser, tried to explain that Everett did not really mean to be attacking the orthodox interpretation. See chapter 12 (pg. 219) for Wheeler's defense of Everett and the later exchange between Everett and Petersen in the discussions following pgs. 236 and 238. See also the discussion of Everett's views in the conceptual introduction, chapter 3 (pg. 32).

[cg] In addition to wanting a theory that satisfies the minimal conditions of being logically consistent and empirically faithful, Everett explains in the second appendix that one might also want a theory that is comprehensive and pictorable. Such a theory would provide models for all physical interactions, including measurements, something the Copenhagen interpretation does not accomplish and explicitly denies as being a virtue.

The ψ-function is therefore regarded as a description of an *ensemble* of systems rather than a single system. Proponents of this interpretation include Einstein,[3] Bohm,[4] Wiener and Siegel.[5,ch]

Einstein hopes that a theory along the lines of his general relativity, where all of physics is reduced to the geometry of space-time could satisfactorily explain quantum effects. In such a theory a particle is no longer a simple object but possesses an enormous amount of structure (i.e., it is thought of as a region of space-time of high curvature). It is conceivable that the interactions of such "particles" would depend in a sensitive way upon the details of this structure, which would then play the role of the "hidden variables."[6] However, these theories are non-linear and it is enormously difficult to obtain any conclusive results. Nevertheless, the possibility cannot be discounted.

Bohm considers ψ to be a real force field acting on a particle which always has a well-defined position and momentum (which are the hidden variables of this theory). The ψ-field satisfying Schrödinger's equation is pictured as somewhat analogous to the electromagnetic field satisfying Maxwell's equations, although for systems of n particles the ψ-field is in a $3n$-dimensional space. With this theory Bohm succeeds in showing that in all actual cases of measurement the best predictions that can be made are those of the usual theory, so that no experiments could ever rule out his interpretation in favor of the ordinary theory. Our main criticism of this view is on the grounds of simplicity—if one desires to hold the view that ψ is a real field then the associated particle is superfluous since, as we have endeavored to illustrate, the pure wave theory is itself satisfactory.

Wiener and Siegel[ci] have developed a theory which is more closely tied to the formalism of quantum mechanics. From the set N of all non-degenerate linear Hermitian operators for a system having a complete set of eigenstates, a subset I is chosen such that no two members of I commute and every element outside I commutes with at least one element of I. The set I therefore contains precisely one operator for every orientation of the

---

[3] Einstein [A. Einstein, in *Albert Einstein, Philosopher-Scientist*. The Library of Living Philosophers, Inc., Vol. 7, p. 665. Evanston: 1949.].

[4] Bohm [D. Bohm, *Phys. Rev.* 84, 166, 1952 and 85, 180, 1952.].

[5] Wiener and Siegel [N. Wiener, A. Siegel, *Nuovo Cimento Suppl.* 2, 982 (1955).].

[6] For an example of this type of theory see Einstein and Rosen [A. Einstein, N. Rosen, *Phys. Rev.* 48, 73, 1935.].

[ch] Everett's reference here originally read "I. E. Siegel." Everett seems to have mistaken Norbert Wiener's collaborator Armand Siegel for the mathematical physicist Irving Segal. Everett's primary argument here against hidden variable theories is that hidden variables are not needed since pure wave mechanics is similarly consistent, empirically faithful, comprehensive, and pictorable, but is also simpler. See appendix II (pg. 168) for Everett's discussion of theoretical virtues and theory selection.

[ci] Text read "Siegal."

principal axes of the Hilbert space for the system. It is postulated that each of the operators of I corresponds to an independent observable which can take any of the real numerical values of the spectrum of the operator. This theory, in its present form, is a theory of infinitely[7] many "hidden variables," since a system is pictured as possessing (at each instant) a value for every one of these "observables" simultaneously, with the changes in these values obeying precise (deterministic) dynamical laws. However, the change of any one of these variables with time depends upon the entire set of observables, so that it is impossible ever to discover by measurement the complete set of values for a system (since only one "observable" at a time can be observed). Therefore, statistical ensembles are introduced, in which the values of all of the observables are related to points in a "differential space," which is a Hilbert space containing a measure for which each (differential space) coordinate has an independent normal distribution. It is then shown that the resulting statistical dynamics is in accord with the usual form of quantum theory.

It cannot be disputed that these theories are often appealing and might conceivably become important should future discoveries indicate serious inadequacies in the present scheme (i.e., they might be more easily modified to encompass new experience). But from our viewpoint they are usually more cumbersome than the conceptually simpler theory based on pure wave mechanics. Nevertheless, these theories are of great theoretical importance because they provide us with examples that "hidden variables" theories are indeed possible.

   d. *The stochastic process interpretation.* This is the point of view which holds that the fundamental processes of nature are stochastic (i.e., probabilistic) processes. According to this picture physical systems are supposed to exist at all times in definite states, but the states are continually undergoing probabilistic changes. The discontinuous probabilistic "quantum-jumps" are not associated with acts of observation but are fundamental to the systems themselves.[cj]

A stochastic theory which emphasizes the particle, rather than wave, aspects of quantum theory has been investigated by Bopp.[8] The particles do not obey deterministic laws of motion, but rather probabilistic laws, and by developing a general "correlation statistics" Bopp shows that his

---

[7] A non-denumerable infinity, in fact, since the set I is uncountable!

[8] Bopp [F. Bopp, *Z. Naturforsch.* 2a(4), 202, 1947; 7a, 82, 1952; 8a, 6, 1953.].

[cj] The GRW formulation of quantum mechanics is a recent example of such a theory (Ghirardi et al., 1986). Everett had no fundamental objection to this strategy. But since he believed that pure wave mechanics formed a satisfactory theory, he took the introduction of a stochastic dynamics to be unnecessary.

quantum scheme is a special case which gives results in accord with the usual theory. (This accord is only approximate, and in principle one could decide between the theories. The approximation is so close, however, that it is hardly conceivable that a decision would be practically feasible.)[ck]

Bopp's theory seems to stem from a desire to have a theory founded upon particles rather than waves, since it is this particle aspect (highly localized phenomena) which is most frequently encountered in present day high-energy experiments (cloud chamber tracks, etc.). However, it seems to us to be much easier to understand particle aspects from a wave picture (concentrated wave packets) than it is to understand wave aspects (diffraction, interference, etc.) from a particle picture.

Nevertheless, there can be no fundamental objection to the idea of a stochastic theory, except on grounds of a naked prejudice for determinism. The question of determinism or indeterminism in nature is obviously forever undecidable in physics, since for any current deterministic [probabilistic] theory one could always postulate that a refinement of the theory would disclose a probabilistic [deterministic] substructure, and that the current deterministic [probabilistic] theory is to be explained in terms of the refined theory on the basis of the law of large numbers [ignorance of hidden variables].[cl] However, it is quite another matter to object to a mixture of the two where the probabilistic processes occur only with acts of observation.

   e. *The wave interpretation.* This is the position proposed in the present thesis, in which the wave function itself is held to be the fundamental entity, obeying at all times a deterministic wave equation.

This view also corresponds most closely with that held by Schrödinger.[9] However, this picture only makes sense when observation processes themselves are treated within the theory. It is only in this manner that the *apparent* existence of definite macroscopic objects, as well as localized phenomena, such as tracks in cloud chambers, can be satisfactorily explained in a wave theory where the waves are continually diffusing. With the deduction in this theory that phenomena will appear to observers to be subject to Process 1, Heisenberg's criticism[10] of Schrödinger's opinion—that continuous wave mechanics could not seem to explain the discontinuities which are everywhere observed—is effectively met. The "quantum-jumps" exist in our theory as *relative* phenomena (i.e., the states of an object-system relative to chosen observer states show this effect), while the absolute states change quite continuously.

---

[9] Schrödinger [E. Schrödinger, *Brit. J. Phil. Sci.* 3, 109, 233, 1952.].
[10] Heisenberg [W. Heisenberg, in *Niels Bohr and the Development of Physics.* McGraw-Hill, p. 12, New York: 1955.].

[ck] See also Everett's discussion of Bopp's theory in his letter to DeWitt in chapter 16 (pg. 256).
[cl] The brackets here are Everett's own.

The wave theory is definitely tenable and forms, we believe, the simplest complete, self-consistent theory.[cm]

We should like now to comment on some views expressed by Einstein. Einstein's[11] criticism of quantum theory (which is actually directed more against what we have called the "popular" view than Bohr's interpretation) is mainly concerned with the drastic changes of state brought about by simple acts of observation (i.e., the infinitely rapid collapse of wave functions), particularly in connection with correlated systems which are widely separated so as to be mechanically uncoupled at the time of observation.[12] At another time he put his feeling colorfully by stating that he could not believe that a mouse could bring about drastic changes in the universe simply by looking at it.[13]

However, from the standpoint of our theory, *it is not so much the system which is affected by an observation as the observer, who becomes correlated to the system.*

In the case of observation of one system of a pair of spatially separated, correlated systems, nothing happens to the remote system to make any of its states more "real" than the rest. It had no independent states to begin with, but a number of states occurring in a superposition with corresponding states for the other (near) system. Observation of the near system simply correlates the observer to this system, a purely local process—but a process which also entails automatic correlation with the remote system. Each state of the remote system still exists with the same amplitude in a superposition, but now a superposition for which element contains, in addition to a remote system state and correlated near system state, an observer state which describes an observer who perceives the state of the near system.[14] From the present viewpoint all elements of this superposition are equally "real." Only the observer state has changed, so as to become correlated with the state of the near system and hence naturally with that of the remote system also. The mouse does not affect the universe—only the mouse is affected.[cn]

---

[11] Einstein [A. Einstein, in *Albert Einstein, Philosopher-Scientist.* The Library of Living Philosophers, Inc., Vol. 7, p. 665. Evanston, Ill.: 1949.].

[12] For example, the paradox of Einstein, Rosen, and Podolsky [A. Einstein, B. Podolsky, N. Rosen, *Phys. Rev.* 47, 777, 1935.].

[13] Address delivered at Palmer Physical Laboratory, Princeton, Spring, 1954.

[14] See in this connection Chapter IV, particularly pp. 205, 206.

[cm] Although Everett took other strategies for addressing the quantum measurement problem seriously, he favored pure wave mechanics on the grounds that it satisfied the minimal conditions of consistency and empirical faithfulness and had the added virtues of being simple and comprehensive. See Appendix II (pg. 168).

[cn] The suggestion is that the mouse's observation does not cause the universe to split. Rather, the observation sets up a correlation between the mouse's measurement record and the world thereby splitting the mouse's *state* into a set of relative states, typically one for each possible measurement outcome. See also pgs. 188–89.

Our theory in a certain sense bridges the positions of Einstein and Bohr, since the complete theory is quite objective and deterministic ("God does not play dice with the universe"), and yet on the subjective level, of assertions relative to observer states, it is probabilistic in the *strong sense* that there is no way for observers to make any predictions better than the limitations imposed by the uncertainty principle.[15]

In conclusion, we have seen that if we wish to adhere to objective descriptions then the principle of the psycho-physical parallelism requires that we should be able to consider some mechanical devices as representing observers. The situation is then that such devices must either cause the probabilistic discontinuities of Process 1 or must be transformed into the superpositions we have discussed. We are forced to abandon the former possibility since it leads to the situation that some physical systems would obey different laws from the rest, with no clear means for distinguishing between these two types of systems. We are thus led to our present theory which results from the complete abandonment of Process 1 as a basic process. Nevertheless, within the context of this theory, which is objectively deterministic, it develops that the probabilistic aspects of Process 1 reappear at the subjective level, as relative phenomena to observers.

One is thus free to build a conceptual model of the universe, which postulates only the existence of a universal wave function which obeys a linear wave equation. One then investigates the internal correlations in this wave function with the aim of deducing laws of physics, which are statements that take the form: Under the conditions C the property A of a subsystem of the universe (subset of the total collection of coordinates for the wave function) is correlated with the property B of another subsystem (with the manner of correlation being specified). For example, the classical mechanics of a system of massive particles becomes a law which expresses the correlation between the positions and momenta (approximate) of the particles at one time with those at another time.[16] All statements about subsystems then become *relative* statements, i.e., statements about the subsystem relative to a prescribed state for the remainder (since this is generally the only way a subsystem even possesses a unique state), and all laws are correlation laws.[co]

The theory based on pure wave mechanics is a conceptually simple causal theory, which fully maintains the principle of the psycho-physical parallelism. It therefore forms a framework in which it is possible to discuss

---

[15] Cf. Chapter V, §2.

[16] Cf. Chapter V, §2.

[co] This is Everett's summary of how pure wave mechanics explains the quasi-classical behavior of macrosystems. See also the earlier extended discussion (pgs. 134–37).

(in addition to ordinary phenomena) observation processes themselves, including the interrelationships of several observers, in a logical, unambiguous fashion. In addition, all of the correlation paradoxes, like that of Einstein, Rosen, and Podolsky,[17] find easy explanation.

While our theory justifies the personal use of the probabilistic interpretation as an aid to making practical predictions, it forms a broader frame in which to understand the consistency of that interpretation. It transcends the probabilistic theory, however, in its ability to deal logically with questions of imperfect observation and approximate measurement.

Since this viewpoint will be applicable to all forms of quantum mechanics which maintain the superposition principle, it may prove a fruitful framework for the interpretation of new quantum formalisms. Field theories, particularly any which might be relativistic in the sense of general relativity, might benefit from this position, since one is free to construct formal (non-probabilistic) theories, and supply any possible statistical interpretations later. (This viewpoint avoids the necessity of considering anomalous probabilistic jumps scattered about space-time, and one can assert that field equations are satisfied everywhere and everywhen, then *deduce* any statistical assertions by the present method).

By focusing attention upon questions of correlations, one may be able to deduce useful relations (correlation laws analogous to those of classical mechanics) for theories which at present do not possess known classical counterparts. Quantized fields do not generally possess pointwise independent field values, the values at one point of space-time being correlated with those at neighboring points of space-time in a manner, it is to be expected, approximating the behavior of their classical counterparts. If correlations are important in systems with only a finite number of degrees of freedom, how much more important they must be for systems of infinitely many coordinates.

Finally, aside from any possible practical advantages of the theory, it remains a matter of intellectual interest that the statistical assertions of the usual interpretation do not have the status of independent hypotheses, but are deducible (in the present sense) from the pure wave mechanics, which results from their omission.

APPENDIX I

We shall now supply the proofs of a number of assertions which have been made in the text.

---

[17] Einstein, Rosen, and Podolsky [A. Einstein, B. Podolsky, N. Rosen, *Phys. Rev.* 47, 777, 1935.].

§1. Proof of Theorem 1

We now show that $\{X, Y, \ldots, Z\} > 0$ unless $X, Y, \ldots, Z$ are independent random variables. Abbreviate $P(x_i, y_j, \ldots, z_k)$ by $P_{ij\ldots k}$, and let

$$Q_{ij\ldots k} = \begin{cases} \frac{P_{ij\ldots k}}{P_i P_j \ldots P_k} & \text{if } P_i P_j \ldots P_k > 0 \\ 1 & \text{if } P_i P_j \ldots P_k = 0 \end{cases} \tag{1.1}$$

(Note that $P_i P_j \ldots P_k = 0$ implies that also $P_{ij\ldots k} = 0$.) Then always

$$P_{ij\ldots k} = Q_{ij\ldots k} P_i P_j \ldots P_k, \tag{1.2}$$

and we have

$$\{X, Y, \ldots, Z\} = \text{Exp}\left[\ln \frac{P_{ij\ldots k}}{P_i P_j \ldots P_k}\right] = \text{Exp}[\ln Q_{ij\ldots k}]$$
$$= \sum_{ij\ldots k} P_i P_j \ldots P_k Q_{ij\ldots k} \ln Q_{ij\ldots k}. \tag{1.3}$$

Applying the inequality for $x \geq 0$:

$$x \ln x > x - 1 \qquad (\text{except for } x = 1) \tag{1.4}$$

(which is easily established by calculating the minimum of $x \ln x - (x - 1)$) to (1.3) we have:

$$P_i P_j \ldots P_k Q_{ij\ldots k} \ln Q_{ij\ldots k} > P_i P_j \ldots P_k (Q_{ij\ldots k} - 1)$$
$$(\text{unless } Q_{ij\ldots k} = 1). \tag{1.5}$$

Therefore we have for the sum:

$$\sum_{ij\ldots k} P_i P_j \ldots P_k Q_{ij\ldots k} \ln Q_{ij\ldots k} > \sum_{ij\ldots k} P_i P_j \ldots P_k Q_{ij\ldots k} - \sum_{ij\ldots k} P_i P_j \ldots P_k, \tag{1.6}$$

unless *all* $Q_{ij\ldots k} = 1$. But $\sum_{ij\ldots k} P_i P_j \ldots P_k Q_{ij\ldots k} = \sum_{ij\ldots k} P_{ij\ldots k} = 1$, and $\sum_{ij\ldots k} P_i P_j \ldots P_k = 1$, so that the right side of (1.6) vanishes. The left side is, by (1.3) the correlation $\{X, Y, \ldots, Z\}$, and the condition that all of the $Q_{ij\ldots k}$ equal 1 is precisely the independence condition that $P_{ij\ldots k} = P_i P_j \ldots P_k$ for all $i, j, \ldots, k$. We have therefore proved that

$$\{X, Y, \ldots, Z\} > 0 \tag{1.7}$$

unless $X, Y, \ldots, Z$ are mutually independent.

§2. Convex function inequalities

We shall now establish some basic inequalities which follow from the convexity of the function $x \ln x$.

LEMMA 1. $x_i \geq 0, \quad P_i \geq 0, \quad \sum_i P_i = 1$

$$\Rightarrow \left(\sum_i P_i x_i \right) \ln \left(\sum_i P_i x_i \right) \leq \sum_i P_i x_i \ln x_i.$$

This property is usually taken as the definition of a convex function,[18] but follows from the fact that the second derivative of $x \ln x$ is positive for all positive x, which is the elementary notion of convexity. There is also an immediate corollary for the continuous case:

COROLLARY 1. $g(x) \geq 0, \quad P(x) \geq 0, \quad \int P(x)\, dx = 1$

$$\Rightarrow \left[\int P(x)g(x)\, dx \right] \ln \left[\int P(x)g(x)\, dx \right] \leq \int P(x)g(x) \ln g(x)\, dx.$$

We can now derive a more general and very useful inequality from Lemma 1:

LEMMA 2. $x_i \geq 0, \quad a_i \geq 0 \quad$ (all i)

$$\Rightarrow \left(\sum_i x_i \right) \ln \left(\frac{\sum_i x_i}{\sum_i a_i} \right) \leq \sum_i x_i \ln \left(\frac{x_i}{a_i} \right).$$

Proof. Let $P_i = a_i / \sum_i a_i$, so that $P_i \geq 0$ and $\sum_i P_i = 1$. Then by Lemma 1:

$$\left[\sum_i P_i \left(\frac{x_i}{a_i} \right) \right] \ln \left[\sum_i P_i \left(\frac{x_i}{a_i} \right) \right] \leq \sum_i P_i \left(\frac{x_i}{a_i} \right) \ln \left(\frac{x_i}{a_i} \right). \qquad (2.1)$$

Substitution for P_i yields:

$$\left[\sum_i \frac{a_i}{(\sum_i a_i)} \left(\frac{x_i}{a_i} \right) \right] \ln \left[\sum_i \frac{a_i}{(\sum_i a_i)} \left(\frac{x_i}{a_i} \right) \right]$$

$$\leq \sum_i \frac{a_i}{(\sum_i a_i)} \left(\frac{x_i}{a_i} \right) \ln \left(\frac{x_i}{a_i} \right), \qquad (2.2)$$

---

[18] See Hardy, Littlewood, and Pólya [G. H. Hardy, J. E. Littlewood, G. Pólya, *Inequalities*. Cambridge University Press: 1952.], p. 70.

which reduces to

$$\left(\sum_i x_i\right) \ln\left(\frac{\sum_i x_i}{\sum_i a_i}\right) \leqq \sum_i x_i \ln\left(\frac{x_i}{a_i}\right), \tag{2.3}$$

and we have proved the lemma. $\qquad\qquad\qquad\qquad\qquad\qquad\square$

We also mention the analogous result for the continuous case:

COROLLARY 2. $f(x) \geqq 0$, $\quad g(x) \geqq 0$ $\quad$ (all x)

$$\Rightarrow \left[\int f(x)\,dx\right] \ln\left[\frac{\int f(x)\,dx}{\int g(x)\,dx}\right] \leqq \int f(x) \ln\left(\frac{f(x)}{g(x)}\right)\,dx.$$

§3. Refinement theorems

We now supply the proof for Theorems 2 and 4 of Chapter II, which concern the behavior of correlation and information upon refinement of the distributions. We suppose that the original (unrefined) distribution is $P_{ij\ldots k} = P(x_i, y_j, \ldots, z_k)$, and that the *refined* distribution is $P_{ij\ldots k}^{\prime\mu_i, \nu_j, \ldots, \eta_k}$, where the original value x_i for X has been resolved into a number of values $x_i^{\mu_i}$, and similarly for $Y, \ldots, Z$. Then:

$$P_{ij\ldots k} = \sum_{\mu_i, \nu_j, \ldots, \eta_k} P_{ij\ldots k}^{\prime\mu_i, \nu_j, \ldots, \eta_k}, \quad P_i = \sum_{\mu_i} P_i^{\prime\mu_i}, \quad \text{etc.} \tag{3.1}$$

Computing the new correlation $\{X, Y, \ldots, Z\}'$ for the refined distribution $P_{ij\ldots k}^{\prime\mu_i, \nu_j, \ldots, \eta_k}$ we find:

$$\{X, Y, \ldots, Z\}' = \sum_{ij\ldots k} \sum_{\mu_i, \nu_j, \ldots, \eta_k} P_{ij\ldots k}^{\prime\mu_i, \nu_j, \ldots, \eta_k} \ln\left(\frac{P_{ij\ldots k}^{\prime\mu_i, \nu_j, \ldots, \eta_k}}{P_i^{\prime\mu_i}, P_j^{\prime\nu_j}, \ldots, P_k^{\prime\eta_k}}\right). \tag{3.2}$$

However, by Lemma 2, §2:

$$\left(\sum_{\mu_i \ldots \eta_k} P_{i\ldots k}^{\prime\mu_i\ldots\eta_k}\right) \ln\left(\frac{\sum_{\mu_i\ldots\eta_k} P_{i\ldots k}^{\prime\mu_i\ldots\eta_k}}{\sum_{\mu_i\ldots\eta_k} P_i^{\prime\mu_i}, P_j^{\prime\nu_j}, \ldots, P_k^{\prime\eta_k}}\right)$$

$$\leqq \sum_{\mu_i\ldots\eta_k} P_{i\ldots k}^{\prime\mu_i\ldots\eta_k} \ln\left(\frac{P_{i\ldots k}^{\prime\mu_i\ldots\eta_k}}{P_i^{\prime\mu_i}, P_j^{\prime\nu_j}, \ldots, P_k^{\prime\eta_k}}\right). \tag{3.3}$$

Substitution of (3.3) into (3.2), noting that $\sum_{\mu_i...\eta_k} P_i^{\prime\mu_i}, P_j^{\prime\nu_j}, \ldots, P_k^{\prime\eta_k}$ is equal to $\left(\sum_{\mu_i} P_i^{\prime\mu_i}\right)\left(\sum_{\nu_j} P_j^{\prime\nu_j}\right)\cdots\left(\sum_{\eta_k} P_k^{\prime\eta_k}\right)$, leads to:

$$\{X, Y, \ldots, Z\}' \geqq \left(\sum_{ij...k}\sum_{\mu_i...\eta_k} P_{ij...k}^{\prime\mu_i...\eta_k}\right)$$

$$\ln\left[\frac{\sum_{\mu_i...\eta_k} P_{ij...k}^{\prime\mu_i...\eta_k}}{\left(\sum_{\mu_i} P_i^{\prime\mu_i}\right)\left(\sum_{\nu_j} P_j^{\prime\nu_j}\right)\cdots\left(\sum_{\eta_k} P_k^{\prime\eta_k}\right)}\right]$$

$$= \sum_{ij...k} P_{ij...k}\ln\frac{P_{ij...k}}{P_i P_j \ldots P_k} = \{X, Y, \ldots, Z\}, \qquad (3.4)$$

and we have completed the proof of Theorem 2 (Chapter II), which asserts that refinement never decreases the correlation.[19]

We now consider the effect of refinement upon the relative information. We shall use the previous notation, and further assume that $a_i^{\prime\mu_i}, b_j^{\prime\nu_j}, \ldots, c_k^{\prime\eta_k}$ are the information measures for which we wish to compute the relative information of $P_{ij...k}^{\prime\mu_i,\nu_j,...,\eta_k}$ and of $P_{ij...k}$. The information measures for the unrefined distribution $P_{ij...k}$ then satisfy the relations:

$$a_i = \sum_{\mu_i} a_i^{\mu_i}, \quad b_j = \sum_{\nu_j} b_j^{\nu_j}, \quad \ldots \quad . \qquad (3.5)$$

The relative information of the refined distribution is

$$I_{XY...Z}' = \sum_{i...j}\sum_{\mu_i...\eta_k} P_{ij...k}^{\prime\mu_i...\eta_k}\ln\left[\frac{P_{ij...k}^{\prime\mu_i...\eta_k}}{a_i^{\prime\mu_i}, b_j^{\prime\nu_j}, \ldots, c_k^{\prime\eta_k}}\right], \qquad (3.6)$$

and by exactly the same procedure as we have just used for the correlation we arrive at the result:

$$I_{XY...Z}' \geqq \sum_{i...k} P_{ij...k}\ln\frac{P_{ij...k}}{a_i b_j \ldots c_k} = I_{XY...Z}, \qquad (3.7)$$

and we have proved that refinement never decreases the relative information (Theorem 4, Chapter II).

It is interesting to note that the relation (3.4) for the behavior of correlation under refinement can be deduced from the behavior of relative information (3.7). This deduction is an immediate consequence of the fact

---

[19] Cf. Shannon [C. E. Shannon, W. Weaver, *The Mathematical Theory of Communication*. University of Illinois Press: 1949.], Appendix 7, where a quite similar theorem is proved.

that the correlation is a relative information—the information of the *joint distribution* relative to the product measure of the *marginal distributions*.

§4. Monotone decrease of information for stochastic processes

We consider a sequence of transition-probability matrices T_{ij}^n ($\sum_j T_{ij}^n = 1$ for all n, i, and $0 \leq T_{ij}^n \leq 1$ for all n, i, j), and a sequence of measures a_i^n ($a_i^n \geq 0$) having the property that

$$a_j^{n+1} = \sum_i a_i^n T_{ij}^n. \tag{4.1}$$

We further suppose that we have a sequence of probability distributions, P_i^n, such that

$$P_j^{n+1} = \sum_i P_i^n T_{ij}^n. \tag{4.2}$$

For each of these probability distributions the relative information I^n (relative to the a_i^n measure) is defined:

$$I^n = \sum_i P_i^n \ln \left(\frac{P_i^n}{a_i^n} \right). \tag{4.3}$$

Under these circumstances we have the following theorem:

THEOREM. $I^{n+1} \leq I^n$.

Proof. Expanding I^{n+1} we get:

$$I^{n+1} = \sum_j P_j^{n+1} \ln \left(\frac{P_j^{n+1}}{a_j^{n+1}} \right) = \sum_j \left(\sum_i P_i^n T_{ij}^n \right) \ln \frac{\left(\sum_i P_i^n T_{ij}^n \right)}{\left(\sum_i a_i^n T_{ij}^n \right)}. \tag{4.4}$$

However, by Lemma 2 (§2, Appendix I) we have the inequality

$$\left(\sum_i P_i^n T_{ij}^n \right) \ln \frac{\left(\sum_i P_i^n T_{ij}^n \right)}{\left(\sum_i a_i^n T_{ij}^n \right)} \leq \sum_i P_i^n T_{ij}^n \ln \frac{P_i^n T_{ij}^n}{a_i^n T_{ij}^n}. \tag{4.5}$$

Substitution of (4.5) into (4.4) yields:

$$I^{n+1} \leq \sum_j \left(\sum_i P_i^n T_{ij}^n \ln \frac{P_i^n}{a_i^n} \right) = \sum_i P_i^n \left(\sum_j T_{ij}^n \right) \ln \left(\frac{P_i^n}{a_i^n} \right)$$

$$= \sum_i P_i^n \ln \left(\frac{P_i^n}{a_i^n} \right) = I^n, \tag{4.6}$$

and the proof is completed. $\square$

This proof can be successively specialized to the case where T is stationary ($T_{ij}^n = T_{ij}$ for all n) and then to the case where T is doubly-stochastic ($\sum_i T_{ij} = 1$ for all j):

COROLLARY 3. T_{ij}^n *is stationary* ($T_{ij}^n = T_{ij}$, *all* n), *and the measure* a_i *is a stationary measure* ($a_j = \sum_i a_i T_{ij}$), *imply that the information,* $I^n = \sum_i P_i^n \ln(P_i^n / a_i^n)$, *is monotone decreasing. (As before,* $P_j^{n+1} = \sum_i P_i^n T_{ij}^n$.)

Proof. Immediate consequence of preceding theorem. $\qquad\qquad\square$

COROLLARY 4. T_{ij} *is doubly-stochastic* ($\sum_i T_{ij} = 1$, *all* j) *implies that the information relative to the uniform measure* ($a_i = 1$, *all* i), $I^n = \sum_i P_i^n \ln P_i^n$, *is monotone decreasing.*

Proof. For $a_i = 1$ (all i) we have that $\sum_i a_i T_{ij} = \sum_i T_{ij} = 1 = a_j$. Therefore the uniform measure is stationary in this case and the result follows from Corollary 1.

These results hold for the continuous case also, and may be easily verified by replacing the above summations by integrations, and by replacing Lemma 2 by its corollary. $\qquad\qquad\square$

§5. Proof of special inequality for Chapter IV (1.7)

LEMMA. *Given probability densities* $P(r)$, $P_1(x)$, $P_2(r)$, *with* $P(r) = \int P_1(x) P_2(r - x\tau)\, dx$. *Then* $I_R \leq I_X - \ln \tau$, *where* $I_X = \int P_1(x) \ln P_1(x)\, dx$ *and* $I_R = \int P(r) \ln P(r)\, dr$.

Proof. We first note that:

$$\int P_2(r - x\tau)\, dx = \int P_2(\omega) \frac{d\omega}{\tau} = \frac{1}{\tau} \qquad \text{(all } r) \tag{5.1}$$

and that furthermore

$$\int P_2(r - x\tau)\, dr = \int P_2(\omega)\, d\omega = 1 \qquad \text{(all } x). \tag{5.2}$$

We now define the density $\tilde{P}^r(x)$:

$$\tilde{P}^r(x) = \tau P_2(r - x\tau), \tag{5.3}$$

which is normalized, by (5.1). Then, according to §2, Corollary 1 of Appendix I, we have the relation:

$$\left(\int \tilde{P}^r(x) P_1(x)\, dx \right) \ln \left(\int \tilde{P}^r(x) P_1(x)\, dx \right) \underset{=}{\leq} \int \tilde{P}^r(x) P_1(x)\, dx. \tag{5.4}$$

Substitution from (5.3) gives

$$\left(\tau \int P_2(r - x\tau) P_1(x) \, dx \right) \ln \left(\tau \int P_2(r - x\tau) P_1(x) \, dx \right)$$
$$\leqq \tau \int P_2(r - x\tau) P_1(x) \ln P_1(x) \, dx. \tag{5.5}$$

The relation $P(r) = \int P_1(x) P_2(r - x\tau) \, dx$, together with (5.5) then implies

$$P(r) \ln \tau P(r) \leqq \int P_2(r - x\tau) P_1(x) \ln P_1(x) \, dx, \tag{5.6}$$

which is the same as:

$$P(r) \ln P(r) \leqq \int P_2(r - x\tau) P_1(x) \ln P_1(x) \, dx - P(r) \ln \tau. \tag{5.7}$$

Integrating with respect to r, and interchanging the order of integration on the right side gives:

$$I_R = \int P(r) \ln P(r) \, dr \leqq \int \left[\int P_2(r - x\tau) \, dr \right] P_1(x) \ln P_1(x) \, dx$$
$$- (\ln \tau) \int P(r) \, dr. \tag{5.8}$$

But using (5.2) and the fact that $\int P(r) \, dr = 1$ this means that

$$I_R \leqq \int P_1(x) \ln P_1(x) \, dx - \ln \tau = I_X - \ln \tau, \tag{5.9}$$

and the proof of the lemma is complete. □

§6. Stationary point of $I_K + I_X$

We shall show that the information sum:

$$I_K + I_X = \int_{-\infty}^{\infty} \phi^* \phi(k) \ln \phi^* \phi(k) \, dk + \int_{-\infty}^{\infty} \psi^* \psi(x) \ln \psi^* \psi(x) \, dx, \tag{6.1}$$

where

$$\phi(k) = (1/\sqrt{2\pi}) \int_{-\infty}^{\infty} e^{-ikx} \psi(x) \, dx$$

is *stationary* for the functions:

$$\psi_0(x) = (1/2\pi\sigma_x^2)^{\frac{1}{4}}e^{-x^2/4\sigma_x^2}, \quad \phi_0(k) = (2\sigma_x^2/\pi)^{\frac{1}{4}}e^{-k^2\sigma_x^2}, \quad (6.2)$$

with respect to variations of ψ, $\delta\psi$, which preserve the normalization:

$$\int_{-\infty}^{\infty} \delta(\psi^*\psi)\,dx = 0. \quad (6.3)$$

The variation $\delta\psi$ gives rise to a variation $\delta\phi$ of $\phi(k)$:

$$\delta\phi = (1/\sqrt{2\pi})\int_{-\infty}^{\infty} e^{-ikx}\delta\psi\,dx. \quad (6.4)$$

To avoid duplication of effort we first calculate variation δI_ξ for an arbitrary wave function $u(\xi)$. By definition,

$$I_\xi = \int_{-\infty}^{\infty} u^*(\xi)u(\xi)\ln u^*(\xi)u(\xi)\,d\xi, \quad (6.5)$$

so that

$$\delta I_\xi = \int_{-\infty}^{\infty} [u^*u\delta(\ln u^*u) + \delta(u^*u)\ln u^*u]\,d\xi$$

$$= \int_{-\infty}^{\infty} (1 + \ln u^*u)(u^*\delta u\,u\delta u^*)\,d\xi. \quad (6.6)$$

We now suppose that u has the *real* form:

$$u(\xi) = ae^{-b\xi^2} = u^*(\xi), \quad (6.7)$$

and from (6.6) we get

$$\delta I_\xi = \int_{-\infty}^{\infty} (1 + \ln a^2 - 2b\xi^2)ae^{-b\xi^2}(\delta u)\,d\xi + \text{complex conjugate}. \quad (6.8)$$

We now compute δI_K for ϕ_0 using (6.8), (6.2), and (6.4):

$$\delta I_K|_{\phi_0} = \int_{-\infty}^{\infty} (1 + \ln a'^2 - 2b'k^2)a'e^{-b'k^2}\frac{1}{\sqrt{2\pi}}\int_{-\infty}^{\infty} e^{ikx}\delta\phi\,dx\,dk + \text{c.c.}, \quad (6.9)$$

where

$$a = (2\sigma_x^2/\pi)^{\frac{1}{4}}, \quad b' = \sigma_x^2.$$

Interchanging the order of integration and performing the definite integration over k we get:

$$\delta I_K|_{\phi_0} = \int_{-\infty}^{\infty} \frac{a'}{\sqrt{2b'}} \left(\ln a'^2 + \frac{x^2}{2b'} \right) e^{-(x^2/4b')} \delta\phi(x)\, dx + \text{c.c.}, \qquad (6.10)$$

while application of (6.8) to ψ_0 gives

$$\delta I_X|_{\psi_0} = \int_{-\infty}^{\infty} (1 + \ln a''^2 - 2b''x^2) a'' e^{-b''x^2} \delta\psi(x)\, dx + \text{c.c.}, \qquad (6.11)$$

where

$$a'' = (1/2\pi\sigma_x^2)^{\frac{1}{4}}, \quad b'' = (1/4\sigma_x^2).$$

Adding (6.10) and (6.11), and substituting for a', b', a'', b'', yields:

$$\delta(I_K + I_X)|_{\psi_0} = (1 - \ln \pi) \int_{-\infty}^{\infty} (1/2\pi\sigma_x^2)^{\frac{1}{4}} e^{-(x^2/4\sigma_x^2)} \delta\psi(x)\, dx + \text{c.c.} \qquad (6.12)$$

But the integrand of (6.12) is simply $\psi_0(x)\delta\psi(x)$, so that

$$\delta(I_K + I_X)|_{\psi_0} = (1 - \ln \pi) \int_{-\infty}^{\infty} \psi_0 \delta\psi\, dx + \text{c.c.} \qquad (6.13)$$

Since ψ_0 is real, $\psi_0\delta\psi + \text{c.c.} = \psi^*\delta\psi + \text{c.c.} = \psi^*\delta\psi + \psi_0\delta\psi^* = \delta(\psi^*\psi)$, so that

$$\delta(I_K + I_X)|_{\psi_0} = (1 - \ln \pi) \int_{-\infty}^{\infty} \delta(\psi^*\psi)\, dx = 0, \qquad (6.14)$$

due to the normality restriction (6.3), and the proof is completed.

APPENDIX II

REMARKS ON THE ROLE OF THEORETICAL PHYSICS[cp]

　　There have been lately a number of new interpretations of quantum mechanics, most of which are equivalent in the sense that they predict the

---

[cp] This is Everett's extended discussion of the nature and cognitive status of physical theories. He explains in the first sentence of the section the primary reason for this discussion. Everett describes his understanding of the physical theories as being essentially the same as Philipp Frank's, chapter 17 (pg. 257). See also Everett's discussion of the material in this section in his correspondence with DeWitt, chapter 16 (pg. 252).

same results for all physical experiments. Since there is therefore no hope of deciding among them on the basis of physical experiments, we must turn elsewhere and inquire into the fundamental question of the nature and purpose of physical theories in general. Only after we have investigated and come to some sort of agreement upon these general questions, i.e., of the role of theories themselves, will we be able to put these alternative interpretations in their proper perspective.

Every theory can be divided into two separate parts: the formal part and the interpretive part. The formal part consists of a purely logico-mathematical structure, i.e., a collection of symbols together with rules for their manipulations, while the interpretive part consists of a set of "associations," which are rules which put some of the elements of the formal part into correspondence with the perceived world. The essential point of a theory, then, is that it is a *mathematical model*, together with an *isomorphism*[1] between the model and the world, of experience (i.e., the sense perceptions of the individual, or the "real world"—depending upon one's choice of epistemology).[cq]

The model nature is quite apparent in the newest theories, as in nuclear physics, and particularly in those fields outside of physics proper, such as

---

[1] By isomorphism we mean a mapping of some elements of the model into elements of the perceived world which has the property that the model is faithful, that is, if in the model a symbol *A* implies a symbol *B*, and *A* corresponds to the happening of an event in the perceived world, then the event corresponding to *B* must also obtain. The word homomorphism would be technically more correct, since there may not be a one-one correspondence between the model and the external world.

[cq] This point concerns the proper relationship between a theory's mathematical model and experience. Everett first describes the relationship as an *isomorphism*. In this description, the mathematical model of experience described by an empirically faithful theory and our actual experience would have precisely the same structure. He then suggests that the relationship between the model and experience is better characterized as a *homomorphism*. Here he seems to have in mind an isomorphism between a proper substructure of the model and a proper substructure of our representation of experience. There are two considerations involved in the homomorphism: (1) the theory is not required to explain all of our experience and (2) there may be parts of the model that are not interpreted as our experience. Consideration (1) allows pure wave mechanics to be a perfectly satisfactory physical theory without capturing all our experience—it need not, for example, say anything concerning the exchange rate between the dollar and the euro. Consideration (2) allows the formal model to contain more than is found in our experience. Everett identifies experience in the correlation model of pure wave mechanics by the memory sequences represented by the terms in an appropriate expansion of the absolute state; but, as he explains to DeWitt, not all of the memory sequences represented in the absolute state are directly relevant to our experience (chapter 16, pgs. 254–55). Rather, it is enough for Everett that one can find our experience represented in a typical term in the absolute state in the norm-squared sense of typical. In this sense pure wave mechanics is taken to be empirically faithful. See also the discussion of the empirical virtues of pure wave mechanics in Everett's letter to DeWitt (chapter 16, pg. 252) and the discussion of Everett's understanding of the status of theories in the conceptual introduction (chapter 3, pgs. 51–54).

the theory of games, various economic models, etc., where the degree of applicability of the models is still a matter of consiaderable doubt. However, when a theory is highly successful and becomes firmly established, the model tends to become identified with "reality" itself, and the model nature of the theory becomes obscured. The rise of classical physics offers an excellent example of this process. The constructs of classical physics are just as much fictions of our own minds as those of any other theory; we simply have a great deal more confidence in them. It must be deemed a mistake, therefore, to attribute any more "reality" here than elsewhere.[cr]

Once we have granted that any physical theory is essentially only a model for the world of experience, we must renounce all hope of finding anything like "*the* correct theory." There is nothing which prevents any number of quite distinct models from being in correspondence with experience (i.e., all "correct"), and furthermore no way of ever verifying that any model is completely correct, simply because the totality of all experience is never accessible to us.

Two types of prediction can be distinguished: the prediction of phenomena already understood, in which the theory plays simply the role of a device for compactly summarizing known results (the aspect of most interest to the engineer), and the prediction of new phenomena and effects, unsuspected before the formulation of the theory. Our experience has shown that a theory often transcends the restricted field in which it was formulated. It is this phenomenon (which might be called the "inertia" of theories) which is of most interest to the theoretical physicist and supplies a greater motive to theory construction than that of aiding the engineer.

From the viewpoint of the first type of prediction we would say that the "best" theory is the one from which the most accurate predictions can be most easily deduced—two not necessarily compatible ideals. Classical physics, for example, permits deductions with far greater ease than the more accurate theories of relativity and quantum mechanics, and in such a case we must retain them all. It would be the worst sort of folly to advocate that the study of classical physics be completely dropped in favor of the newer theories. It can even happen that several quite distinct models can exist which are completely equivalent in their predictions, such that different ones are most applicable in different cases, a situation which seems to be realized in quantum mechanics today. It would seem foolish to attempt to

---

[cr] Everett held that the nonempiricical entities and structures of our best physical theories are to be regarded as fictions and that even the long-term success of a theory is not an indication of its descriptive truth. It was, consequently, the logical, empirical, and pragmatic virtues of theories that formed the proper basis for their evaluation. This view agrees well with Everett's identification of his position with Philipp Frank's operational view in chapter 17 (pg. 257). Everett did not opt for a simple-minded version of positivism since he held that taking a theory to be nothing more than a representation of experience missed the picturing and forward-looking aspects of empirical inquiry. See pg. 171.

reject all but one in such a situation, where it might be profitable to retain them all.

Nevertheless, we have a strong desire to construct a single all-embracing theory which would be applicable to the entire universe. From what stems this desire? The answer lies in the second type of prediction—the discovery of new phenomena—and involves the consideration of inductive inference and the factors which influence our *confidence* in a given theory (to be applicable outside of the field of its formulation). This is a difficult subject and one which is only beginning to be studied seriously. Certain main points are clear, however, for example, that our confidence increases with the number of successes of a theory. If a new theory replaces several older theories which deal with separate phenomena, i.e., a comprehensive theory of the previously diverse fields, then our confidence in the new theory is very much greater than the confidence in either of the older theories, since the range of success of the new theory is much greater than any of the older ones. It is therefore this factor of confidence which seems to be at the root of the desire for comprehensive theories.

A closely related criterion is *simplicity*—by which we refer to conceptual simplicity rather than ease in use, which is of paramount interest to the engineer. A good example of the distinction is the theory of general relativity which is conceptually quite simple, while enormously cumbersome in actual calculations. Conceptual simplicity, like comprehensiveness, has the property of increasing confidence in a theory. A theory containing many *ad hoc* constants and restrictions, or many independent hypotheses, in no way impresses us as much as one which is largely free of arbitrariness.

It is necessary to say a few words about a view which is sometimes expressed, the idea that a physical theory should contain no elements which do not correspond directly to observables. This position seems to be founded on the notion that the only purpose of a theory is to serve as a summary of known data, and overlooks the second major purpose, the discovery of totally new phenomena. The major motivation of this viewpoint appears to be the desire to construct perfectly "safe" theories which will never be open to contradiction. Strict adherence to such a philosophy would probably seriously stifle the progress of physics.

The critical examination of just what quantities are observable in a theory does, however, play a useful role, since it gives an insight into ways of modification of a theory when it becomes necessary. A good example of this process is the development of Special Relativity. Such successes of the positivist viewpoint, when used merely as a tool for deciding which modifications of a theory are possible, in no way justify its universal adoption as a general principle which all theories must satisfy.[cs]

[cs] A simple-minded positivist view does not acknowledge the broad collection of theoretical virtues that are properly relevant to the acceptance of a theory as summarized below.

In summary, a physical theory is a logical construct (model), consisting of symbols and rules for their manipulation, some of whose elements are associated with elements of the perceived world. The fundamental requirements of a theory are logical consistency and correctness. There is no reason why there cannot be any number of different theories satisfying these requirements, and further criteria such as usefulness, simplicity, comprehensiveness, pictorability, etc., must be resorted to in such cases to further restrict the number. Even so, it may be impossible to give a total ordering of the theories according to "goodness," since different ones may rate highest according to the different criteria, and it may be most advantageous to retain more than one.^ct

As a final note, we might comment upon the concept of *causality*. It should be clearly recognized that causality is a property of a model and not a property of the world of experience. The concept of causality only makes sense with reference to a theory in which there are logical dependencies among the elements. A theory contains relations of the form "*A* implies *B*," which can be read as "*A* causes *B*," while our experience, uninterpreted by any theory, gives nothing of the sort, but only a *correlation* between the event corresponding to *B* and that corresponding to *A*.

Further, as Everett has argued, a theory should take the risk of predicting future experience of new unexperienced types.

^ct The picture is one of theory selection by means of a pragmatic and forward-looking cost–benefit analysis where one should be willing to take epistemic risks but where the ultimate descriptive truth of the theory and the metaphysical nature of the world are largely irrelevant. One consequently may wish to keep more than one theory out of a set of strictly incompatible theories. More specifically, whereas Everett took the relative-state interpretation to be the best option, he did not argue that one should simply reject other interpretations.

CHAPTER 9

Short Thesis: "Relative State" Formulation of Quantum Mechanics (1957)

Largely due to the criticism of the long thesis by the Copenhagen colleagues, Everett and his advisor John Wheeler rewrote Everett's thesis in the winter of 1957 to produce a much shorter version, which Everett subsequently defended for his Ph.D. under the title "On the Foundations of Quantum Mechanics." Whereas the long thesis was organized around the quantum measurement problem and how it is best solved by pure wave mechanics, the short thesis presented Everett's relative-state formulation of pure wave mechanics more as a suitable theory for the development of quantum gravity, cosmology, and field theory. The short thesis no longer contains Everett's chapter on information theory and correlation, his survey of possible solutions to the measurement problem, or his extended discussion of the nature of physical theories. What remains is a distilled presentation of pure wave mechanics, his principle of the fundamental relativity of states, and his derivation of the standard quantum statistics.[cu] The short thesis was retitled for publication in Reviews of Modern Physics *in July 1957 as "The 'Relative State' Formulation of Quantum Mechanics." This article was followed in the journal issue with a supporting paper by Wheeler. Wheeler's paper similarly follows Everett's paper in the present volume.[cv]*

[cu] This last was, for Everett, the most significant feature of his thesis: the sketch of the proof that measure-one of the sequences of relative records, in the norm-squared coefficient measure, exhibits the standard quantum probabilities in the limit as the same measurement is repeated on identically prepared systems. See for example pgs. 274–75.

[cv] See chapter 1 (pg. 3) and chapter 2 (pg. 17) for further discussion of the short thesis and how it related to Everett's earlier work.

"Relative State" Formulation of Quantum Mechanics[*, cw, cx]

Hugh Everett, III[†]

Palmer Physical Laboratory, Princeton University, Princeton, New Jersey

[*] Thesis submitted to Princeton University March 1, 1957, in partial fulfillment of the requirements for the Ph.D. degree. An earlier draft dated January, 1956, was circulated to several physicists whose comments were helpful. Professor Niels Bohr, Dr. H. J. Groenewold, Dr. Aage Petersen, Dr. A. Stern, and Professor L. Rosenfeld are free of any responsibility, but they are warmly thanked for the useful objections that they raised.[cy] Most particular thanks are due to Professor John A. Wheeler for his continued guidance and encouragement. Appreciation is also expressed to the National Science Foundation for fellowship support.

[†] Present address: Weapons Systems Evaluation Group, The Pentagon, Washington, D. C.[cz]

[cw] The following is a list of people who checked the short thesis out from the Princeton University library before the end of 1982:

Abner Shimony—Aug. 6–14, 1957 (a graduate student in physics at Princeton)
William Faris—Oct. 20–Dec. 12, 1960 (a graduate student in math at Princeton)
Stephen H. Gimber, Jr.—May 15, 1963
Jonathan L. Rosner—Jan. 25, 1965 (a graduate student in physics at Princeton)
David K. Lewis—Jan. 25, 1966 (a graduate student at Harvard)
Eric Hannah—Mar. 9–10, 1970 (a graduate student in physics at Princeton)
Lee Smolin—Dec. 28, 1982–Feb. 3, 1983 (at the Institute for Advanced Study).

[cx] The short thesis that Everett submitted for his Ph.D. (entitled "On the Foundations of Quantum Mechanics") began with the following abstract:

Quantum Mechanics is reformulated in a way which eliminates its present dependence on the special treatment of observation of a system by an external observer. The result is believed to be a more suitable formulation for application to field theories, particularly general relativity. The new formulation does not deny or contradict the conventional formulation but is a more general and complete formulation from which the conventional interpretation can be *deduced* within its own realm of applicability. In this sense the new theory plays the role of a metatheory for the older theory, that is, it is an underlying theory in which the nature and consistency of the conventional theory can be investigated and clarified. The new theory results from the conventional formulation by *omitting* the special postulates concerned with external observation. In their place a concept of "relativity of states" is developed for treating and interpreting the quantum description of isolated systems within which observation processes can occur. Abstract models for observers are formulated that can be treated within the theory as physical systems subject at all times to the same laws as all other physical systems. Isolated systems containing these model observers in interaction with other subsystems are investigated, and certain changes that occur in an observer as a consequence of the interaction with the surrounding systems are deduced. When these changes are interpreted as the experience of the observer this experience is found to be in accord with the statistical predictions of the conventional "external observation" formulation of quantum mechanics.

[cy] In the original, Everett misspelled Petersen and Groenewold's names.
[cz] Rather than pursue a traditional academic position, Everett accepted a job as a defense analyst. This gave him the opportunity to apply his interests in game theory, decision theory, and computer modeling to concrete problems. He never served in a traditional academic position.

1. Introduction

The task of quantizing general relativity raises serious questions about the meaning of the present formulation and interpretation of quantum mechanics when applied to so fundamental a structure as the space-time geometry itself. This paper seeks to clarify the foundations of quantum mechanics. It presents a reformulation of quantum theory in a form believed suitable for application to general relativity.

The aim is not to deny or contradict the conventional formulation of quantum theory, which has demonstrated its usefulness in an overwhelming variety of problems, but rather to supply a new, more general and complete formulation, from which the conventional interpretation can be *deduced*.

The relationship of this new formulation to the older formulation is therefore that of a metatheory to a theory, that is, it is an underlying theory in which the nature and consistency, as well as the realm of applicability, of the older theory can be investigated and clarified.

The new theory is not based on any radical departure from the conventional one. The special postulates in the old theory which deal with observation are omitted in the new theory. The altered theory thereby acquires a new character. It has to be analyzed in and for itself before any identification becomes possible between the quantities of the theory and the properties of the world of experience. The identification, when made, leads back to the omitted postulates of the conventional theory that deal with observation, but in a manner which clarifies their role and logical position.

We begin with a brief discussion of the conventional formulation and some of the reasons which motivate one to seek a modification.

2. Realm of Applicability of the Conventional or "External Observation" Formulation of Quantum Mechanics

We take the conventional or "external observation" formulation of quantum mechanics to be essentially the following[1]: A physical system is completely described by a state function ψ, which is an element of a Hilbert space, and which furthermore gives information only to the extent of specifying the probabilities of the results of various observations which can be made *on* the system *by* external observers. There are two fundamentally different ways in which the state function can change:

Process 1: The discontinuous change brought about by the observation of a quantity with eigenstates $\phi_1, \phi_2, \ldots$, in which the state ψ will be changed to the state ϕ_j with probability $|(\psi, \phi_j)|^2$.

---

[1] We use the terminology and notation of J. von Neumann, *Mathematical Foundations of Quantum Mechanics*, translated by R. T. Beyer (Princeton University Press, Princeton, 1955).

Process 2: The continuous, deterministic change of state of an isolated system with time according to a wave equation $\partial\psi/\partial t = A\psi$, where A is a linear operator.

This formulation describes a wealth of experience. No experimental evidence is known which contradicts it.[da]

Not all conceivable situations fit the framework of this mathematical formulation. Consider for example an isolated system consisting of an observer or measuring apparatus, plus an object system. Can the change with time of the state of the *total* system be described by Process 2?[db] If so, then it would appear that no discontinuous probabilistic process like Process 1 can take place. If not, we are forced to admit that systems which contain observers are not subject to the same kind of quantum-mechanical description as we admit for all other physical systems. The question cannot be ruled out as lying in the domain of psychology. Much of the discussion of "observers" in quantum mechanics has to do with photoelectric cells, photographic plates, and similar devices where a mechanistic attitude can hardly be contested. For the following one can *limit himself to this class of problems*, if he is unwilling to consider observers in the more familiar sense on the same mechanistic level of analysis.

What mixture of Processes 1 and 2 of the conventional formulation is to be applied to the case where only an approximate measurement is effected; that is, where an apparatus or observer interacts only weakly and for a limited time with an object system? In this case of an approximate measurement a proper theory must specify (1) the new state of the object system that corresponds to any particular reading of the apparatus and (2) the probability with which this reading will occur. von Neumann showed how to treat a special class of approximate measurements by the method of projection operators.[2] However, a general treatment of all approximate measurements by the method of projection operators can be shown (Sec. 4) to be impossible.

How is one to apply the conventional formulation of quantum mechanics to the space-time geometry itself? The issue becomes especially acute in the

---

[2] Reference 1 [(von Neumann, 1955), Chap. 4, Sec. 4.]

[da] See the conceptual introduction, chapter 3 (pg. 28) for a discussion of von Neumann's theory. Everett gives a more detailed description of the theory in the long version of his thesis, chapter 8 (pg. 73).

[db] This is a version of what is now known as the Wigner's Friend story. Eugene Wigner would later use such a story to argue that, by an implicit principle of charity, the observer inside the system must be the cause of the collapse (Wigner, 1961). Everett uses it here to argue that the standard collapse theory is ultimately inconsistent. The problem is that the orthodox formulations of quantum mechanics cannot consistently describe the evolution of systems that themselves contain observers. It is this that the relative-state formulation is meant to address. See also the discussion in the long thesis, chapter 8 (pgs. 73–75).

case of a closed universe.[3,dc] There is no place to stand outside the system to observe it. There is nothing outside it to produce transitions from one state to another. Even the familiar concept of a proper state of the energy is completely inapplicable. In the derivation of the law of conservation of energy, one defines the total energy by way of an integral extended over a surface large enough to include all parts of the system and their interactions.[4] But in a closed space, when a surface is made to include more and more of the volume, it ultimately disappears into nothingness. Attempts to define the total energy for a closed space collapse to the vacuous statement, zero equals zero.[de]

How are a quantum description of a closed universe, of approximate measurements, and of a system that contains an observer to be made? These three questions have one feature in common, that they all inquire about the *quantum mechanics* that is *internal to an isolated system*.

No way is evident to apply the conventional formulation of quantum mechanics to a system that is not subject to *external* observation. The whole interpretive scheme of that formalism rests upon the notion of external observation. The probabilities of the various possible outcomes of the observation are prescribed exclusively by Process 1. Without that part of the formalism there is no means whatever to ascribe a physical interpretation to the conventional machinery. But

---

[3] See A. Einstein, *The Meaning of Relativity* (Princeton University Press, Princeton, 1950), third edition, p. 107.

[4] L. Landau and E. Lifshitz, *The Classical Theory of Fields*, translated by M. Hamermesh (Addison-Wesley Press, Cambridge, 1951), p. 343.[dd]

[dc] The term "closed universe" can be ambiguous in this context. Cosmologists often use the expression to refer to a universe with positive curvature (such as a spherical universe), but it is sometimes used for a universe represented by a closed manifold (i.e., a compact manifold without boundary). A closed universe in the first sense is necessarily closed in the second, whereas an open universe in the first sense may be either open or closed in the second sense. Given the reference to Einstein, however, it seems that Everett has the first meaning in mind.

[dd] This is one of several footnotes that Everett added to the text while reworking the thesis under Wheeler's direction.

[de] Everett's remarks here are cryptic, but the reference to Landau and Lifshitz (1951) gives some indication of what he has in mind. Everett is thinking of the total energy associated with a region of space-time as an integral calculated over the region's volume. By Stokes' theorem, an integral over a volume can be reinterpreted as a flux integral calculated over the surface of the region. In a closed universe, however, one has a choice about what to interpret as the "inside" of any surface and what to interpret as the "outside," which means that (using Stokes' theorem again) one can show that the integral of a quantity over a volume of space-time is equal up to a sign to the integral of that quantity over all of space-time except for that volume (i.e., everything "outside" the region's surface). This leads one to the conclusion that the total energy of a closed universe must be zero, since there can be nothing "outside" the entire universe. Hence it is meaningless to talk of the energy state of a closed universe. Such considerations may in fact generate problems for the notion of a universal wave function.

Process 1 is out of the question for systems not subject to external observation.[5]

3. QUANTUM MECHANICS INTERNAL TO AN ISOLATED SYSTEM

This paper proposes to regard pure wave mechanics (Process 2 only) as a complete theory. It postulates that a wave function that obeys a linear wave equation everywhere and at all times supplies a complete mathematical model for every isolated physical system without exception. It further postulates that every system that is subject to external observation can be regarded as part of a larger isolated system.

The wave function is taken as the basic physical entity with *no a priori interpretation*. Interpretation only comes *after* an investigation of the logical structure of the theory. Here as always the theory itself sets the framework for its interpretation.[5]

For any interpretation it is necessary to put the mathematical model of the theory into correspondence with experience.[df] For this purpose it is necessary to formulate abstract models for observers that can be treated within the theory itself as physical systems, to consider isolated systems containing such model observers in interaction with other subsystems, to deduce the changes that occur in an observer as a consequence of interaction with the surrounding subsystems, and to interpret the changes in the familiar language of experience.

Section 4 investigates representations of the state of a composite system in terms of states of constituent subsystems. The mathematics leads one to recognize the concept of the *relativity of states*, in the following sense: a constituent subsystem cannot be said to be in any single well-defined state, independently of the remainder of the composite system. To any arbitrarily chosen state for one subsystem there will correspond a unique *relative state* for the remainder of the composite system. This relative state will usually depend upon the choice of state for the first subsystem. Thus the state of one subsystem does not have an independent existence, but is fixed only by the state of the remaining subsystem. In other words, the states occupied by the subsystems are not independent, but *correlated*. Such correlations between systems arise whenever systems interact. In the present formulation all measurements and observation processes are to be regarded simply as interactions between the physical systems involved—interactions

---

[5] See in particular the discussion of this point by N. Bohr and L. Rosenfeld, Kgl. Danske Videnskab. Selskab, Mat.-fys. Medd. **12**, No. 8 (1933).

[df] This is a methodological position that Everett discusses in some detail in the second appendix to the long thesis, chapter 8 (pg. 168).

which produce strong correlations. A simple model for a measurement, due to von Neumann, is analyzed from this viewpoint.

Section 5 gives an abstract treatment of the problem of observation. This uses only the superposition principle and general rules by which composite system states are formed of subsystem states, in order that the results shall have the greatest generality and be applicable to any form of quantum theory for which these principles hold. Deductions are drawn about the state of the observer relative to the state of the object system. It is found that experiences of the observer (magnetic tape memory, counter system, etc.) are in full accord with predictions of the conventional "external observer" formulation of quantum mechanics, based on Process 1.

Section 6 recapitulates the "relative state" formulation of quantum mechanics.

4. CONCEPT OF RELATIVE STATE

We now investigate some consequences of the wave mechanical formalism of composite systems. If a composite system S, is composed of two subsystems S_1 and S_2, with associated Hilbert spaces H_1 and H_2, then, according to the usual formalism of composite systems, the Hilbert space for S is taken to be the *tensor product* of H_1 and H_2 (written $H = H_1 \otimes H_2$).[dg] This has the consequence that if the sets $\{\xi_i^{S_1}\}$ and $\{\eta_j^{S_2}\}$ are complete orthonormal sets of states for S_1 and S_2, respectively, then the general state of S can be written as a superposition:

$$\psi^S = \sum_{i,j} a_{ij} \xi_i^{S_1} \eta_j^{S_2}. \tag{1}$$

From (3.1)[dh] although S is in a definite state ψ^S, the subsystems S_1 and S_2 do not possess anything like definite states independently of one another (except in the special case where all but one of the a_{ij} are zero).[di]

We *can*, however, for any choice of a state in one subsystem, *uniquely* assign a corresponding *relative* state in the other subsystem. For example, if we choose ξ_k as the state for S_1, while the composite system S is in the state ψ^S given by (3.1), then the corresponding *relative state* in S_2,

---

[dg] Pure wave mechanics adopts the standard rule for representing composite systems and quantum mechanically independent physical properties. This is *rule 5* in the conceptual introduction, chapter 3 (pg. 29).

[dh] In the original text, some of Everett's references to equations do not match with his equation numbers.

[di] Insofar as Everett holds that a system in an entangled state does not have anything like a definite physical state of its own, he is expressing a commitment to one of the consequences of the standard eigenvalue–eigenstate link. This is *rule 3* in the conceptual introduction, chapter 3 (pg. 28).

$\psi(S_2; \text{rel } \xi_k, S_1)$, will be:

$$\psi(S_2; \text{rel } \xi_k, S_1) = N_k \sum_j a_{kj} \eta_j^{S_2} \qquad (2)$$

where N_k is a normalization constant. This relative state for ξ_k is *independent* of the choice of basis $\{\xi_i\}$ $(i \neq k)$ for the orthogonal complement of ξ_k, and is hence determined uniquely by ξ_k alone. To find the relative state in S_2 for an arbitrary state of S_1 therefore, one simply carries out the above procedure using any pair of bases for S_1 and S_2 which contains the desired state as one element of the basis for S_1. To find states in S_1 relative to states in S_2, interchange S_1 and S_2 in the procedure.

In the conventional or "external observation" formulation, the relative state in S_2, $\psi(S_2; \text{rel } \phi, S_1)$, for a state ϕ^{S_1} in S_1, gives the conditional probability distributions for the results of all measurements in S_2, given that S_1 has been measured and found to be in state ϕ^{S_1}—i.e., that ϕ^{S_1} is the eigenfunction of the measurement in S_1 corresponding to the observed eigenvalue.

For any choice of basis in S_1, $\{\xi_i\}$, it is always possible to represent the state of S, (1), as a *single* superposition of pairs of states, each consisting of a state from the basis $\{\xi_i\}$ in S_1 and its relative state in S_2. Thus, from (2), (1) can be written in the form:

$$\psi^S = \sum_i \frac{1}{N_i} \xi_i^{S_1} \psi(S_2; \text{rel } \xi_i, S_1). \qquad (3)$$

This is an important representation used frequently.

Summarizing: There does not, in general, exist anything like a single state for one subsystem of a composite system. Subsystems do not possess states that are independent of the states of the remainder of the system, so that the subsystem states are generally correlated *with one another. One can arbitrarily choose a state for one subsystem, and be led to the relative state for the remainder. Thus we are faced with a* fundamental *relativity of states, which is implied by the formalism of composite systems. It is meaningless to ask the absolute state of a subsystem—one can only ask the state relative to a given state of the remainder of the subsystem.*[dj]

At this point we consider a simple example, due to von Neumann, which serves as a model of a measurement process. Discussion of this example prepares the ground for the analysis of "observation." We start with a

---

[dj] The procedure for determining relative states does not in any way involve the selection of a physically preferred basis. Everett held that branching is always relative to a *choice of basis*. He seems to have held this view throughout his life. See for example fn. gf on pg. 226 and fn. la on pg. 287.

system of only one coordinate, q (such as position of a particle), and an apparatus of one coordinate r (for example the position of a meter needle). Further suppose that they are initially independent, so that the combined wave function is $\psi_0^{S+A} = \phi(q)\eta(r)$ where $\phi(q)$ is the initial system wave function, and $\eta(r)$ is the initial apparatus function. The Hamiltonian is such that the two systems do not interact except during the interval $t = 0$ to $t = T$, during which time the total Hamiltonian consists only of a simple interaction,

$$H_I = -i\hbar q(\partial/\partial r). \tag{4}$$

Then the state

$$\psi_t^{S+A}(q, r) = \phi(q)\eta(r - qt) \tag{5}$$

is a solution of the Schrödinger equation,

$$i\hbar(\partial\psi_t^{S+A}/\partial t) = H_I\psi_t^{S+A}, \tag{6}$$

for the specified initial conditions at time $t = 0$.

From (5) at time $t = T$ (at which time interaction stops) there is no longer any definite independent apparatus state, nor any independent system state. The apparatus therefore does not indicate any definite object-system value, and nothing like Process 1 has occurred.

Nevertheless, we *can* look upon the total wave function (5) as a *superposition* of pairs of subsystem states, each element of which has a definite q value and a correspondingly displaced apparatus state. Thus after the interaction the state (5) has the form:

$$\psi_T^{S+A} = \int \phi(q')\delta(q - q')\eta(r - qT)dq', \tag{7}$$

which is a superposition of states $\psi_{q'} = \delta(q - q')\eta(r - qT)$. Each of these elements, $\psi_{q'}$, of the superposition describes a state in which the system has the definite value $q = q'$, and in which the apparatus has a state that is displaced from its original state by the amount $q'T$. These elements $\psi_{q'}$ are then superposed with coefficients $\phi(q')$ to form the total state (7).

Conversely, if we transform to the representation where the *apparatus* coordinate is definite, we write (5) as

$$\psi_T^{S+A} = \int (1/N_{r'})\xi^{r'}(q)\delta(r - r')dr',$$

where

$$\xi^{r'}(q) = N_{r'}\phi(q)\eta(r' - qT) \tag{8}$$

and

$$(1/N_{r'})^2 = \int \phi^*(q)\phi(q)\eta^*(r' - qT)\eta(r' - qT)dq.$$

Then the $\xi^{r'}(q)$ are the relative system state functions[6] for the apparatus states $\delta(r - r')$ of definite value $r = r'$.

If T is sufficiently large, or $\eta(r)$ sufficiently sharp (near $\delta(r)$), then $\xi^{r'}(q)$ is nearly $\delta(q - r'/T)$ and the relative system states $\xi^{r'}(q)$ are nearly eigenstates for the values $q = r'/T$.

We have seen that (8) is a superposition of states $\psi_{r'}$, *for each of which* the apparatus has recorded a definite value r', and the system is left in approximately the eigenstate of the measurement corresponding to $q = r'/T$. The discontinuous "jump" into an eigenstate is thus only a relative proposition, dependent upon the mode of decomposition of the total wave function into the superposition, and relative to a particularly chosen apparatus-coordinate value. So far as the complete theory is concerned all elements of the superposition exist simultaneously, and the entire process is quite continuous.

von Neumann's example is only a special case of a more general situation. Consider any measuring apparatus interacting with any object system. As a result of the interaction the state of the measuring apparatus is no longer capable of independent definition. It can be defined only *relative* to the state of the object system. In other words, there exists only a correlation between the two states of the two systems. It seems as if nothing can ever be settled by such a measurement.

This indefinite behavior seems to be quite at variance with our observations, since physical objects always appear to us to have definite positions. Can we reconcile this feature wave mechanical theory built purely on Process 2 with experience, or must the theory be abandoned as untenable? In order to answer this question we consider the problem of observation itself within the framework of the theory.

---

[6] This example provides a model of an approximate measurement. However, the relative system states after the interaction $\xi^{r'}(q)$ cannot ordinarily be generated from the original system state ϕ by the application of *any* projection operator, E. Proof: Suppose on the contrary that $\xi^{r'}(q) = NE\phi(q) = N'\phi(q)\eta(r' - qt)$, where N, N' are normalization constants. Then

$$E(NE\phi(q)) = NE^2\phi(q) = N''\phi(q)\eta^2(r' - qt)$$

and $E^2\phi(q) = (N''/N)\phi(q)\eta^2(r' - qt)$. But the condition $E^2 = E$, which is necessary for E to be a projection, implies that $N'/N'' \eta(q) = \eta^2(q)$, which is generally false.

5. OBSERVATION

We have the task of making deductions about the appearance of phenomena to observers which are considered as purely physical systems and are treated within the theory. To accomplish this it is necessary to identify some present properties of such an observer with features of the past experience of the observer. Thus, in order to say that an observer 0 has observed the event α, it is necessary that the state of 0 has become changed from its former state to a new state which is dependent upon α.

It will suffice for our purposes to consider the observers to possess memories (i.e., parts of a relatively permanent nature whose states are in correspondence with past experience of the observers). In order to make deductions about the past experience of an observer it is sufficient to deduce the present contents of the memory as it appears within the mathematical model.

As models for observers we can, if we wish, consider automatically functioning machines, possessing sensory apparatus and coupled to recording devices capable of registering past sensory data and machine configurations. We can further suppose that the machine is so constructed that its present actions shall be determined not only by its present sensory data, but by the contents of its memory as well. Such a machine will then be capable of performing a sequence of observations (measurements), and furthermore of deciding upon its future experiments on the basis of past results. If we consider that current sensory data, as well as machine configuration, is immediately recorded in the memory, then the actions of the machine at a given instant can be regarded as a function of the memory contents only, and all relevant experience of the machine is contained in the memory.

For such machines we are justified in using such phrases as "the machine has perceived A" or "the machine is aware of A" if the occurrence of A is represented in the memory, since the future behavior of the machine will be based upon the occurrence of A. In fact, all of the customary language of subjective experience is quite applicable to such machines and forms the most natural and useful mode of expression when dealing with their behavior, as is well known to individuals who work with complex automata.

When dealing with a system representing an observer quantum mechanically we ascribe a state function, ψ^0, to it. When the state ψ^0 describes an observer whose memory contains representations of the events $A, B, \ldots, C$ we denote this fact by appending the memory sequence in brackets as a subscript, writing:

$$\psi^0_{[A, B, \ldots, C]}. \tag{9}$$

The symbols $A, B, \ldots, C$, which we assume to be ordered time-wise, therefore stand for memory configurations which are in correspondence with the past experience of the observer. These configurations can be regarded as punches in a paper tape, impressions on a magnetic reel, configurations of a relay switching circuit, or even configurations of brain cells. We require only that they be capable of the interpretation "The observer has experienced the succession of events $A, B, \ldots, C$." (We sometimes write dots in a memory sequence, $\ldots A, B, \ldots, C$, to indicate the possible presence of previous memories which are irrelevant to the case being considered.)

The mathematical model seeks to treat the interaction of such observer systems with other physical systems (observations), within the framework of Process 2 wave mechanics, and to deduce the resulting memory configurations, which are then to be interpreted as records of the past experiences of the observers.

We begin by defining what constitutes a "good" observation. A good observation of a quantity A, with eigenfunctions ϕ_i, for a system S, by an observer whose initial state is ψ^0, consists of an interaction which, in a specified period of time, transforms each (total) state

$$\psi^{S+0} = \phi_i \psi^0_{[\ldots]} \tag{10}$$

into a new state

$$\psi^{S+0'} = \phi_i \psi^0_{[\ldots \alpha_i]} \tag{11}$$

where α_i characterizes[7] the state ϕ_i. (The symbol, α_i, might stand for a recording of the eigenvalue, for example.) That is, we require that the system state, *if it is an eigenstate*, shall be unchanged, and (2) that the observer state shall change so as to describe an observer that is "aware" of which eigenfunction it is; that is, some property is recorded in the memory of the observer which characterizes ϕ_i, such as the eigenvalue. The requirement that the eigenstates for the system be unchanged is necessary if the observation is to be significant (repeatable), and the requirement that the observer state change in a manner which is different for each eigenfunction is necessary if we are to be able to call the interaction an observation at all. How closely a general interaction satisfies the definition of a good observation depends upon (1) the way in which the interaction depends upon the dynamical variables of the observer system—including memory variables—and upon the dynamical variables of the object system and (2) the initial state of the observer system. Given (1) and (2), one can for example solve the wave equation, deduce the state of the composite system after the end of the interaction, and check whether an object system that

---

[7] It should be understood that $\psi^0_{[\ldots \alpha_i]}$ is a *different* state for each i. A more precise notation would write $\psi^0_{i[\ldots \alpha_i]}$, but no confusion can arise if we simply let the ψ^0_i be indexed only by the index of the memory configuration symbol.

was originally in an eigenstate is left in an eigenstate, as demanded by the repeatability postulate. This postulate is satisfied, for example, by the model of von Neumann that has already been discussed.

From the definition of a good observation we first deduce the result of an observation upon a system which is *not* in an eigenstate of the observation. We know from our definition that the interaction transforms states $\phi_i \psi^0_{[\cdots]}$ into states $\phi_i \psi^0_{[\cdots \alpha_i]}$. Consequently these solutions of the wave equation can be superposed to give the final state for the case of an arbitrary initial system state. Thus if the initial system state is not an eigenstate, but a general state $\sum_i a_i \phi_i$, the final total state will have the form:

$$\psi^{S+0'} = \sum_i a_i \phi_i \psi^0_{[\cdots \alpha_i]}. \quad ^{\text{dk}} \tag{12}$$

This superposition principle continues to apply in the presence of further systems which do not interact during the measurement. Thus, if systems $S_1, S_2, \ldots, S_n$ are present as well as 0, with original states $\psi^{S_1}, \psi^{S_2}, \ldots, \psi^{S_n}$, and the only interaction during the time of measurement takes place between S_1 and 0, the measurement will transform the initial total state:

$$\psi^{S_1+S_2+\cdots+S_n+0} = \psi^{S_1} \psi^{S_2} \cdots \psi^{S_n} \psi^0_{[\cdots]} \tag{13}$$

into the final state:

$$\psi'^{S_1+S_2+\cdots+S_n+0} = \sum_i a_i \phi_i^{S_1} \psi^{S_2} \cdots \psi^{S_n} \psi^0_{[\cdots \alpha_i]} \tag{14}$$

where $a_i = (\phi_i^{S_1}, \psi^{S_1})$ and $\phi_i^{S_1}$ are the eigenfunctions of the observation.

Thus we arrive at the general rule for the transformation of total state functions which describe systems within which observation processes occur:

Rule 1: The observation of a quantity A, with eigenfunctions $\phi_i^{S_1}$, in a system S_1 by the observer 0, transforms the total state according to:

$$\psi^{S_1} \psi^{S_2} \cdots \psi^{S_n} \psi^0_{[\cdots]} \rightarrow \sum_i a_i \phi_i^{S_1} \psi^{S_2} \cdots \psi^{S_n} \psi^0_{[\cdots \alpha_i]} \tag{15}$$

---

dk This follows from (1) the dispositions of a good measuring device in those cases where the initial state of the object system is an eigenstate of the observable being measured and (2) the fact that Process 2 is linear. The result of the linear dynamics on a superposition of eigenstates is the superposition of the result of the dynamics on each element of the superposition. More generally, the linearity of the dynamics provides a very simple way to calculate the final state if one knows what the dynamics does to each element of a set of states that spans the state space. The eigenstates of a physical observable form an orthonormal basis that spans the relevant state space for describing a measurement of the observable.

where

$$a_i = (\phi_i^{S_1}, \psi^{S_1}).$$

If we next consider a *second* observation to be made, where our total state is now a superposition, we can apply Rule 1 separately to each element of the superposition, since each element separately obeys the wave equation and behaves independently of the remaining elements, and then superpose the results to obtain the final solution. We formulate this as:

Rule 2: Rule 1 may be applied separately to each element of a superposition of total system states, the results being superposed to obtain the final total state. Thus, a determination of B, with eigenfunctions $\eta_j^{S_2}$, on S_2 by the observer 0 transforms the total state

$$\sum_i a_i \phi_i^{S_1} \psi^{S_2} \cdots \psi^{S_n} \psi_{[\cdots\alpha_i]}^0 \tag{16}$$

into the state

$$\sum_{i,j} a_i b_j \phi_i^{S_1} \eta_j^{S_2} \psi^{S_3} \cdots \psi^{S_n} \psi_{[\cdots\alpha_i,\beta_j]}^0 \tag{17}$$

where $b_j = (\eta_j^{S_2}, \psi^{S_2})$, which follows from the application of Rule 1 to each element $\phi_i^{S_1} \psi^{S_2} \cdots \psi^{S_n} \psi_{[\cdots\alpha_i]}^0$, and then superposing the results with the coefficients a_i.[dl]

These two rules, which follow directly from the superposition principle, give a convenient method for determining final total states for any number of observation processes in any combinations. We now seek the *interpretation* of such final total states.

Let us consider the simple case of a single observation of a quantity A, with eigenfunctions ϕ_i, in the system S with initial state ψ^S, by an observer 0 whose initial state is $\psi_{[\cdots]}^0$. The final result is, as we have seen, the superposition

$$\psi'^{S+0} = \sum_i a_i \phi_i \psi_{[\cdots\alpha_i]}^0. \tag{18}$$

There is no longer any independent system state or observer state, although the two have become correlated in a one-one manner.[dm] However, in each *element* of the superposition, $\phi_i \psi_{[\cdots\alpha_i]}^0$, the object-system state is a particular

---

[dl] Everett's Rule 1 and Rule 2 hold because Process 2 is linear.

[dm] As entailed by the standard interpretation of states, *Rule 3* in the conceptual introduction, chapter 3 (pg. 28), neither physical system has an *absolute* physical state of its own in this entangled state. They do, however, have well-defined *relative* states as Everett explained on pg. 180 and below.

eigenstate of the observation, and *furthermore the observer-system state describes the observer as definitely perceiving that particular system state.* This correlation is what allows one to maintain the interpretation that a measurement has been performed.[dn]

We now consider a situation where the observer system comes into interaction with the object system for a second time. According to Rule 2 we arrive at the total state after the second observation:

$$\psi''^{S+0} = \sum_i a_i \phi_i \psi^0_{[\cdots\alpha_i,\alpha_i]}. \tag{19}$$

Again, each element $\phi_i \psi^0_{[\cdots\alpha_i,\alpha_i]}$ describes a system eigenstate, but this time also describes the observer as having obtained the *same result* for each of the two observations.[do] Thus for every separate state of the observer in the final superposition the result of the observation was repeatable, even though different for different states. The repeatability is a consequence of the fact that after an observation the *relative* system state for a particular observer state is the corresponding eigenstate.

Consider now a different situation. An observer-system 0, with initial state $\psi^0_{[\cdots]}$, measures the *same* quantity A in a number of separate, identical, systems which are initially in the same state, $\psi^{S_1} = \psi^{S_2} = \cdots = \psi^{S_n} = \sum_i a_i \phi_i$ (where the ϕ_i are, as usual, eigenfunctions of A). The initial total state function is then

$$\psi_0^{S_1+S_2+\cdots+S_n+0} = \psi^{S_1} \psi^{S_2} \cdots \psi^{S_n} \psi^0_{[\cdots]}. \tag{20}$$

We assume that the measurements are performed on the systems in the order $S_1, S_2, \ldots, S_n$. Then the total state after the first measurement is by Rule 1,

$$\psi_1^{S_1+S_2+\cdots+S_n+0} = \sum_i a_i \phi_i^{S_1} \psi^{S_2} \cdots \psi^{S_n} \psi^0_{[\cdots\alpha_i^1]} \tag{21}$$

(where α_i^1 refers to the first system, S_1).

After the second measurement it is, by Rule 2,

$$\psi_2^{S_1+S_2+\cdots+S_n+0} = \sum_{i,j} a_i a_j \phi_i^{S_1} \phi_j^{S_2} \psi^{S_3} \cdots \psi^{S_n} \psi^0_{[\cdots\alpha_i^1,\alpha_j^2]} \tag{22}$$

[dn] See chapter 8 (pg. 121) for slightly more detail from Everett regarding how such correlations are supposed to explain determinate perceptions. See also the discussion of empirical faithfulness in the conceptual introduction, chapter 3 (pg. 52).

[do] This follows from the linearity of the dynamics and the fact Everett supposes that the correlations induced by a good measurement are perfect.

and in general, after r measurements have taken place ($r \leq n$), Rule 2 gives the result:

$$\psi_r = \sum_{i,j,\cdots k} a_i a_j \cdots a_k \phi_i^{S_1} \phi_j^{S_2} \cdots \phi_k^{S_r} \psi^{S_{r+1}} \cdots \psi^{S_n} \psi^0_{[\cdots \alpha_i^1, \alpha_j^2, \cdots \alpha_k^r]} \qquad (23)$$

We can give this state, ψ_r, the following interpretation. It consists of a superposition of states:

$$\psi'_{ij\cdots k} = \phi_i^{S_1} \phi_j^{S_2} \cdots \phi_k^{S_r} \times \psi^{S_{r+1}} \cdots \psi^{S_n} \psi^0_{[\alpha_i^1, \alpha_j^2, \cdots \alpha_k^r]} \qquad (24)$$

each of which describes the observer with a definite memory sequence $[\alpha_i^1, \alpha_j^2, \dots \alpha_k^r]$. Relative to him the (observed) system states are the corresponding eigenfunctions $\phi_i^{S_1}, \phi_j^{S_2}, \dots, \phi_k^{S_r}$, the remaining systems, $S_{r+1}, \dots, S_n$, being unaltered.

A typical element $\psi'_{ij\cdots k}$ of the final superposition describes a state of affairs wherein the observer has perceived an apparently random sequence of definite results for the observations.[dp] Furthermore the object systems have been left in the corresponding eigenstates of the observation. At this stage suppose that a redetermination of an earlier system observation (S_l) takes place. Then it follows that every element of the resulting final superposition will describe the observer with a memory configuration of the form $[\alpha_i^1, \cdots \alpha_j^l, \cdots \alpha_k^r, \alpha_j^l]$ in which the earlier memory coincides with the later—i.e., the memory states are *correlated*. It will thus *appear* to the observer, as described by a typical element of the superposition, that each initial observation on a system caused the system to "jump" into an eigenstate in a random fashion and thereafter remain there for subsequent measurements on the same system.[dq] Therefore—disregarding for the moment quantitative questions of relative frequencies—the probabilistic assertions of Process 1 *appear* to be valid to the observer described by a typical element of the final superposition.

We thus arrive at the following picture: Throughout all of a sequence of observation processes there is only one physical system representing the

---

[dp] Everett here introduces the notion of a *typical* element of the superposition. The notion of a typical element of a set S is characterized by specifying a measure over the set. A probability measure μ is a function that assigns a number between zero and one to each subset of S such that for $Q, R \subset S$ (1) $\mu(S) = 1$, (2) $\mu(R) = 1 - \mu(\bar{R})$, and (3) if $R \cap Q =$ then $\mu(R \cup Q) = \mu(R) + \mu(Q)$. It is also usually supposed that μ is similarly additive for any countable union of disjoint subsets of S. What Everett says here is true for a broad selection of typicality measures and notions of what it might mean for a sequence to be random. Everett explains what notion of typicality he has in mind later (pg. 190) and in the long thesis, chapter 8 (pgs. 123–27). Importantly, it is not a measure where he means *most* elements by count.

[dq] This is just the earlier repeatability property in the context of a more complicated experiment.

observer, yet there is no single unique *state* of the observer (which follows from the representations of interacting systems). Nevertheless, there is a representation in terms of a *superposition*, each element of which contains a definite observer state and a corresponding system state. Thus with each succeeding observation (or interaction), the observer state "branches" into a number of different states. Each branch represents a different outcome of the measurement and the *corresponding* eigenstate for the object-system state. All branches exist simultaneously in the superposition after any given sequence of observations.[‡]

The "trajectory" of the memory configuration of an observer performing a sequence of measurements is thus not a linear sequence of memory configurations, but a branching tree, with all possible outcomes existing simultaneously in a final superposition with various coefficients in the mathematical model. In any familiar memory device the branching does not continue indefinitely, but must stop at a point limited by the capacity of the memory.

[‡] *Note added in proof.*—In reply to a preprint of this article some correspondents have raised the question of the "transition from possible to actual," arguing that in "reality" there is—as our experience testifies—no such splitting of observer states, so that only one branch can ever actually exist. Since this point may occur to other readers the following is offered in explanation.[dr]

The whole issue of the transition from "possible" to "actual" is taken care of in the theory in a very simple way—there is no such transition, nor is such a transition necessary for the theory to be in accord with our experience. From the viewpoint of the theory *all* elements of a superposition (all "branches") are "actual," none any more "real" than the rest. It is unnecessary to suppose that all but one are somehow destroyed, since all the separate elements of a superposition individually obey the wave equation with complete indifference to the presence or absence ("actuality" or not) of any other elements. This total lack of effect of one branch on another also implies that no observer will ever be aware of any "splitting" process.[dt]

Arguments that the world picture presented by this theory is contradicted by experience, because we are unaware of any branching process, are like the criticism of the Copernican theory that the mobility of the Earth as a real physical fact is incompatible with the common sense interpretation of nature because we feel no such motion. In both cases the argument fails when it is shown that the theory itself predicts that our experience will be what it in fact is. (In the Copernican case the addition of Newtonian physics was required to be able to show that the Earth's inhabitants would be unaware of any motion of the Earth.)[du]

[dr] Both Wiener and DeWitt worried about the transition from possible to actual. Everett explained his position here, in his notes on his copy of the letter from Wiener, in his reply to the letter from Wiener (fn. hg, pg. 233), and in his reply to the letter to DeWitt (pg. 254).

[dt] See Everett's discussion of this point in his letter to DeWitt (chapter 16, pg. 254).

[du] Everett discussed this analogy between pure wave mechanics and Copernican astronomy in his letter to DeWitt (pg. 254). Everett held that pure wave mechanics never makes such statements as "outcome A is actually realized" except *relative* to some specified state of a correlated system. All possibilities then are actually realized *as relative states*. See, for example, Everett's note on his copy of the letter from Wiener (hg).

In order to establish quantitative results, we must put some sort of measure (weighting) on the elements of a final superposition.[ds] This is necessary to be able to make assertions which hold for almost all of the observer states described by elements of a superposition. We wish to make quantitative statements about the relative frequencies of the different possible results of observation—which are recorded in the memory—for a typical observer state; but to accomplish this we must have a method for selecting a typical element from a superposition of orthogonal states.

We therefore seek a general scheme to assign a measure to the elements of a superposition of orthogonal states $\sum_i a_i \phi_i$. We require a positive function m of the complex coefficients of the elements of the superposition, so that $m(a_i)$ shall be the measure assigned to the element ϕ_i. In order that this general scheme be unambiguous we must first require that the states themselves always be normalized, so that we can distinguish the coefficients from the states. However, we can still only determine the *coefficients*, in distinction to the states, up to an arbitrary phase factor. In order to avoid ambiguities the function m must therefore be a function of the amplitudes of the coefficients alone, $m(a_i) = m(|a_i|)$.[dv]

We now impose an additivity requirement. We can regard a subset of the superposition, say $\sum_{i=1}^{n} a_i \phi_i$, as a single element $\alpha \phi'$:

$$\alpha \phi' = \sum_{i=1}^{n} a_i \phi_i. \tag{25}$$

We then demand that the measure assigned to ϕ' shall be the sum of the measures assigned to the ϕ_i (i from 1 to n):[dw]

$$m(\alpha) = \sum_{i=1}^{n} m(a_i). \tag{26}$$

[ds] This is the beginning of Everett's argument for the special status of the norm-squared measure of typicality in pure wave mechanics. A somewhat longer version of this argument is found the long thesis, chapter 8 (pgs. 123–27). Everett considered his reflections regarding typicality in pure wave mechanics to be his most significant accomplishment. As examples of its status for Everett, see pgs. 273, 274–75, and 295–96. See also the discussion of empirical faithfulness in the conceptual introduction (pgs. 51–53).

[dv] This is a further background assumption. The grounds for this constraint involve what Everett believes one might know regarding the renormalized coefficients on the relative state of a subsystem of a composite system. If one could only know the coefficients up to a phase factor, one might take only their squared amplitudes to be significant. There are, however, quantum-mechanical predictions, such as the Aharonov–Bohm effect, that depend on the precise relative phase of each element of the superposition (Aharonov and Bohm, 1959).

[dw] This additivity condition determines a relationship between the measures over elements in different expansions of the absolute state. This provides related notions of typicality for relative states in different bases. Everett needs a general notion of typicality that works for any basis since he denies that there is any physically preferred basis for the specification of relative states. See for example pg. 180 and chapter 21 (fn. la on pg. 287).

Then we have already restricted the choice of m to the square amplitude alone; in other words, we have $m(a_i) = a_i^* a_i$, apart from a multiplicative constant.

To see this, note that the normality of ϕ' requires that $|\alpha| = (\sum a_i^* a_i)^{\frac{1}{2}}$. From our remarks about the dependence of m upon the amplitude alone, we replace the a_i by their amplitudes $u_i = |a_i|$. Equation (26) then imposes the requirement,

$$m(\alpha) = m \left(\sum a_i^* a_i \right)^{\frac{1}{2}} = m \left(\sum u_i^2 \right)^{\frac{1}{2}} = \sum m(u_i) = \sum m(u_i^2)^{\frac{1}{2}}. \quad (27)$$

Defining a new function $g(x)$

$$g(x) = m(\sqrt{x}) \quad (28)$$

we see that (27) requires that

$$g \left(\sum u_i^2 \right) = \sum g(u_i^2). \quad (29)$$

Thus g is restricted to be linear and necessarily has the form:

$$g(x) = cx \quad (c \text{ constant}). \quad (30)$$

Therefore $g(x^2) = cx^2 = m(\sqrt{x^2}) = m(x)$ and we have deduced that m is restricted to the form

$$m(a_i) = m(u_i) = cu_i^2 = ca_i^* a_i. \quad (31)$$

We have thus shown that the only choice of measure consistent with our additivity requirement is the square amplitude measure, apart from an arbitrary multiplicative constant which may be fixed, if desired, by normalization requirements. (The requirement that the total measure be unity implies that this constant is 1.)[dx]

The situation here is fully analogous to that of classical statistical mechanics, where one puts a measure on trajectories of systems in the phase space by placing a measure on the phase space itself, and then making assertions

[dx] This is the argument for the uniqueness of the measure m satisfying Everett's specified constraints. Note that m is unique up to the constant c. The constant c might be thought of as a renormalization factor and set so that m satisfies the mathematical conditions for a probability measure over the elements of the superposition. That does not, however, mean that m is a probability. There are no postulated chance events, and Everett does not argue here that m represents the epistemic uncertainty of an agent. In order that m not be confused with probability, Everett insisted that probability talk in the theory be translated back to talk in terms of the typicality measure m. See for example (pg. 193) and (pg. 127) in chapter 8, the long thesis. Compare the discussion here to pgs. 123–27.

(such as ergodicity, quasi-ergodicity, etc.) which hold for "almost all" trajectories. This notion of "almost all" depends here also upon the choice of measure, which is in this case taken to be the Lebesgue measure on the phase space. One could contradict the statements of classical statistical mechanics by choosing a measure for which only the exceptional trajectories had nonzero measure. Nevertheless the choice of Lebesgue measure on the phase space can be justified by the fact that it is the only choice for which the "conservation of probability" holds, (Liouville's theorem) and hence the only choice which makes possible any reasonable statistical deductions at all.

In our case, we wish to make statements about "trajectories" of observers. However, for us a trajectory is constantly branching (transforming from state to superposition) with each successive measurement. To have a requirement analogous to the "conservation of probability" in the classical case, we demand that the measure assigned to a trajectory at one time shall equal the sum of the measures of its separate branches at a later time.[dy] This is precisely the additivity requirement which we imposed and which leads uniquely to the choice of square-amplitude measure. Our procedure is therefore quite as justified as that of classical statistical mechanics.[dz]

Having deduced that there is a unique measure which will satisfy our requirements, the square-amplitude measure, we continue our deduction. This measure then assigns to the $i, j, \ldots k$th element of. the superposition (24),

$$\phi_i^{S_1} \phi_j^{S_2} \cdots \phi_k^{S_r} \psi^{S_{r+1}} \cdots \psi^{S_n} \psi^0_{[\alpha_i^1, \alpha_j^2, \cdots \alpha_k^r]} \tag{32}$$

the measure (weight)

$$M_{ij\ldots k} = (a_i a_j \cdots a_k)^* (a_i a_j \cdots a_k) \tag{33}$$

so that the observer state with memory configuration $[\alpha_i^1, \alpha_j^2, \ldots, \alpha_k^r]$ is assigned the measure $a_i^* a_i a_j^* a_j \cdots a_k^* a_k = M_{ij\ldots k}$. We see immediately that this is a product measure, namely,

$$M_{ij\ldots k} = M_i M_j \cdots M_k \tag{34}$$

where

$$M_l = a_l^* a_l$$

[dy] While Everett's earlier statement of the condition involved the additivity of the measure associated with linearly related elements in a superposition *at a time*, this statement involves conservation of measure over branches *at different times*. Everett wants to tighten the analogy between his justification of the norm-squared measure for pure wave mechanics and standard justifications of the choice of typicality measure in classical thermodynamics.

[dz] See also the discussion in the long thesis, chapter 8 (pg. 126). Everett further explained the parallel with statistical mechanics in his comments at the Xavier conference in 1961, chapter 19 (pgs. 274–75).

so that the measure assigned to a particular memory sequence $[\alpha_i^1, \alpha_j^2, \ldots, \alpha_k^r]$ is simply the product of the measures for the individual components of the memory sequence.

There is a direct correspondence of our measure structure to the probability theory of random sequences. *If we regard* the $M_{ij\ldots k}$ as probabilities for the sequences then the sequences are equivalent to the random sequences which are generated by ascribing to each term the *independent* probabilities $M_l = a_l^* a_l$. Now probability theory is equivalent to measure theory mathematically, so that we can make use of it, while keeping in mind that all results should be translated back to measure theoretic language.[ea]

Thus, in particular, if we consider the sequences to become longer and longer (more and more observations performed) *each* memory sequence of the final superposition will satisfy any given criterion for a randomly generated sequence, generated by the independent probabilities $a_l^* a_l$, except for a set of total measure which tends toward zero as the number of observations becomes unlimited.[eb] Hence all averages of functions over *any* memory sequence, including the special case of frequencies, can be computed from the probabilities $a_i^* a_i$, except for a set of memory sequences of measure zero. We have therefore shown that the statistical assertions of Process 1 will appear to be valid to the observer, *in almost all* elements of the superposition (24), in the limit as the number of observations goes to infinity.[ec]

While we have so far considered only sequences of observations of the same quantity upon identical systems, the result is equally true for arbitrary sequences of observations, as may be verified by writing more general sequences of measurements, and applying Rules 1 and 2 in the same manner as presented here.

We can therefore summarize the situation when the sequence of observations is arbitrary, when these observations are made upon the same or different systems in any order, and when the number of observations of

[ea] As he explains here, Everett did not interpret this measure as providing probabilities. This is understandable since pure wave mechanics is deterministic and since Everett describes no systematic uncertainty concerning the state that would yield the standard quantum probabilities. See also chapter 8 (pg. 127). Rather, Everett understood the norm-squared measure as providing a notion of branch typicality. He felt that he had accounted for the appearance of a stochastic process to observers treated within the system by showing that a typical branch would exhibit the standard quantum statistics. For Everett's discussions of typicality in pure wave mechanics see (pgs. 77, 189, 123 and 295). See also the discussion of empirical faithfulness in the conceptual introduction (pg. 51).

[eb] This is true for any standard of *random sequence* for which the limit of the ratio of length-n random memory sequences to the total number of length-n memory sequences goes to zero as n gets large since the product measure associated with each term goes to zero. See Barrett (1999, limiting properties of the bare theory) for a detailed discussion of this issue.

[ec] More precisely, as the number of measurements gets large, the limit of the sum of the norm-squared of the coefficients associated with those elements of the superposition where the relative frequencies differ from those predicted by the standard collapse theory by more than ϵ goes to zero for all $\epsilon > 0$. See Barrett (1999).

each quantity in each system is very large, with the following result:

> Except for a set of memory sequences of measure nearly zero, the averages of any functions over a memory sequence can be calculated approximately by the use of the independent probabilities given by Process 1 for each initial observation, on a system, and by the use of the usual transition probabilities for succeeding observations upon the same system. In the limit, as the number of all types of observations goes to infinity the calculation is exact, and the exceptional set has measure zero.

This prescription for the calculation of averages over memory sequences by probabilities assigned to individual elements is precisely that of the conventional "external observation" theory (Process 1). Moreover, these predictions hold for almost all memory sequences. Therefore all predictions of the usual theory will appear to be valid to the observer in almost all observer states.

In particular, the uncertainty principle is never violated since the latest measurement upon a system supplies all possible information about the relative system state, so that there is no direct correlation between any earlier results of observation on the system and the succeeding observation. Any observation of a quantity B, between two successive observations of quantity A (all on the same system) will destroy the one-one correspondence between the earlier and later memory states for the result of A. Thus for alternating observations of different quantities there are fundamental limitations upon the correlations between memory states for the same observed quantity, these limitations expressing the content of the uncertainty principle.[ed]

As a final step one may investigate the consequences of allowing several observer systems to interact with (observe) the same object system, as well as to interact with one another (communicate). The latter interaction can be treated simply as an interaction which correlates parts of the memory configuration of one observer with another. When these observer systems are investigated, in the same manner as we have already presented in this section using Rules 1 and 2, one finds that in *all elements* of the final superposition:[ee]

[ed] This short paragraph refers to Everett's extended discussion of approximate measurement, uncertainty, and information in the long thesis. See, in particular, the discussion in chapter 8 following pg. 145 for approximate measurement, and the discussions following pgs. 80 and 110 for information theory and the proposed relationship to the uncertainty principle, respectively. Most of the discussion of information theory was dropped in the short thesis.

[ee] More precisely, the following properties hold for all elements of the superposition as Everett suggests if and only if the correlation between each observer's records and the corresponding object systems are perfect. See chapter 8 (pg. 130) for Everett's extended treatment of multiple observers. Since pure wave mechanics allows one to describe nested observations consistently, it solves the measurement problem as understood by Everett (see pgs. 176 and 73–75).

1. When several observers have separately observed the same quantity in the object system and then communicated the results to one another they find that they are in agreement. This agreement persists even when an observer performs his observation *after* the result has been communicated to him by another observer who has performed the observation.

2. Let one observer perform an observation of a quantity *A* in the object system, then let a second perform an observation of a quantity *B* in this object system which does not commute with *A*, and finally let the first observer repeat his observation of *A*. Then the memory system of the first observer will *not* in general show the same result for both observations. The intervening observation by the other observer of the non-commuting quantity *B* prevents the possibility of any one-to-one correlation between the two observations of *A*.

3. Consider the case where the states of two object systems are correlated but where the two systems do not interact. Let one observer perform a specified observation on the first system, then let another observer perform an observation on the second system, and finally let the first observer repeat his observation. Then it is found that the first observer always gets the same result both times, and the observation by the second observer has no effect whatsoever on the outcome of the first's observations. Fictitious paradoxes like that of Einstein, Podolsky, and Rosen[8] which are concerned with such correlated, noninteracting systems are easily investigated and clarified in the present scheme.

Many further combinations of several observers and systems can be studied within the present framework. The results of the present "relative state" formalism agree with those of the conventional "external observation" formalism in all those cases where that familiar machinery is applicable.

In conclusion, the continuous evolution of the state function of a composite system with time gives a complete mathematical model for processes that involve an idealized observer. When interaction occurs, the result of the evolution in time is a superposition of states, each element of which assigns a different state to the memory of the observer. Judged by the state of the memory in almost all of the observer states, the probabilistic conclusions of the usual "external observation" formulation of quantum theory are valid.[ef] In other words, pure Process 2 wave mechanics, without any initial probability assertions, leads to all the probability concepts of the familiar formalism.

[8] Einstein, Podolsky, and Rosen, Phys. Rev. **47**, 777 (1935). For a thorough discussion of the physics of observation, see the chapter by N. Bohr in *Albert Einstein, Philosopher-Scientist* (The Library of Living Philosophers, Inc., Evanston, 1949).

[ef] This is true insofar as such probabilistic conclusions are understood as assertions concerning the statistical properties of typical sequences of measurement records in the memory of an idealized observer at a time in the norm-squared sense of typicality that Everett proposes.

6. Discussion

The theory based on pure wave mechanics is a conceptually simple, causal theory, which gives predictions in accord with experience. It constitutes a framework in which one can investigate in detail, mathematically, and in a logically consistent manner a number of sometimes puzzling subjects, such as the measuring process itself and the interrelationship of several observers. Objections have been raised in the past to the conventional or "external observation" formulation of quantum theory on the grounds that its probabilistic features are postulated in advance instead of being derived from the theory itself. We believe that the present "relative-state" formulation meets this objection, while retaining all of the content of the standard formulation.

While our theory ultimately justifies the use of the probabilistic interpretation as an aid to making practical predications, it forms a broader frame in which to understand the consistency of that interpretation. In this respect it can be said to form a *metatheory* for the standard theory. It transcends the usual "external observation" formulation, however, in its ability to deal logically with questions of imperfect observation and approximate measurement.[cg]

The "relative state" formulation will apply to all forms of quantum mechanics which maintain the superposition principle.[ch] It may therefore prove a fruitful framework for the quantization of general relativity. The formalism invites one to construct the formal theory first, and to supply the statistical interpretation later. This method should be particularly useful for interpreting quantized unified field theories where there is no question of ever isolating observers and object systems. They all are represented in a *single* structure, the field. Any interpretative rules can probably only be deduced in and through the theory itself.

Aside from any possible practical advantages of the theory, it remains a matter of intellectual interest that the statistical assertions of the usual interpretation do not have the status of independent hypotheses, but are deducible (in the present sense) from the pure wave mechanics that starts completely free of statistical postulates.

[cg] While he does not explain in any detail here how the theory handles approximate measurement, this is a main theme in the long thesis (chapter 8, see the discussion following (pg. 145).

[ch] That the relative-state formulation of pure wave mechanics is comprehensive was, for Everett, one of its most significant virtues. See the discussion of theoretical virtues in the long thesis, chapter 8 (pgs. 168–72). It made pure wave mechanics a suitable foundation for general relativity and future field theories in particular.

CHAPTER 10

Wheeler Article: Assessment of Everett's "Relative State" Formulation of Quantum Theory (1957)

John Wheeler's paper was published in Reviews of Modern Physics *immediately following Everett's short thesis.[ei] In his paper Wheeler sought both to endorse Everett's relative-state formulation of pure wave mechanics and to put it in a context where it might be more acceptable to his colleagues. Although Wheeler's support for Everett's proposal faded over time, Everett remained steadfast in his commitment to the relative-state formulation.[ej]*

Assessment of Everett's "Relative State" Formulation of Quantum Theory

John A. Wheeler
Palmer Physical Laboratory, Princeton University, Princeton, New Jersey

The preceding paper[ek] puts the principles of quantum mechanics in a new form.[1] Observations are treated as a special case of normal interactions that occur within a system, not as a new and different kind of process that takes place from without. The conventional mathematical formulation with its well-known postulates about probabilities of observations is derived as a

---

[1] Hugh Everett, III, Revs. Modern Phys. 29, 454 (1957).

[ei] The short thesis is reprinted in this volume as chapter 9 (pg. 173).

[ej] See Everett's letter to Raub in this volume (chapter 25, pg. 315) for an example of Everett's own late views and his sense of Wheeler's position.

[ek] Wheeler much later recalled that the "paper by my student Hugh Everett[...] was so impenetrable that I was moved to publish next to it a short paper." He recalled Everett as "an independent, intense, driven young man." Wheeler reported: "When he brought me the draft of his thesis, I could sense its depth and see that he was grappling with some very basic problems, yet I found the draft barely comprehensible. I knew that if I had that much trouble with it, other faculty members on his committee would have even more trouble. They not only would find it incomprehensible; they might find it without merit. So Hugh and I worked long hours at night in my office to revise the draft. Even after that effort, I decided the thesis needed a companion piece, which I prepared for publication with this paper. My real intent was to make his thesis more digestible to his other committee members" (Wheeler and Ford, 1998, pg. 268).

consequence of the new or *"meta"* quantum mechanics. Both formulations apply as well to complex systems as to simple ones, and as well to particles as to fields. Both supply mathematical models for the physical world. In the new or "relative state" formalism this model associates with an isolated system a state function that obeys a linear wave equation. The theory deals with the totality of all the possible ways in which this state function can be decomposed into the sum of products of state functions for subsystems of the over-all system—and nothing more. For example, in a system endowed with four degrees of freedom x_1, x_2, x_3, x_4, and a time coordinate, t, the general state can be written $\psi(x_1, x_2, x_3, x_4, t)$. However, there is *no* way in which ψ defines any unique state for any subsystem (subset of x_1, x_2, x_3, x_4). The subsystem consisting of x_1 and x_3, say, *cannot* be assigned a state $u(x_1, x_3, t)$ independent of the state assigned to the subsystem x_2 and x_4. In other words, there is ordinarily *no* choice of f or u which will allow ψ to be written in the form $\psi = u(x_1, x_3, t) f(x_2, x_4, t)$. The most that can be done is to associate a *relative* state to the subsystem, $u_{\mathrm{rel}}(x_1, x_3, t)$, relative to some *specified* state $f(x_2, x_4, t)$ for the remainder of the system. The method of assigning relative states $u_{\mathrm{rel}}(x_1, x_3, t)$ in one subsystem to specific states $f(x_2, x_4, t)$ for the remainder, permits one to decompose ψ into a superposition of products, each consisting of one member of an orthonormal set for one subsystem and its corresponding relative state in the other subsystem:

$$\psi = \sum_i a_i\, f_i(x_2, x_4, t) u_{\mathrm{rel}}\, f_i(x_1, x_3, t),$$

where $\{f\}$ is an orthonormal set. According as the functions f_n constitute one or another family of orthonormal functions, the relative state functions $u_{\mathrm{rel}\, f_n}$ have one or another dependence upon the variables of the remaining subsystem.

Another way of phrasing this unique association of relative state in one subsystem to states in the remainder is to say that the states are correlated. The totality of these correlations which can arise from all possible decompositions into states and relative states is all that can be read out of the mathematical model.

The model has a place for observations only insofar as they take place within the isolated system. The theory of observation becomes a special case of the theory of correlations between subsystems.

How does this mathematical model for nature relate to the present conceptual scheme of physics? Our conclusions can be stated very briefly: (1) The conceptual scheme of "relative state" quantum mechanics is completely different from the conceptual scheme of the conventional "external observation" form of quantum mechanics and (2) The conclusions from the new treatment correspond completely in familiar cases to the conclusions from the usual analysis. The rest of this note seeks to stress this *correspondence in conclusions* but also this *complete difference in concept*.

The "external observation" formulation of quantum mechanics has the great merit that it is dualistic.[el] It associates a state function with the system under study—as for example a particle—but not with the *ultimate* observing equipment. The system under study can be enlarged to include the original object as a subsystem and also a piece of observing equipment—such as a Geiger counter—as another subsystem. At the same time the number of variables in the state function has to be enlarged accordingly. However, the *ultimate* observing equipment still lies outside the system that is treated by a wave equation. As Bohr[2] so clearly emphasizes, we always interpret the wave amplitude by way of observations of a classical character made from outside the quantum system. The conventional formalism admits no other way of interpreting the wave amplitude; it is logically self-consistent; and it rightly rules out any classical description of the internal dynamics of the system. With the help of the principle of complementarity the "external observation" formulation nevertheless keeps all it consistently can of classical concepts. Without this possibility of classical measuring equipment the mathematical machinery of quantum mechanics would seem at first sight to admit no correlation with the physical world.

Instead of founding quantum mechanics upon classical physics, the "relative state" formulation uses a completely different kind of model for physics. This new model has a character all of its own; is conceptually self-contained; defines its own possibilities for interpretation; and does not require for its formulation any reference to classical concepts. It is difficult to make clear how decisively the "relative state" formulation drops classical concepts. One's initial unhappiness at this step can be matched but few times in history:[3] when Newton described gravity by anything so preposterous as action at a distance; when Maxwell described anything so natural as action at a distance in terms as unnatural as field theory; when Einstein denied a privileged character to any coordinate system, and the whole foundations of physical measurement at first sight seemed to collapse. How can one consider seriously a model for nature that

---

[2] Chapter by Niels Bohr in *Albert Einstein, Philosopher-Scientist*, edited by P.A. Schilpp (The Library of Living Philosophers, Inc., Evanston, Illinois, 1949).

[3] See, for example, Philipp Frank's *Modern Science and Its Philosophy* (George Braziller, New York, 1955), Chap. 12.

[el] Referring to this "great merit" starkly contrasts with Wheeler's later thoughts concerning the orthodox formulation of quantum mechanics: "A difficulty with this 'Copenhagen interpretation,' a difficulty that still deeply troubles me and many others, is that it splits the world in two: a quantum world in which probabilities play themselves out, and a classical world, in which actual measurements are made. How can one clearly draw a line between the two?" (Wheeler and Ford, 1998, pg. 269). It also disagrees with Everett's views; see for example pgs. 238–40.

follows neither the Newtonian scheme, in which coordinates are functions of time, nor the 'external observation' description, where probabilities are ascribed to the possible outcomes of a measurement? Merely to analyze the alternative decompositions of a state function, as in (1), without saying what the decomposition means or how to interpret it, is apparently to define a theoretical structure almost as poorly as possible! Nothing quite comparable can be cited from the rest of physics except the principle in general relativity that all relative coordinate systems are equally justified. As in general relativity, so in the relative-state formulation of quantum mechanics the analysis of observation is the key to the physical interpretation.

Observations are not made from outside the system by some super-observer. There is no observer on hand to use the conventional "external observation" theory. Instead, the whole of the observer apparatus is treated in the mathematical model as part of an isolated system. All that the model will say or ever can say about observers is contained in the interrelations of eigenfunctions for the object part of this isolated system and relative state functions of the remaining part of the system. Every attempt to ascribe probabilities to observables is as out of place in the relative state formalism as it would be in any kind of quantum physics to ascribe coordinate and momentum to a particle at the same time. The word "probability" implies the notion of observation from outside with equipment that will be described typically in classical terms. Neither these classical terms, nor observation from outside, nor a priori probability considerations come into the *foundations* of the relative state form of quantum theory.

So much for the conceptual differences between the new and old formulations. Now for their correspondence. The preceding paper shows that this correspondence is detailed and close. The tracing out of the correspondence demands that the system include something that can be called an observing subsystem. This subsystem can be as simple as a particle which is to collide with a particle that is under study. In this case the correspondence occurs at a primitive level between the relative state formalism where the system consists of two particles, and the external observation theory where the system consists of only one particle. The correlations between the eigenfunctions of the observer particle in the one scheme are closely related in the other scheme to the familiar statements about the relative probabilities for various possible outcomes of a measurement on the object particle.

A more detailed correspondence can be traced between the two forms of quantum theory when the observing system is sufficiently complex to have what can be described as memory states. In this case one can see the complementary aspects of the usual external observation theory coming into evidence in another way in the relative state theory. They are expressed in terms of limitations on the degree of correlation between the memory states for successive observations on a system of the same quantity, when there has been an intervening observation of a noncommuting quantity. In this sense one has in the relative state formalism for the first time the possibility of a closed mathematical model for complementarity.

In physics it is not enough for a single observer or apparatus to make measurements. Different pieces of equipment that make the same type of measurement on the same object system must show a pattern of consistency if the concept of measurement is to make sense. Does not such consistency demand the external observation formulation of quantum theory? There the results of the measurements can be spelled out in classical language. Is not such "language" a prerequisite for comparing the measurements made by different observing systems?

The analysis of multiple observers in the preceding paper by the theory of relative states indicates that the necessary consistency between measurements is already obtained without going to the external observer formulation. To describe this situation one can use if he will the words "communication in clear terms always demands classical concepts." However, the kind of physics that goes on does not adjust itself to the available terminology; the terminology has to adjust itself in accordance with the kind of physics that goes on. In brief, the problem of multiple observers solves itself within the theory of relative states, not by adding the conventional theory of measurement to that theory.[em]

It would be too much to hope that this brief survey should put the relative state formulation of quantum theory into completely clear focus. One can at any rate end by saying what it does not do. It does not seek to supplant the conventional external observer formalism, but to give a new and independent foundation for that formalism.[en] It does not introduce the idea of a super-observer; it rejects that concept from the start. It does not supply a prescription to say what is the correct functional form of the Hamiltonian of any given system. Neither does it supply any prediction as to the functional dependence of the over-all state function of the isolated system upon the variables of the system. But neither does the classical universe of Laplace supply any prescription for the original positions and velocities of all the particles whose future behavior Laplace stood ready to predict. In other words, the relative state theory does not pretend to answer all the questions of physics. The concept of relative state does demand

---

[em] The problem of multiple observers is the quantum measurement problem for Everett (see pg. 176 in chapter 8 and pgs. 73–75 in chapter 9). He believed that he had solved this problem by his treatment of multiple observers in pure wave mechanics (chapter 8, pg. 130).

[en] The Copenhagen colleagues strongly denied that quantum mechanics, at least as they understood it, needed a new foundation; rather, they took Everett simply to be confused about how quantum mechanics works when properly interpreted. See, for example, Alexander Stern's letter to Wheeler (chapter 12, pg. 215). See also (chapter 15, pg. 238) and the following for Everett's response to the charge of having misunderstood.

a totally new view of the foundational character of physics. No escape seems possible from this relative state formulation if one wants to have a complete mathematical model for the quantum mechanics that is internal to an isolated system. Apart from Everett's concept of relative states, no self-consistent system of ideas is at hand to explain what one shall mean by quantizing[4] a closed system like the universe of general relativity.

[4] C. W. Misner, Revs. Modern Phys. 29, 497 (1957).

PART III

The Copenhagen Debate

CHAPTER 11

Correspondence: Wheeler and Everett (1956)

In the spring of 1956, Wheeler was in residence at the University of Leiden, Netherlands. A copy of the long thesis was mailed to Bohr's Institute for Theoretical Physics in April 1956. In May, Wheeler traveled to Copenhagen to discuss Everett's theory with Bohr and Petersen. As Wheeler knew, Everett had already accepted a position as a weapons system analyst with WSEG, and he was eager to graduate. But the professor was reluctant to sign off on Everett's dissertation unless his own mentor, Bohr, accepted at least the possibility that Everett was onto a potentially viable idea, suitable at least for quantizing gravity. Wheeler took notes of his discussion with Petersen, who argued that Everett's theory failed in light of Bohr's view that the quantum world is epistemologically inaccessible. Everett annotated his copy of Wheeler's notes. Everett's annotations are included here in the lettered footnotes.

WHEELER TO EVERETT, MAY 22, 1956

PALMER PHYSICAL LABORATORY
PRINCETON UNIVERSITY
PRINCETON, NEW JERSEY

Dictated from Holland
Received by MBP 22 May, 1956

Hugh Everett
Palmer Laboratory

Dear Hugh:

Professor Hamilton tells me that you wish—or that it appears that you may wish—to remain registered as a graduate student until your 26th birthday, November 11. He asks me my understanding of this point. I think you had better check this important issue with him directly, and let me know the outcome. I didn't realize you had this in mind.

In our discussions in Princeton on April 9–10, and on the train to the MATS terminal in Washington April 13, I understood you prefer to go with the operations research group in the fall rather than take an academic position then (a) because of draft problems (b) because of desire anyway to do your bit for national defense (c) because equipment in Washington would allow you to do some special projects you are interested in. I respect all three wishes, but feel that in the following year you ought to start working towards a first class academic position that will allow you to stand with full freedom for a cause and subject of your own. You have something original and important to contribute and I feel you ought not to let yourself be distracted from it.

The bound copy of the second version of your thesis I sent to Aage Petersen in Copenhagen shortly before I went there two weeks ago, for review with Niels Bohr. After my arrival the three of us had three long and strong discussions about it. I will send you separately notes about specific points. Stating conclusions briefly, your beautiful wave function formalism of course remains unshaken; but all of us feel that the real issue is the words that are to be attached to the quantities of the formalism. We feel that complete misinterpretation of what physics is about will result unless the words that go with the formalism are drastically revised.[eo]

I hope you will accept the points that I will list in a separate note; also those that Bohr promised to write me about this week.[ep] (He was arranging when I left for Stern to give last week a seminar report on your thesis, so it could be thoroughly reviewed before he wrote.)

However, I feel that the issues are too complex to be cleared up by correspondence. I told Bohr I'd arrange to pay from the Elementary Particle Research Fund half your minimum rate steamship fare New York to Copenhagen; I think there's an appreciable chance Bohr would take care of the other half, according to what he said. He would welcome very much a several weeks' visit from you to thrash this out. You ought not to go of course except when he signifies to you that you are picking a time when he can spend a lot of time with you. Unless and until you have fought out the issues of interpretation one by one with Bohr, I won't feel happy about the conclusions to be drawn from a piece of work as far reaching as yours. *Please* go (and see me too each way if you can!)

So in one way your thesis is all done; in another way, the hardest part of the work is just beginning.

I would like to feel happier than I do with the final product; then I would like to see it published in the Danish Academy in full—that's the perfect place for it.

[eo] This is the central problem of choosing the right words to describe the theory again. Among other things, the interpretation of the theory depends on this choice. For a few examples of the problem see pgs. 68, 121, 209–10, and 223.

[ep] Wheeler's notes from these conversations begin on pg. 207. Bohr himself may have never produced notes for Wheeler.

How soon can you come?

Sincerely,

John Wheeler

P. S. Let me know the result of your discussion with Professor Hamilton. I would like to see your final exam scheduled for Friday afternoon September 14 or September 28, but I think it is foolish of me to talk about dates until this whole issue of words is straightened out.

cc: Professor Hamilton
      Professor Shenstone

WHEELER NOTES ON CONVERSATION WITH PETERSEN, MAY 3, 1956

3 May '56

A. Petersen[eq] —Paradox outlined by Everett.

Distinction between Bohr way & the 2 postulate way to do q. mech.[er]

How analyze paradox à la Copen? See Bohrs papers. How extend system under investigation?

If QM description of measuring tool prevents its use as a meas. tool

Complexity of human being—exclude by psycho phys parallelism

A. P. says more cautious to deal with spots on plate, forget the brain side. Then do sharpen up to everyday issues—more immediate. Lucky thing that how we see spot doesn't matter. Everett ought to reform so.[es]

What do

See Warsaw ~ 1938 ± 1, New Theories of Physics Bohr's new Definition of "Phenom. in Qu. Physics". cf also Einstein volume. Have a wholeness that is foreign to class—stone throwing. Q. phenom. are indivisible. To subdivide, have introduce an apparatus which is incompatible with the phenom one wanted to subdivide. (Slits; which hole did electron go through—specify history more closely; change phenom.—Bad for interf;

[eq] Aage Petersen was Bohr's assistant from 1952 to 1962 and Everett's friend. Everett was a student and Petersen was visiting Princeton when they met. The lengths to which Wheeler was willing to go in defense of Everett's thesis suggest his own commitment to the project at this time. Petersen is referred to in these notes by his initials, A. P.

[er] The complaint is that Everett needs to better distinguish between how the standard von Neumann–Dirac collapse formulation of quantum mechanics works and how Bohr's Copenhagen formulation of the theory works. The following notes provide a sketch of how Bohr's theory works. While he clearly understood the difference, Everett did at times equivocate between the two in his thesis. Everett later reported that the standard von Neumann–Dirac collapse theory was the target of the thesis but that the Copenhagen formulation of the theory was not preferable (pg. 239).

[es] Measurement records for Everett's automata model of observers are, however, no more subtle than spots on a photographic plate.

if we want to retain diff. between phys. & math. Have to pick one exptl arrang or other—cf Bohr-Einstein example. When loosen diaph. to measure recoil & tell slit, then it becomes part of meas. system. Then apply QM to slit, uncertainty principle. $B + A$ working on this—B says A you don't have good set up to meas mom—but then A's app. no more suitable for A's purpose. But then say not every system has a ψ fn does not pertain to a phys system in same way as a dynamical variable (Cant meas $\hbar$ class—symbolizes a wholeness foreign to class phys.) (Consider this for our Lorentz notes). ψ fn for elec doesn't have sense until we get something like a prob dist of spots.[et] Only a coord sys can give a vector a meaning. Have to know ψ <u>plus</u> expt'l apparatus to make predictions.[eu]

With photon of known mom., can get distb'n. of spots in one expt; or can say through which slit goes.

So, Q. M. formalism not well defined appl'n. without expt'l arrangement.[ev]

Wholeness.

2 formalisms: Must & do give same result.

Objection: Does not appreciate the change brought into physics by the quantum. E Calls Bohrs too conserv.—but AP: no more conserve. than demanded by exist of q.[ew]

Bohr would say Everett much too class,[ex] not in math but in recognize

---

[et] Everett writes in the margin: "<u>Nonsense!</u>"

[eu] Everett is instructed to read Bohr's new papers to better understand Bohr's most recent description of how phenomena work in quantum mechanics. Perhaps the most influential of these is (Bohr, 1949). The worry here is that by trying to provide an account of measurement as a simple physical process involving just a correlation between the measuring device and the object system, Everett is not properly recognizing that the description of any quantum phenomena is always contingent on one's characterization of a measuring apparatus that is compatible with the phenomena, and no one such characterization will serve all purposes. This is largely just a restatement of the Copenhagen interpretation as Bohr described it at this point. See pg. 152 for Everett's understanding of the theory. Insofar as Everett took his most important contribution to be showing how to treat all quantum measurements in terms of simple correlation interactions within pure wave mechanics, the complaint of the Copenhagen colleagues on this point would have represented for Everett a fundamental and irreconcilable disagreement. Everett cannot agree that one must know both ψ *and* an explicit apparatus to make predictions since in pure wave mechanics the wave function provides a complete physical description of the world.

[ev] The Copenhagen interpretation holds that consistent talk of quantum phenomena requires both (1) a specification of the quantum mechanical states in the language of the quantum mechanical formalism and (2) a specification on the measurement apparatus and exactly how it would be used over the course of the experiment in the language of classical mechanics. See pg. 152 for Everett's description of the Copenhagen interpretation in the long thesis, chapter 8.

[ew] Everett took Bohr's theory to be too conservative because it insisted on keeping a fully classical description of the measuring apparatus. Everett held that the apparatus itself should be treated quantum mechanically like another system (pg. 239).

[ex] Wheeler writes in the margin: "This would make a good introductory statement early in thesis."

new features. Just as in past formalism, the whole problem the tough one was to find the right words to express the content of the formalism in acceptable form.[ey]

2. Mech not about the things Everett thinks it is about; only for domain where $\frac{e^2}{\hbar c}$ small, where can meas. mom or pos. with ∞ accuracy; not a matter of our only having tested business so far; only in H atom as long as small radn damping; if no longer
Surprising we can have a
JAW[ez] position always QM, complex words for $e^2/\hbar c$; much opposed by AP—no poss to connect with ordinary words.

JAW evolution in a strongly coupled world. Help each other out. Need words.[fa]

AP Math can never be used in phys until have words. Aren't comparing selves with servo mechanisms. What mean by physics is what can be expressed unambig in ordinary language.[fb] Spots on plate have meaning but not in Everett—he talks of correlation but can never build that up by ψ fn's.[fc] —H atom stability

Bohr (ac to AP) need non-rel. way to live self into rel. world—have to sep. between space & time—consider watch; <u>entrance into</u> complex nos only via real nos; hence entrance into <u>rel</u> via <u>non rel</u>.[fd]

JAW—in world of big assemblies of points $\frac{e^2}{\hbar c}$ evolution occurs—Help each other fight way thru to a language.—language <u>second</u>. Very contrary to Bohr say A.P.

What do <u>re</u> HE III thesis. (1) Knowledge limited (2) Class phys didnt help learn about <u>obsv'n</u> (3) In QM learned more.

[ey] Which, on the Copenhagen interpretation, requires that one always distinguish between a classical measuring apparatus and the quantum-mechanical system one is observing.

[ez] John Archibald Wheeler.

[fa] There are two reasons why Everett's precise choice of words mattered. The words Everett chose to describe the structure of pure wave mechanics, for better or worse, provided an implicit interpretation of the theory. For his part, Everett did not adhere to any fixed vocabulary for describing the correlation structure of pure wave mechanics. See, for example, pg. 121 for an example of Everett's shifting language. Everett's flexible choice of descriptive language has always posed a difficulty for readers. See for examples pgs. 68, 206, 210, and 223. The second consideration here is pragmatic. Insofar as Wheeler wanted to preserve the possibility of Everett getting an academic job, he wanted Everett and the Copenhagen colleagues to reach some sort of agreement where Everett was not seen as attacking the orthodox view. Wheeler believed that this would be possible if he could find the right words for Everett to use in the description of his theory.

[fb] Everett writes in the margin: "Def of Physics?"

[fc] Everett took himself to have captured both the standard quantum statistics and ordinary descriptions of classical systems in the long thesis. Everett writes in the margin: "obviously hasn't completed reading of thesis! <u>it does just that.</u>"

[fd] At this point in the notes, Wheeler includes a diagram that he labels "For Book." The significance of the diagram is unclear, but it can be seen online at the UCIspace archive (http://hdl.handle.net/10575/1143).

New thing in Q Mech. Concept of wholeness.

I̲f use space & time (c numbers etc)

S̲el things on QM (and objections) based on this c-number idea (hence Bohrs great emphasis on class obvn).

Great divergence not on formalism but on words.[fe]

In this realm of discussion there is no <u>problem of several observers</u>.[ff]

JAW—need to describe detailed QM of slit, not merely $\Delta p \Delta q \sim \hbar$—hence have to analyze in detail.

AP cites Bohr. Rosenfeld (cf my vol Pergamon)—

[Says Von N & Wig all nonsense; their stuff beside the point; doesn't come into; could overlook test body atomicity—Von N & Wig—mess up by including meas tool in system].[fg]

Silly to say † ‡ ‡ apparatus has a ψ-function. Example chosen for corresp. with classical.[fh]

But more typical—meas. system has a certain amt of dynamics.

If want to predict prob distributions of class variable[fi]

[fe] The difference in words reflects the difference in the respective interpretations of the same formalism between Everett and the Copenhagen colleagues. Not even the Copenhagen colleagues would want to be seen as denying that pure wave mechanics holds at some level of description.

[ff] Everett held that both the standard collapse formulation and Bohr's formulation encounter serious problems when one considers more than one observer. See pgs. 176 and 73–75 for detailed discussions.

[fg] Both the standard von Neumann–Dirac and the Wigner interpretations are taken by the Copenhagen colleagues to "mess up" precisely because they seek to do what Everett does: to include the description of the measurement device in the physical description of the world. They held that Bohr's interpretation had avoided this fatal mistake.

[fh] The diagram that Wheeler drew at this point seems to represent a standard two-path experiment. The thought is that the experimental apparatus itself is stipulated to be classical in the very description of the experiment and hence cannot be described quantum mechanically. But this is precisely what Everett denied in treating all physical systems quantum mechanically.

[fi] Insofar as quantum mechanics is supposed to predict probability distributions for classical variables, the Copenhagen colleagues held that measurement records, in addition to the experimental apparatus itself, must be describable in terms of classical variables, not by a quantum mechanical wave function as Everett, von Neumann, and Wigner require.

WHEELER TO EVERETT, MAY 26, 1956

INSTITUUT-LORENTZ[fj]   LEIDEN, SAT AM
VOOR THEORETISCHE NATUURKUNDE   26 MAY
    PROF. DR. S. R. DE GROOT
    LANGEBRUG 111, LEIDEN
      TEL. K 1710-31725

Dear Hugh:

I am sending in a few minutes this cable:

Niels Bohr
    Physicum
    Copenhagen
    Everett now Princeton phoned asking confer with you. Hopes fly almost immediately but must return midjune. Could you cable him if inconvenient. My great hope thesis suitable Danish Academy publication after revision Have answered Stern.[fk] Regards

John

It is well for you to tell me and Bohr you have to return 15 June but I hope to goodness you keep open the necessity[fl] to stay until the battle is won. I would prepare WSEG for this.[fm]

I hope very much you will stop here coming and going as you can without extra cost if you travel KLM. I am quite prepared to believe SAS will do this much for you at no extra cost:[fn]

My dates:

Free coming week except for lecture Tues am May 29 and seminar Wed night and a 1 hr appointment 11–12 am Sat. Leave Sun night June 3 for 4 days England back Fri am June 8. From Sat. afternoon June 9 to Sun. afternoon June 10 am attending a physics meeting. Back here late Sun June 10. Here until Sat June 16 when am going on weeks sailing trip with children newly arrived from U.S. Then here continuously.

[fj] At the top of the letter, Wheeler included the phone numbers for the office he was using and a coffee house near the home in which he was staying.

[fk] Everett's thesis was not published by the Danish Academy. The long thesis was published only much later, in DeWitt and Graham (1973), and the short thesis ended up in *Reviews of Modern Physics*.

[fl] Wheeler originally wrote "the possibility," but crossed it out.

[fm] WSEG stands for Weapons Systems Evaluation Group, the Pentagon research group that Everett had planned to work for after defending his thesis.

[fn] Wheeler included hand drawn diagram showing proposed flight paths between New York, Amsterdam, and Copenhagen.

Here is a note to give Mrs. Pratt:[fo]
(MBP—Could you please handle as promptly as E. needs)
Mrs. Pratt—This is to authorize and request the university to disburse $260 from the Elementary Particle Research Fund for post payment of Hugh Everett's expenses to come to Amsterdam and Copenhagen to confer with Bohr and me on his work.—John A. Wheeler.   26 May 1956

WHEELER TO EVERETT, SEPTEMBER 17, 1956

PALMER PHYSICAL LABORATORY
PRINCETON UNIVERSITY
PRINCETON, NEW JERSEY

17 September 1956

Mr. Hugh Everett
22 Glen Drive
Alexandria, Va.

Dear Hugh:

Many thanks for your letter. I think your thesis is going to prove very important, but first it has to be made javelin proof. I look forward eagerly to seeing you. I may be down the weekend of October 6 or October 13 or both, and will write more in detail shortly.

Best regards on the eve of departure for Seattle.

John Wheeler

Enclosure

[fo] Wheeler writes in the margin: "cut here ↓".

Copy[fp]

Dear Professor Wheeler,

As for the moment I have no more time to spend on it, I have provisionally finished my remarks on quantum measurement and I have sent you a copy by sea mail one of these days. I have avoided polemics with the many divergent papers on the subject and after having read Everett's paper I found (apart from a technical reference) no special reason to go into that one. I cant agree with his interpretation and in those things which have a direct physical meaning I have (apart from the extensive treatment of correlation) found no particular new points. I hope that in spite of their sketchy and incomplete character my remarks sufficiently make clear my point of view.

Although you may now have abandoned the idea of interaction-at-a-distance, I still should like to ask you whether you have a spare copy of in particular your and Feynman's first paper on that subject.

With many thanks for the loan of the Everett paper,

Yours sincerely,
H. J. Groenewold

[fp] What follows is the enclosure Wheeler mentions above. See also Groenewold's subsequent letter in this volume (chapter 13, pg. 226).

CHAPTER 12

Correspondence: Wheeler, Everett, and Stern (1956)

After Wheeler left Copenhagen, Alexander Stern, an American physicist in residence at Bohr's Institute, led a seminar on Everett's theory. He then wrote to Wheeler, critiquing the theory forcefully from the point of view of complementarity, labeling it "theology." Interestingly, Stern referred to a prior correspondence in which Wheeler had compared the mathematics of game theory to Everett's formulation of quantum mechanical formulation, since in both cases observers are included as a critical part of the theory itself. Although we have not found a copy of this letter, from Everett's handwritten notes, it is clear that his approach to quantum mechanics was influenced by his work in game theory in which concepts of probability, utility, and information interrelate.[1]

Everett penciled in his own critique of Stern's critique, concentrating on the issue of irreversibility. He concluded that Stern had not tried hard enough to comprehend his argument about including observers in the wave function.

Spurned by Bohr and his circle of physicists, Wheeler was very keen that Everett struggle directly with Bohr. He informed him that unless the language in which his formalism was expressed was made less controversial, he was not going to sign off on his degree. And he wrote a conciliatory letter to Stern claiming that Everett did not intend to disparage Bohr's philosophy of complementarity and his correspondence model. That was not a true statement, however, as Everett disliked both the von Neumann–Dirac wave function collapse postulate and the Copenhagen interpretation, which he considered to be "hopelessly incomplete" as well as a "philosophic monstrosity with a 'reality' concept for the macroscopic world and denial of the same for the microcosm" (pg. 255).

---

[1] Information theory was a starting point for Everett, and it is not surprising that physicists of Bohr's and Stern's generation tended to think of information in terms of "meaning", whereas Everett thought of information as a formal notion that might be represented in the state of almost any physical system—in keeping with his background in game theory and the new science of "cybernetics." That is perhaps why Everett could easily conceive of an observer as a servomechanism, whereas Bohr (a neo-Kantian) and Stern (a Bohrian) could not separate measurement from human agency.

STERN TO WHEELER, MAY 20, 1956

COPY

UNIVERSITETETS INSTITUT   Blegdamsvej 17
                    FOR   Copenhagen, Denmark
            TEORETISK FYSIK   Telegrams, Physicum, Copenhagen

May 20, 1956

Prof. John A. Wheeler
Instituut-Lorentz
Leiden, Nederland

Dear Prof. Wheeler:

Many thanks for your nice letter. I gave a seminar on Everett's paper on Monday, May 14. Prof. Bohr was kind enough to make a few introductory remarks and open the discussion. Prof. Mercier and Prof. Møller made some comments.

In my opinion, there are some notions of Everett's that seem to lack meaningful content, as, for example, his universal wave function. Moreover, he employs the concept of observer to mean different things at different times—the measuring apparatus, a servo-mechanism for registering experimental results, and its dictionary meaning, that is, common usage. Also, at times, he talks about observers in the sense of a statistical ensemble of classical statistical mechanics, and reminds one of other attempts at a causal, deterministic interpretation of quantum mechanics.[fq]

I do not follow him when he claims that, according to his theory, one can view the accepted probabilistic interpretation of quantum theory as representing the subjective appearances of observers. At times he gives one the impression that he believes that, were it not for the interference of physicists (observers) quantum theory would be a continuous, deterministic, and elegant theory. Significantly enough, his second chapter on Probability, Information, and Correlation is the best in his book.

But, to my mind, the basic shortcoming in his method of approach of his erudite, but inconclusive and indefinite paper is his lack of an adequate understanding of the measuring process. Everett does not seem to appreciate the FUNDAMENTALLY irreversible character and the FINALITY of a macroscopic measurement.[fr]

[fq] In his comments concerning this letter Everett says that for him the "statistical ensemble of observers is within the context of the theory, a *real*, in distinction to a *virtual*, ensemble" (chapter 12, pg. 224). This is the real ensemble of the relative states of an observer. See Everett's discussion of conditional expectations being given by relative states (chapter 8, pgs. 99–100) and the discussion of Everett's notion of relative states in the conceptual introduction (chapter 3, pgs. 34–35).

[fr] Everett considered the apparent irreversibility of measurement in the long thesis (see for example pgs. 143–44). Everett thus concluded that Stern's remarks concerning his

One cannot follow through, nor can one trace the interaction between the apparatus and the atomic system under observation. It is *not* an "uncontrollable interaction", a phrase often used in the literature. Rather, it is an INDEFINABLE interaction.[fs] Such a connotation would be more in accord with the fact of the indivisibility, the wholeness of the quantum phenomenon as embodied in the experimental arrangement.

Interestingly enough, now that molecular causes are being discovered for various serious ailments as well as for basic biological phenomena, a problem similar to the observation problem has arisen in the biological sciences. Let us take a concrete example that has recently been in the news. There has been good evidence accumulating in recent years that schizophrenia is due to some "abnormal" molecular process in the blood. However, to trace the schizophrenic phenomenon from the basic molecular level to the observational level of its psychological symptomatic manifestations is an aspect of the observation problem. It cannot be traced in the detail of a space-time description. Moreover, the space-time coordinates, such as, position and momentum, are not relevant for the description of such phenomena. All this may seem far from physics, but the transition from the microscopic to the macroscopic observational level presents a problem in biology similar to that in physics. In biology, of course, the event is infinitely more complex, but it is likely that a common basic principle is operative in both cases.

Then there is the concept of a state in quantum theory. An elementary system does not come with a "ready-made" state. It does not possess a state in the sense of classical physics. Its state (usually the initial state) is prepared. The probability distribution in quantum theory implies more than a mere information content. Rather, the experiment is so designed as to give a meaningful information PATTERN. One can no more exclude meaning and understanding from physics than one can substitute servo-mechanisms for physicists. Wave mechanics without probability excludes physicists.[ft]

---

misunderstanding the fundamental irreversibility of the measurement process "indicate rather clearly that he has had insufficient time to read the entire work." Everett continued that "several rereadings on his part seem to be called for" (pg. 224). For Everett the question was how measurements can appear to be reversible when the dynamics are not. He believed that pure wave mechanics provided an explanation in terms of the subjective appearance of irreversibility to observers treated within the theory.

[fs] Everett would have taken this to be an unnecessarily mysterious position to take given the entirely satisfactory nature of pure wave mechanics. In his notes, Everett replied "Technically, 'observer' can be applied to *any* physical system capable of changing its state to new state with some fairly permanent characteristics which depend upon the object system (with which it interacts)." He held that the criteria for a measuring apparatus and for an observer were precisely the same (pg. 224).

[ft] It is implicit in the claim that one cannot substitute servo-mechanisms for physicists, that an observer is somehow more than just a physical system. Everett, however, believed that the principle of psychophysical parallelism ("PPP") required that it be possible to model observers as ordinary physical systems (pg. 224).

In your letter you ask, "Do we need mathematical models, like those of game theory, that will include the observers, in order to put across to the mathematically minded what is meant by these ideas?" (I take it you mean complementarity and other ideas of quantum theory "as distinct from the mere formalism.") But the mathematical model of game theory has different ideational implications from those of quantum theory.

The role and behavior of the observer in game theory are fundamentally different from those of the observer in quantum theory. The observer in quantum theory prepares the initial state of the system, but this is certainly not a legitimate function of the observer in game theory. For example, the preparation of a deck of cards in a certain initial state in a game of poker would violate the rules of the game, if not the ethics of the observer who has prepared the system. In physics, the meaningful pattern is associated with the probability distribution of the system other than that of the observer, while in game theory it is associated with the behavior of the observer (strategy). There is no objective feature concerning the rules of a game. They are man-made and are completely determined by man.

The unobservables in a theory should have observable consequences. The unobservables and the observables together form the theoretical structure, and they must be logically connected. If Everett's universal wave equation demands a universal observer, an idealized observer, then this becomes a matter of theology. If a complete knowledge of the state of the composite system (apparatus plus atomic subsystems) involves practically an infinite number of observers which cannot communicate with one another, then we are talking metaphysics. One may invoke the image of a large number of mystics in different "resonant" states. Heisenberg's recent attempt at a theory of elementary particles is a good example of what I mean. Heisenberg has conceived of a Hilbert Space II in which the rules of quantum theory do not apply. It is in some sense a symbolic receptacle wherein he can deposit the undesirable features of present-day field theory, *and there is no logical thread of connectivity between his symbolic limbo and the rest of his theory.*

I cannot help believing that, if Everett went further and carried out his mathematical ideas, forgetting his preconceived model of the universe, which guided, channeled, and concluded his mathematical investigation, he would have come across a contradiction in his work. His claim that process 1 and process II are inconsistent when one treats the apparatus system and the atomic object system under observation as a single composite system and if one allows for more than one observer is, to my mind, not tenable.[fu]

[fu] The thought here is that Everett's theory must itself involve a contradiction since Stern cannot believe that there is a logical conflict between Process 1 and Process 2 in the standard collapse formulation of quantum mechanics. Stern is simply denying that the standard collapse theory encounters a measurement problem.

The subjective aspect of physics, which some scholars and philosophers have claimed to detect but have not understood, has its origin in the fact that physics must make contact with reality which is, after all, the way the world appears to us, and can be understood by us. Mathematics and the abstract concepts of physics which it inspires are the connecting links between our all-too narrow and little world and the infinite pageant that is nature. In the pursuit of truth the concepts of objectivity and subjectivity are no longer antithetical, but take on a complementarity aspect. Our formalism must be in terms of possible or idealized experiments whose interpretations thereby involve the use of concepts intimately connected with our own sphere of experience which we choose to call reality. The epistemological nature of our experiments and the objective nature of the abstract mathematical formalism TOGETHER form the body and spirit of science.

I am sorry I have not read Bush's book on "Stochastic Models for Learning." I went to the University library the day after receiving your letter, but they did not have the book in their collection. I intend to read it when I get back to the States.

Finally, I hope I have not bored you by telling you many things which you know better than I do. Also, I realize that, in places, my remarks may have been too cryptic, and I should have liked to expatiate, but I felt I should keep my comments within the confines of a letter.

My wife joins me in sending our best regards to you and Mrs. Wheeler.

Sincerely,
Alexander W. Stern

WHEELER TO STERN, MAY 25, 1956

C O P Y

25 May 1956
Dr. Alexander W. Stern
Blegdamsvej 17
København Ø

Dear Dr. Stern:

Many thanks for your kind letter of May 20, and for your extensive comments on Everett's paper. Also I am grateful for the time you and Professor Bohr and Aage Petersen and other friends have spent in reviewing Everett's ideas, both in the special seminar and otherwise.

I fully recognize that there are many places in Everett's presentation that are open to heavy objection, and still more that are subject to misinterpretation. To make the whole discussion consistent at every point with what we know about the measuring problem is going to be a very heavy task. On it I would like to make sure that Everett has the benefit of a number of weeks in Copenhagen. I have already written him to this effect.

I would not have imposed upon my friends the burden of analyzing Everett's ideas, nor given so much time to past discussions of these ideas myself, if I did not feel that the concept of "universal wave function" offers an illuminating and satisfactory way to present the content of quantum theory. I do not in any way question the self consistency and correctness of the present quantum mechanical formalism when I say this. On the contrary, I have vigorously supported and expect to support in the future the current and inescapable approach to the measurement problem. To be sure, Everett may have felt some questions on this point in the past, but I do not. Moreover, I think I may say that this very fine and able and independently thinking young man has gradually come to accept the present approach to the measurement problem as correct and self consistent, despite a few traces that remain in the present thesis, draft of a past dubious attitude.[fv] So, to avoid any possible misunderstanding, let me say that Everett's thesis is not meant to *question* the present approach to the measurement problem, but to accept it and *generalize* it.[fw]

[fv] This is too much to say on behalf of Everett. While he was eventually willing to produce a redacted version of his thesis where he did not repeat his complaints concerning the Copenhagen interpretation (pgs. 152–53), Everett continued to insist that his formulation of quantum mechanics was strongly preferable to both the standard collapse theory and the Copenhagen interpretation. See for example chapter 15 (pgs. 238–40) and the discussion of Everett's position in the conceptual introduction (pg. 32).

[fw] The problem with Wheeler's defense here is that Everett did explicitly criticize as inadequate the Copenhagen interpretation in his long thesis (pgs. 152–53), and Everett

Realizing that the exchange of a few letters is not going to be enough to clear up many deep issues, let me nevertheless outline the point of view of Everett's thesis in my own words:

(1) WAVE FUNCTION IN THE CONVENTIONAL FORMALISM

(a) One recalls the familiar distinction in quantum mechanics between the system under study and the many varieties of apparatus which can be used to study it. (b) A complete observation with one choice of equipment leaves the system in a well defined quantum state. (c) This state, and its free evolution with time, are described by a wave function that depends upon the system coordinates alone. (d) This wave function, and the wave equation that it obeys, together form a machinery to predict in a way free of all contradictions the outcome of one or another type of measurement made with one or another of the complementary varieties of apparatus at the disposal of the observer.

(2) WAVE FUNCTION IN EVERETT'S FORMALISM

(a) Nothing prevents one from *considering* a wave function and its time evolution in abstracto; that is, without ever talking about the equipment which originally prepared the system in that state, or even mentioning the many alternative pieces of apparatus that might be used to study that state. (b) A state function as used in this sense has absolutely nothing to do with the state function as used in the customary discussion of the measurement problem, for now *no means of external observation are admitted to the discussion.* (c) Why in the world talk of a wave function under such conditions, for it in no way measures up to the role of the wave function in the customary formulation, that we accept without question? (d) Because it is proposed, of Everett's free volition, to formulate *a new physical theory*, as a step of free creation. (e) In this theory, as in every new theory, the quantities that enter have roles and positions that will be defined and determined *by the logical structure of the theory itself.* (f) The greatest possible confusion will result if the mathematical quantities in Everett's theory, such as the wave function, are thought of as having the purpose that the wave function fulfills in the customary formulation. (g) The present draft of the thesis by no means keeps clear throughout this distinction between the idea of wave function in the two formulations, let alone the ideas of observer and measurement. (h) Everett's "universal wave function" is not to be thought of as subject to external observation. Every idea of an observation from outside is to be rejected. (i) If the universal wave function is not subject to external observation, is it not as you put it "a matter of theology"? To this question I should be frank in saying I have no complete answer, nor am I sure that it is necessary to give one. Is there any

continued to hold that the Copenhagen interpretation was unacceptable, as indicated in his letter to Petersen (chapter 15, pgs. 238–40).

difference on this score between Everett's universe and Laplace's universe? No one seriously believed that it would be a practical possibility ever to know at one moment the position and velocity of every particle, but it was convenient to *postulate* that these quantities nevertheless had well defined values. Likewise Everett *postulates* that the "universal wave function" had at one moment a well defined dependence upon the state variables, and therefore also a well defined dependence upon these variables at every other moment. (j) This postulate is a creative act, beyond any step by step pre-justification. (k) The justification must instead lie in the internal consistency of the resulting theory and in its *consequences*. (l) The very meaning of the word "consequences" has to be defined within the framework of the theory itself, not by applying to Everett's concept of wave function epistemological considerations that refer to "wave function" in the *completely different* context of the usual formalism.

(3) EVERETT'S SYSTEM

(a) Everett develops his theory in the following *spirit*: (this is not the theory, but how he goes about constructing it): Follow the evolution of the wave function with time, and look for connections between regions of maximum in the wave function and the kind of happenings we encounter. (b) The connection is meant to supply the possibility of a complete *model* for our world. (c) For this reason the system must contain a sufficiently large number of degrees of freedom to admit subsystems that can be compared to the observers of our world. (d) The idea is accepted, and is essential to Everett's analysis, that these subsystems are sufficiently complex to simulate all that human beings do, including acting, observing, and recording. (e) Thinking and communication between these model observers and also postulated on a basis that I should like to express in the following way. (f) These subsystems or model observers have arisen within the overall system by organic evolution: reproduction and destruction or survival.[fx] (g) Mutual assistance being indispensable to survival, these subsystems have learned to communicate with each other about their experiences. (h) The kind of physics that occurs does not adjust itself to the available words; the words evolve in accordance with the kind of physics that goes on. (i) Thus thinking, experimentation and communication—or psychological duplicates thereof—are all taken by Everett as going on *within* the model universe. (j) As thinking can take place on various levels, from the most rudimentary and mechanical to the most highly complex, so the system that forms the model universe can be so primitive that in it there is no place for anything that we would call reason, or so complex that it admits all the irreversibility of measurements as we know them.

[fx] Everett himself never tells a story quite like this, but there is no reason to believe that he would have disagreed with this.

(4) THE PHYSICS OF EVERETT'S SYSTEM

(a) Whether simple or complex, the model universe has properties that are fixed by the universal wave function. (b) The nature of the statements that can be made within Everett's theory is not fixed by the pattern of the usual quantum formalism at all. (c) Instead, it is determined by the postulates and rules of calculation that he lays down in his thesis, and which I will not attempt to recapitulate here. (d) For clarity, every deduction that comes out at this level of discussion ought to bear the phrase "in the model universe". Such words as "correlation" and "observer" ought to bear quite different names, to emphasize the absolutely fundamental distinction between the model universe and the real physical world.[fy] (e) Everett's theory receives its test number one, the test of logic, by its internal self consistency down to this point.

(5) RELATION BETWEEN EVERETT'S SYSTEM AND PHYSICS AS WE DO IT.

(a) None at all? No, because Everett traces out a correspondence between the "correlations" in his model universe on the one hand, and on the other hand what we observe when we go about making measurements. (b) The closeness and clarity of this correspondence is test number two of his theory. (c) Has the closeness of the correspondence been proven in the second draft of Everett's thesis, that you have? In large measure, in my opinion; but there are still logical loopholes left. I believe Everett can fill in the missing steps in this material. (d) Has the nature of the correspondence been made clear, including the sharp distinction between the model universe and the ideas we use in everyday quantum physics? Far from it, in my opinion. What I have written in this letter is a very feeble attempt to go a very small ways beyond Everett in clarifying this issue. The full job will be a heavy task, but I have confidence that Everett can accomplish it *if* he can have the benefit of some weeks in Copenhagen to struggle out these problems. So I hope to hear from him that he will find it possible to come.

I am taking the liberty to send copies of this letter to (1) Niels Bohr and Aage Petersen and to (2) Hugh Everett, as well as to forward to Everett a copy of your letter to me. If further discussion of these problems takes place at Copenhagen, I shall appreciate learning how far the points that I have numbered in this letter for convenience of reference might be found acceptable.

Thank you again for your kindness in writing. Please accept best regards for yourself and your wife and daughter.

Sincerely yours,
John A. Wheeler

[fy] Wheeler thought of Everett's theory as describing a model universe that may or may not be isomorphic to the real world. Everett himself held a similar but somewhat weaker view. Everett believed that one could only ever check the correspondence between a physical theory and the real universe up to empirical homomorphism (pg. 169). See also pg. 253.

WHEELER TO EVERETT, MAY 25, 1956

LEIDEN, May 25, 1956.

Mr. Hugh Everett
Palmer Physical Laboratory
Princeton, New Jersey.

Dear Hugh:

I will appreciate it if you will see if you agree with my enclosed reply to Stern's letter, which is also enclosed.[fz] In case you write more, I would appreciate it if you would send one carbon to Niels Bohr care of Aage Petersen and another carbon to me. But this long distance writing I for one find a very inefficient way of using my time to forward your work; discussion would be much more to the point. Hence my great hope that you will arrange to come.

From the tenor of the present discussions by others and from what I have written myself you will see that I have no escape from one sad but important conclusion: that your thesis must receive heavy revision of words and discussion, very little of mathematics, before I can rightfully take the responsibility to recommend it for acceptance. Moreover, I think it will be humanly impossible to come to agreement on all issues unless you and I are in the same place for several weeks, or unless you and Bohr and associates are in the same place for several weeks, or both. If you agree with this conclusion, I would like to hear from you what you think should be the next step.

As I have said before, I feel that your work is most interesting and am sure that it will receive discussion of a scope comparable to what has attended Bohm's publications. But in your case I must ask that the bugs be got out and the sources of misunderstanding be clarified *before* the job is made public, not *afterwards*. I hope you will realize that I mean this as what is called here your "promoter," and one actively interested in your reputation and promising future.

May I suggest you show this correspondence (Stern to me, me to Stern and me to you) to Professor Dicke or Hamilton or Treimen or Wightman, preferably to two out of the four, and get further advice from them. I believe you should struggle this out in Copenhagen now, not wait and let things cool until my return in the fall. You cannot expect to meet all objections and clarify all obscurities without the fullest opportunity to meet strong and well informed opposition.

Sincerely,

John Wheeler

[fz] Stern's letter begins on pg. 215.

Comment on Stern's Letter:[ga]

1. True, word observer given many connotations—but <u>all</u> these connotations can be used consistently within the scheme of Chap II (abstract disc. of observation). Such usage <u>deliberate</u> to show how general the treatment is. Technically, "observer" can be applied to <u>any</u> physical system capable of changing its state to new state with some fairly permanent characteristics which depend upon the object system (with which it interacts).—i.e. any measuring apparatus could be called observer—and criteria for observer same as for meas. app. Only possible distinction is often one might reserve word "observer" for more complex automatic mechanism, capable of carrying out actions according to its past experience. Dictionary meaning <u>must</u> also be included, since otherwise human observer not ordinary phys. system and PPP violated.

2. Statistical ensemble of observers is, within the context of the theory, a <u>real</u>, in distinction to a <u>virtual</u>, ensemble!

3. Sterns remarks about misunderstand. of <u>fundamental</u> view. of measurement proc. indicate rather clearly that he has had insufficient time to read the entire work.

   Several rereadings on his part seem to be called for

   Also—Stern is quite guilty in these remarks of begging the question—one of the fundamental motivations of the paper is the question of <u>how can it be</u> that mac. measurements are "irreversible," the answer to which <u>is</u> contained in my theory (see remarks Chap V), but is a serious lacuna in the other theory.

[ga] Everett wrote the following comments in response to Stern's letter under the label "Comment on Stern's Letter." There is no evidence that he sent these to Stern or Wheeler. See the footnotes on pgs. 215–16 for explanations of Everett's comments.

CHAPTER 13

Correspondence: Groenewold to Everett (1957)

After losing the initial round of struggle with Bohr and his circle, Everett left Princeton in the spring of 1956, moving to the Washington, D.C., metropolitan area to commence working at WSEG, quickly establishing himself as WSEG's chief mathematician and computer expert. He married Nancy Gore, who was pregnant with their first child, Elizabeth. He did not attend the January 1957 conference of gravitation in Chapel Hill, but shortly afterwards, he and Wheeler redacted and rewrote his dissertation (cutting it by 75 percent). He received his doctoral degree in April 1957, and his dissertation (with minor changes) was printed in Reviews of Modern Physics *in July 1957 as part of the published proceedings of the Conference on the Role of Gravitation in Physics (see DeWitt (1957), Cecile DeWitt's synopsis of the conference).*

Before that, Wheeler had sent preprints of Everett's and his own article in Reviews of Modern Physics *to Bohr and other prominent physicists. In April, H. J. Groenewold of Natuurkundig Laboratorium der Rijks-Universiteit te Groningen, Germany, wrote to Everett and Wheeler with his criticism of the revised dissertation. He had previously read the long dissertation sent to Copenhagen. He thought that the preprint was a "much improved" abstract of a (rewritten) longer thesis, although he confessed to not being able to follow many of the arguments. Everett furiously annotated his copy of Groenewold's letter, answering his criticisms point by point and concluding that Groenewold, too, had not thoroughly studied his work. Everett's annotations are reproduced here in the lettered footnotes. We have attempted to connect these annotations to the text appropriately, but for more precise detail one may wish to also consult the original document.*

GROENEWOLD TO EVERETT AND WHEELER, APRIL 11, 1957

Natuurkundig Laboratorium
der Rijks-Universiteit
te GRONINGEN
WESTERSINGEL 34
TEL. 22578 (K 5900)

Onze ref. HG854-57.

Groningen, April 11th 1957

Dr. Hugh Everett, III and Professor John A. Wheeler[gb]
Palmer Physical Laboratory
PRINCETON New Jersey
U.S.A.

Dear Dr Everett and Professor Wheeler,

Just returning from the Colston Symposium at Bristol on "Observation and interpretation," where during 4 days from early in the morning until late at night we have been fighting on the foundations of quantum mechanics,[gc] I find your 2 papers dealing with just the same subject. Thank you for sending them. If I were going to give my full comments criticisms, I could do no better than completely rewriting and enlarging my MS on "Quantum Measurements," which I provisionally wrote down last year in Copenhagen and of which I have sent you a copy. Until now I only have rewritten a few sections of it (in particular sections 2 and 4) and I am afraid that in the near future I will not find time to revise it in the way I have in mind. Therefore I now have to restrict myself to some very crude general remarks without the detailed arguments and reasoning and which I only make quite hastily and with much hesitation and reservation.

Compared with the MS on "Wave mechanics without probability," which I borrowed last summer, I find the present abstract much improved,[gd] but with regard to the fundamental physical and epistemological aspects I must say that I still profoundly disagree. Let me just briefly outline some of my main objections and comments in a number of short points (which are not independent).

1) All physical observable quantities may ultimately be expressed in statistical relations between results of various measurements. These relations may be expressed with merely the unitary operators representing the motion

---

[gb] Wheeler writes at the top of the letter: "MBP—Please make 1 photo copy for me; send this to HEIII JAW 15 Apr" and "Hugh—Do you want to draft a reply that I could review before it is sent off? See you soon!—John Wheeler"

[gc] Everett writes in the margin: "very inter. are proceedings to be available?"

[gd] The abstract referred to here is a version of the short thesis.

and the projection operators[ge] representing the various possible measuring results and without wave functions (or more general statistical operators). The statistical relations may be expressed in the form of conditional probabilities of certain measuring results (which e.g. have not yet been read, so that no information has yet been obtained about them) with respect to other measuring results (which e.g. already have been read, so that information has been obtained about them). Now one can introduce the statistical operator, which just represents in a very efficient way all the information which already has been obtained and which may be used to calculate the conditional probability (with respect to this information) of other information which still may be obtained or used. Thus also the statistical operator is conditional and depends on the standpoint from which the system is described. It is relative like the coordinate frame in relativity theory. It seems to me that this conditional character has been overlooked in your papers (as well as in many others).[gf]

2) I fully sympathize with the idea of describing the measuring process on purely physical systems without including living observers. So the "measuring chain" has to be cut off.[gg] But it is extremely fundamental that [blank space in source document] off is made after the measuring result has been recorded in a more permanent way, so that it no longer can be essentially changed if it is observed on its turn (i.e., if the chain is set forth).[gh] This recording has to be more or less irreversible and can only take place in a macrophysical (recording) system.[gi] This macrophysical character of the later part of the measuring chain is decisive for the measuring process. I do not think that it can be left out of consideration in its description. It does not seem to act an essential part in your considerations.[gj]

[ge] Everett writes in the margin: "X Wrong! Won't work for approx. meas." See the discussion in chapter 8 following pg. 145 for Everett's argument that approximate measurements cannot be represented by projection operators.

[gf] Everett writes in the margin: "On contrary, there is full discussion in my paper. In fact is at *very* basis of 'relativity of states.'" Everett allows for the description of a physical system from any standpoint whatsoever by insisting that relative states make sense with respect to *any* basis and any quantum observable. See for example Everett's discussion of relative states (pg. 180) and his discussions of how conditional expectations are given by relative states (pg. 100 and fn. la on pg. 287). See also the discussion of relative states in the conceptual introduction (pg. 34).

[gg] Everett writes in the margin: "Nonsense. Whole idea not to cut off till after final observ."

[gh] Everett writes in the margin: "No meaning in Q.M. Q.M. says it effected just like microsystems." Everett notes again that genuine physical irreversibility is flatly incompatible with the linear dynamics of quantum mechanics. This is why irreversibility for Everett is only *apparent irreversibility* to observers themselves treated within the system. See for examples pgs. 143–44, 224, 240, and fn. kz on pg. 287.

[gi] Everett writes in the margin: "Whence this magic irrev.?"

[gj] Everett writes in the margin: "A very basic idea is to avoid this necessity, which is a serious difficulty in ordin. view." The very basic idea for Everett is that macroscopic systems behave in precisely the same way as microscopic systems.

3) "Process 2" is only an approximation in as far as the influence of other (external) systems is represented by a classical external field. In many cases of elementary quantum mechanics this is a satisfactory approximation.[gk] (It is corrected in field theory). The influence of the measuring instrument can obviously not be represented in this way. I agree that in a certain way it could be represented by applying process 2 to the combined object and measuring systems. But this is not sufficient, because one has also to account for the destruction of the coherence between those Hilbert subspaces, which are to be distinguished by the measurement.[gl] (It should be observed that coherence only has a statistical meaning). There are various effects which lead to such a destruction, one of them being the macrophysical character of the recording system.[gm] I do not see that this destruction of coherence, which is very essential in the measuring process, has been dealt with in your theory.[gn]

4) On one hand the combined object and measuring systems are considered from the microphysical quantum mechanical point of view. So far one could not even speak of a measurement. On the other hand the later part of the measuring chain and in particular the recording system is regarded from the macrophysical classical point of view. A satisfactory theory of measurement has to relate these two aspects to each other.[go] Only some very

---

[gk] Everett writes in the margin: "? What process then?" If Process 2 only approximately describes the behavior of a measuring device, then Everett is puzzled as to what physical process could be taken to provide the actual behavior. Is a *new* physical law being proposed here? For Everett, environmental interactions do not make Process 2 only approximate; rather, they are themselves also fully described by Process 2.

[gl] Everett writes in the margin: "Yes, in ordinary interp. It is failure of ord. interp. to acc. for this which led to present view which doesn't need it."

[gm] Everett writes in the margin: "Not so. To assert this is to be exposed to the 'For what n [?]' embarrassment." Taken with his earlier comments, the suggestion seems to be that such an explanation is not essential, and that, even if it were, the standard theory does not provide a consistent explanation since it would not explain why Process 2 is only approximate. That said, it would have been enough to explain the destruction of coherence for Everett to explain why one should not expect to observe branching. The explanation for this is that there would be no physical record of any branching process in the memory sequence of an idealized observer unless the observer made a macroscopic interference measurement, but such measurements would be difficult to perform simply because of the degree of isolation and control required. It may be that the "for what n" embarrassment here is the embarrassment of having to say how many basic physical systems one must aggregate in order to expect the destruction of interference effects.

[gn] Everett writes: "Nor in the usual theory either in a consistent manner!" Interestingly, Everett does not claim that he has fully dealt with this; rather, he points out that the standard theory cannot handle this problem either. That said, pure wave mechanics is at least a thoroughly self-consistent theory. This is a place where decoherence considerations might have helped Everett. Here, however, he takes the position that the theory need not explain the destruction of interference effects. That said, Everett does discuss that massive systems should be expected to exhibit classical behavior. See for example pgs. 136 and 158.

[go] Everett writes in the margin: "Precisely what is done in the theory (in detail in first version)." There Everett's explanation that one will typically not observe coherence is

preliminary attempts of a microphysical quantum mechanical foundation of a macrophysical classical description of macrosystems have as yet been made. In such a theory destruction of coherence has to play a fundamental part.[gp]

5) After the measuring result has been recorded it may be read by an observer outside the system considered so far.[gq] The recording is not essentially changed by the reading. On the ground of the information obtained from the reading the observer (or his associate theoretician) may assign a new (conditional) statistical operator to the object system. On the other hand he may for a great number of similar measurements statistically test the conditional probabilities, which had been calculated for these measuring results.

6) Now I guess that your "memory of the observer" corresponds somehow with the ensemble of all the recording systems,[gr] but I do not see how your automatical observer included in the described combined systems also could be used for describing the activities of reading the recorded measuring result and of assigning statistical operators to the object system on the ground of the obtained information.[gs] In a certain way the "disturbance" of the object system under the influence of the measuring instrument could be described in the combined systems, but I do not see how that could also be done for the description of the object system, which does not "disturb" it at all.[gt]

7) It seems to me that the relation between the object system and the observer (who not only "observers" the object system, but also describes it with some theory and "interprets" if you like) is a rather delicate one, which should be analyzed very carefully and that neither of them could even be formulated without the other one.

8) Because all observable quantities may ultimately be expressed in statistical relations between measuring results and the latter are represented by essentially macrophysical recordings, the former ones may ultimately be expressed in macrophysical language.[gu] That does of course not mean that the formalism, which serves as a tool for calculating these statistical

ultimately given by arguing that the memory sequences of typical observers exhibit what we in fact see both quantum mechanically and classically. See pgs. 134–37 and pg. 158.

[gp] Everett writes in the margin: "Not necessarily. Only with your preconceived ideas."

[gq] Everett writes in the margin: "Wave mechanics doesn't give this answer. Where does it come from?" This is a subtle point, and Everett gets it just right. If the internal observer communicates his result to the external observer, the internal observer is no longer in an eigenstate of having a determinate result. See Albert (1992, Ch. 8).

[gr] Everett writes in the margin: "Misunderstand. What ensemble?"

[gs] Everett writes in the margin: "Why not?"

[gt] Everett writes in the margin: "?"

[gu] Everett writes in margin: "Epistemologically garbage. Lack of understanding of nature of phys. theory." Everett held that one must have a clear understanding of the proper cognitive status of physical theories in order to properly appreciate his formulation of quantum mechanics. He describes his views most completely in the second appendix to the long thesis (chapter 8, pg. 168).

relations could also be expressed in macrophysical language. On the contrary in this field the macrophysical language is liable to lose its original more or less unambiguous meaning.[gv]

9) A similar loss of the meaning of the language and of our notions might occur under the extension of the described physical system to the entire universe and the inclusion of the observer, who not only observes, but also theoretically describes the system.

10) In many existing considerations of the foundations of quantum mechanics the non-recognition of the conditional character of the statistical operator (or wave function) and also of the statistical character of notions like interference a. o. and the neglect of the macrophysical features of the later part of the measuring chain (in particular the recording) introduces all kinds of paradoxes (e.g. cat paradox, Einstein-Rosen-Podolsky paradox a. o.) I do not see how in your theory these paradoxes could be avoided.[gw]

I am quite clear that this haphazard hasty collection of points with the necessary arguments cannot be clear at all. All I could do in these few lines was perhaps to point out that in various aspects I cannot follow your arguments and that I fail to see that your theory would give a deeper insight into quantum mechanics. I regret that I have no opportunity to discuss these things with you in all detail. I am really much interested in your points of view and in case copies with the full text of Dr Everett's text might be available, I should be glad to read it. My impression that problems in this field recently regained interest and highly puzzle various people was strongly fortified at the Bristol symposium last week.[gx] Also that in various aspects our understanding of them is still far from satisfactory.[gy]

Yours sincerely,

H. J. Groenewold

[gv] Everett writes in the margin: "Why base concept of *reality* on Classical Macrophysical realm?"

[gw] Everett writes in the margin: "Didn't even read my paper. Conditional character fully discussed, as well as the paradoxes, which are more easily explained than usual." Everett concludes that Groenewold has not read carefully because he in fact directly considers such paradoxes. Wigner's friend and the considerations of nested observers that forms the conceptual starting point of Everett's project is a general version of the Schrödinger cat paradox. See pgs. 176 and 73–75. Everett also considers the EPR. See for example pgs. 131–32.

[gx] Among other topics discussed at the symposium were the virtues and vices of Bohmian mechanics.

[gy] Everett brackets the last three sentences of Groenewold's letter and writes in the margin: "Ha!"

CHAPTER 14

Correspondence: Everett and Wiener (1957)

Everett had long been an admirer of Massachusetts Institute of Technology Professor Norbert Wiener, whose path-breaking book Cybernetics or Control and Communications in the Animal and the Machine *(Wiener, 1948) was a seminal influence on his own thinking about the role of information and probability in game theory and physics. So he must have been thrilled to receive a letter from Wiener in early April 1957, responding to the Everett and Wheeler preprints. It turned out that although Wiener liked Everett's inclusion of the observer in the wave function of the system observed, he found fault with Everett's derivation of quantum probabilities. In his comments on Wiener's paper and his reply to Wiener, Everett pointed out that Wiener had misread his argument. Everett's annotations are reproduced here as lettered footnotes.*

WIENER TO WHEELER, APRIL 9, 1957

MASSACHUSETTS INSTITUTE OF TECHNOLOGY
CAMBRIDGE 39, MASS.

DEPARTMENT OF MATHEMATICS

April 9, 1957

Professor John A. Wheeler [gz]
Mr. Hugh Everett
Palmer Physical Laboratory
Princeton
N. J.

Gentlemen,

I have received the two papers you have sent me and intend to give you a more detailed report on them by May 1st.[ha] In the meantime I want to give

---

[gz] Wheeler writes at the top of the page: "MBP—Pls. photocopy for me; original send to HE III JAW 15 Apr.

[ha] These papers were Everett's short thesis and Wheeler's companion paper.

you my first impressions. These are that the inclusion of the observer as an intrinsic part of the observed system is absolutely sound.[hb] Namely, I see no way of bringing into quantum theory the concept that something particular actually happens without introducing an observer as part of the quantum-theoretical system. Matter of fact quantum theory as a partly probabilistic theory without actual occurrence does not seem to me to make sense. Perhaps I may say as you do: the probability in quantum theory must have the properties of true Lebesgue measures.[hc] Nevertheless, the particular things that happen in the universe are mostly not directly accessible to us, and the individual occurrences of probability, where directly observed, must be those in a subsystem which we call the observer. In my opinion it is essential for a usable theory that this observer must represent a very thin selection of the larger universe which he is observing, so that under certain circumstances we are justified in treating the observation which he makes, which may theoretically not be fully independent, as if it had a full property of independence. Roughly speaking, the observer must be something like a human retina which over one very short interval of time only receives impressions of a relatively small part of the universe. Eventually the small parts and their memories add up to a very good representation of the universe.

In other words, I am sympathetic to your point of view, while I think that your discussion of it meets one essential difficulty which I believe is removable. It is this that the whole basis of your quantum theory is Hilbert space, and in Hilbert space, as it is universally understood, there is no true Lebesgue measure.[hd] It is just because of this that I have been forced to introduce the notion of differential-space into quantum theory. (A New Form for the Statistical Postulate of Quantum Mechanics, N. Wiener and A. Siegel, Phys. Rev., Vol. 91, No. 6, September 15, 1953, p. 1551.—The Differential-Space Theory of Quantum Systems, N. Wiener and A. Siegel, No. 4 del Supplemento al Vol. 2, Ser. X. del Nuovo Cimento, pp. 982–1003, 1955.—Fourier Transforms in the Complex Domain, R. E. A. C. Paley and N. Wiener, American Mathematical Society, New York, 1934,

---

[hb] Wiener is here endorsing what Everett himself took to be the most important aspect of the project.

[hc] Everett writes in the margin: "I never said that!" It is unclear what Wiener had in mind here.

[hd] Everett writes in the margin: "I do *not* need Leb. measure in Hilbert space. Whole problem neatly avoided by my treatment. My measure on *trajectories*, i.e., sup of orthog. states, not entire H. space." Everett notes that his measure is over orthogonal trajectories, more specifically, memory sequences, and hence that he does not encounter any technical problems on this view.

Chap. 9.—N. Wiener, Acta Math. 55, 117–258, 1930, Sec. 13.—N. Wiener, J. Math. and Phys. 2, 131, 1923)

This is a space in which there is a true volume and which therefore is clearly distinct from Hilbert space, but is so related to Hilbert space that every unitary transformation of Hilbert space generates a measure-preserving point transformation with the associated differential-space.

Another point where your theory needs amplification (although I imagine that you are fully aware of that, since your theory is a meta-theory; as such, amplification is necessary) is that I do not find an adequate discussion of what it means to say that a certain fact or a certain group of facts is actually realized.[he] Dr. Siegel and I have been working in this field as you will see by the articles. However, as you proceed at your work, I am convinced that you yourself will have to attack this problem.[hf]

In short, I think you have made a real contribution to a possible future quantum theory, particularly in your insisting that the observer be an intrinsic part of the quantum system. However, you will have to go much further before you have rendered this suggestion into something concrete and usable. If I may be allowed to coin a word: I do not think that you have even yet formulated a meta-theory, but simply a set of prolegomena to a future meta-theory.[hg]

I hope that you will go on with your work, and I wish you all success. I certainly think that your paper should be published, but more as comments on the present intellectual situation than as a definitive result.

Sincerely yours,

Norbert Wiener.

NW/ls
Enclosure: 1 reprint[hh]

[he] Everett writes in margin:" In *theory* the universal state function is *the* realized fact. In superposition after measurement *all* elements actually realized."

[hf] Wiener's worry here is natural and was shared at about the same time by DeWitt (chapter 16, pg. 250). Everett's marginal note gives his response: all elements of the universal state are actually realized as relative states in the superposition after the measurement. See also pgs. 254–55 and Everett's note added in proof (pg. 189).

[hg] Everett writes in margin: "I am fully aware that this question of 'actualization' is a serious difficulty for convent. Q.M., and is in fact one of main motives for present formulation. No problem in present form, however. no such statements *ever made* in theory like 'case A actually realized,' except *relative* to some other state! All possibilities 'actually realized,' with corresp observer states."

[hh] Everett writes in the margin: "where?"

EVERETT TO WIENER, MAY 31, 1957

Hugh Everett, III
Arlington Towers, T-438
Arlington, Virginia
May 31, 1957

Professor Norbert Wiener
Department of Mathematics
Massachusetts Institute of Technology
Cambridge 39, Massachusetts

Dear Professor Wiener:

I am writing in reply to your letter of April 9 concerning my paper, "On the Foundations of Quantum Mechanics." I regret that pressing matters have prevented a more prompt response.

First, I would like to correct any impression that my theory requires a Lebesgue measure on Hilbert space. The only measure which I introduced was a measure on the orthogonal states which are superposed to form another state—a measure which presents no mathematical difficulties—and *not* a measure on the Hilbert space itself, the difficulties of which I am fully aware. Perhaps you were misled by my analogy with the case of classical statistical mechanics.[hi] Although in classical statistical mechanics the measure on trajectories does derive from the Lebesgue measure on the phase space itself, my measure on trajectories does *not* derive from a measure on Hilbert space itself.

You also raise the question of what it means to say that a fact or a group of facts is actually realized. Now I realize that this question poses a serious difficulty for the conventional formulation of quantum mechanics, and was in fact one of the main motives for my reformulation. The difficulty is removed in the new formulation, however, since it is quite unnecessary in this theory ever to say anything like "Case A is actually realized."[hj]

Since I have discussed this point of the transition from "possible to actual" with Dr. Bryce DeWitt, who also raised the question, I am enclosing a copy of our correspondence in lieu of a fuller discussion here.[hk]

[hi] Everett argued that probabilities worked exactly the same way in the relative state formulation as in classical statistical mechanics. He used this analogy to argue for the norm-squared measure as the standard of typicality in the relative state formulation. See pg. 125.

[hj] Everett's marginal notes on Wiener's letter explain this point in more detail. To paraphrase, one does not say "Case A is actually realized"; but rather, "Case A is realized relative to a specified state of a correlated system." See Everett's note (fn. hg on pg. 233).

[hk] See in particular pgs. 254–55.

I shall be grateful to receive any further comments or criticisms that may occur to you.

Sincerely yours,

Hugh Everett, III

Note: Address after August 1:

607 Pelham Street
Alexandria, Virginia

HE:nge

Copy sent to Wheeler

Enclosures: Copy of DeWitt's letter & reply

CHAPTER 15

Correspondence: Everett and Petersen (1957)

Toward the end of April 1957, Everett received a letter from Petersen, reporting on how the two preprints had been received in Copenhagen. The gist of Petersen's renewed critique was that Everett's theory did not conform to Bohr's principle of complementarity or his insistence on the special status of observation. Everett was, however, not at all interested in working within what he took to be a deeply flawed approach to quantum mechanics.

PETERSEN TO EVERETT, APRIL 24, 1957

24. 4. 57[hl]

Dear Hugh,

Thanks for sending the two papers.[hm] It was good to see that your ideas are now going to be published. During the past months I have often been wondering if we might see you here as we had hoped, and I trust that you have not quite given up coming and seeing us.

As you can imagine, the papers have given rise to much discussion at the Institute, especially with Bohr, Källén[hn] and Rosenfeld who was here for a few days. Koman also read the manuscripts. Of course, I am not going to report in detail about these discussions, but I think that most of us here look differently upon the problems and don't feel those difficulties in quantum mechanics which your paper sets out to remove. Accordingly, we cannot agree with you and Wheeler that the relative state formulation entails a further clarification of the foundations of quantum mechanics.

I don't think you can find anything in Bohr's papers which conforms with what you call the external observation interpretation.[ho] Rather, his

---

[hl] Wheeler writes at the top of the page: "MBP—Please make a photocopy for me and send this original to Hugh Everett."

[hm] Given what Petersen says below, he had just read Everett's short thesis and Wheeler's companion paper.

[hn] Gunnar Källén was a Swedish physicist working at CERN.

[ho] While Everett sometimes was not perfectly clear in distinguishing between the external observation interpretation of quantum mechanics and the Copenhagen interpretation, there

analysis follows the line of the correspondence argument which was the basic guide in establishing the formalism as a mathematical generalization of classical theories. As emphasized in the Einstein–article there can on this view be no special observational problem in quantum mechanics—in accordance with the fact that the very idea of observation belongs to the frame of classical concepts.[hp] The aim of the analysis is only to make explicit what the formalism implies about the application of the elementary physical concepts. The requirement that these concepts are indispensable for an unambiguous account of the observations is met without further assumptions and is directly reflected by the way in which c- and q-numbers appear in the formalism. There is no arbitrary distinction between the use of classical concepts and the formalism since the large mass of the apparatus compared with that of the individual atomic objects permits that neglect of quantum effects which is demanded for the account of the experimental arrangement.[hq] There may be cases, e.g. in the treatments of the γ-ray microscope, where the placing of the separation is to some extent a matter of taste, but the free choice is limited to a region where quantum description is equivalent with the classical one. In the recording of observations, like a mark on a photographic plate, we are also concerned only with measurements performed on *heavy* bodies. Such recordings may of course be witnessed by any number of observers, and also as regards approximate measurements I can see no new problems.

   I do not understand what you mean by quantized observers. Obviously, one can treat any interaction quantum-mechanically, including the interaction between an electron and a photographic plate, but when utilized as an "observer" the definition of the "state" (position) of the plate excludes considerations of quantum effects.[hr] It seems to me that as far as your treatment of many-body systems is consistent with the proper use of the formalism it has nothing to do with the measuring problem.

is good reason to suppose that he understood the distinction while working on his thesis since he discusses the Copenhagen interpretation separately in the long thesis, and in his subsequent letter to Petersen, Everett explicitly distinguishes between the Copenhagen and von Neumann–Dirac formulations (pg. 239). See pg. 153 for Everett's description and criticisms of the Copenhagen interpretation in the long thesis and pgs. 32–34 in the conceptual introduction for a discussion of Everett's understanding of the distinction.

[hp] The argument is that since measurement must necessarily be understood classically, there is a sense in which there can be no *quantum* measurement problem. Everett, of course, disagreed with the assumption that measurement must necessarily be understood classically insofar as he held that he had provided a thoroughgoing quantum mechanical account of measurement.

[hq] Petersen's appeal to quantum mechanical considerations to justify treating the experimental apparatus classically is inconsistent with the view that observation is *only conceivable* as involving a classical apparatus. Everett notes this inconsistency in his reply. Insofar as one simply stipulates that a measuring apparatus cannot be thought of as quantum mechanical, there can be no justification for the stipulation from the quantum mechanical properties one would expect of the apparatus (pg. 240).

[hr] This is the orthodox Copenhagen line that an observer *as an observer* can only be conceived of classically.

Of course, I am aware that from the point of view of your model-philosophy most of these remarks are beside the point.[hs] However, to my mind this philosophy is not suited for approaching the measuring problem. I would not like to make it a universal principle that ordinary language is indispensable for definition or communication of physical experience, but for the elucidation of the measuring problems hitherto met with in physics the correspondence approach has been quite successful. How radically this approach will have to be modified in order to cope with the many unsolved difficulties remains to be seen.

Betty and I hope to see you sometime when you have finished your work in the Pentagon. As you may know we have got a son who is now learning some elementary communication.

Please give our greetings to the Wheelers.

Yours

Aage

EVERETT TO PETERSEN, MAY 31, 1957

Hugh Everett, III
Arlington Towers, T-438
Arlington, Virginia
May 31, 1957

Dr. Aage Petersen
Blegdamsvej 17
Copenhagen, Denmark

Dear Aage:

It was very good to hear from you again. Perhaps we will be able to talk together again sometime soon. There is a good chance that I will be sent to Europe in the fall on business, and I could probably take a few weeks off and come to Copenhagen. Please let me know what the best times to come are so that I can arrange things (to the extent that I am able) to be most convenient.

In the meantime, lest the discussion of my paper die completely, let me add some fuel to the fire with a number of random comments and criticisms of the "Copenhagen interpretation."[ht]

[hs] The Copenhagen interpretation did not seek to provide a single, consistent model of all physical systems including measuring devices. Rather, it sought to find a place for classical description in quantum mechanics, and hence what the Copenhagen colleagues took as a precondition for genuine understanding. See pgs. 152–53.

[ht] This does not mesh well with Wheeler's report to the Copenhagen colleagues that Everett did not mean to criticize the Copenhagen interpretation. See for example pg. 219. See pg. 236

First of all, the particular difficulties with quantum mechanics that are discussed in my paper have mostly to do with the more common (at least in this country) form of quantum theory, as expressed for example by von Neumann, and not so much with the Bohr (Copenhagen) interpretation. The Bohr interpretation is to me even more unsatisfactory, and on quite different grounds. Primarily my main objections are the complete reliance on classical physics from the outset (which precludes *even in principle* any deduction at all of classical physics from quantum mechanics, as well as any adequate study of measuring processes), and the strange duality of adhering to a "reality" concept for macroscopic physics and denying the same for the microcosm.

Now I do not think you can dismiss my viewpoint as simply a misunderstanding of Bohr's position. I am willing to admit that Bohr's complementarity principle, which expresses limitations on the unrestricted use of *classical concepts*, is a valid principle. I even am prepared to admit that in the initial stages of formulation of quantum theory this principle was very useful in clarifying the theory and showing that it does not lead to any of the more obvious kinds of contradictions. The trouble goes much deeper than this however. I believe that the basing of quantum mechanics upon classical physics was a necessary provisional step, but that the time has come to proceed to something more fundamental.

There is a good analogy in mathematics. The complex numbers were first defined only in terms of the real numbers. However, with sufficient experience and familiarity with their properties, it became possible and indeed more natural to define them first *in their own right* without reference to the real numbers, and to derive from them the special case of the reals. I would suggest that the time has come to do the same for quantum mechanics—to treat it in its own right as a fundamental theory without any dependence on classical physics, and to derive classical physics from it. While it is true that initially the classical concepts were required for its formulation, we now have sufficient familiarity to formulate it without classical physics, as in the case of the complex numbers. I am sure that you will recognize this as Bohr's own example turned against him.

The analogy goes further yet. Just as we no longer regard complex numbers as mere appendages tacked on to the reals to cover annoying inabilities to solve certain equations, we should no longer regard quantum mechanics as a mere appendage to classical physics tacked on to cover annoying discrepancies in the behavior of microscopic systems.

for Petersen's discussion of the Copenhagen interpretation and the conceptual introduction (pgs. 32–34) for a discussion of Everett's dissatisfaction with the Copenhagen interpretation. Everett believed that the Copenhagen interpretation, properly understood, would be subsumed within his relative state formulation.

Let me now mention a few more irritating features of the Copenhagen interpretation. You talk of the massiveness of macrosystems allowing one to neglect further quantum effects (in discussions of breaking the measuring chain), but never give any justification for this flatly asserted dogma. Is this an independent postulate? It most certainly does *not* follow from wave mechanics which leads to quite strange superposition states even for macrosystems when applied to any measuring processes! In fact, by the formulation of your viewpoint you are totally incapable of any justification and *must* make it an independent postulate—that macrosystems are relatively immune to quantum effects.

Another inconsistency: you vigorously state that when an apparatus can be used as measuring apparatus then one cannot simultaneously give consideration to quantum effects—but proceed blithely to apply the formula $\Delta x \Delta p \geq \hbar/2$ to such devices, tacitly admitting quantum effects.[hu]

You say you see no further difficulties with approximate measurements. I have yet to see any adequate account of the phenomena and would appreciate any references you can supply.

Just one final point. I am getting weary of hearing on the one hand that it is the fundamental irreversibility of the measuring process which allows the destruction of phase relations and makes possible the probability interpretation of quantum mechanics, and on the other hand that the fundamentally probabilistic processes of quantum mechanics allow truly reversible processes and for the first time make a satisfactory thermodynamics possible. As a matter of fact, there is nowhere to be found any consistent explanation of this "irreversibility" of the measuring process. It is again certainly not implied by wave mechanics, nor-classical mechanics either. Another independent postulate?[hv]

I am sure that these points (by no means exhaustive) are poorly and inadequately expressed here, but hope you will think about them until we can have a full discussion. I look forward very much to renewing our always enjoyable arguments. Please give my regards to Betty.

Sincerely,

Hugh Everett, III

Note: Address after August 1:
  607 Pelham Street
  Alexandria, Virginia
HE:nge[hw]

[hu] Here Everett notes that Petersen is violating that Copenhagen dogma by treating measuring devices quantum mechanically at all. See (pg. 237).

[hv] Everett held that pure wave mechanics was fully reversible. Any appearance of irreversibility was just that—an appearance. See for examples pgs. 143–44, 224, and fn. kz on pg. 287.

[hw] Written at the bottom of the page: "Copy sent to Wheeler".

CHAPTER 16

Correspondence: Everett and DeWitt (1957)

As an organizer of the Chapel Hill conference on gravitation, DeWitt was invited to guest-edit the July 1957 issue of Reviews of Modern Physics, *which published the proceedings of the conference, including Everett's short thesis and Wheeler's companion paper extolling its virtues. DeWitt did not read the long thesis until many years later. In May 1957, DeWitt wrote a long and thoughtful letter to Wheeler about how Everett's theory had conceptual parallels with Einstein's theory of general relativity. DeWitt was unhappy with both wave function collapse and the Copenhagen interpretation. He wrote that Everett's formalistic argument for pure wave mechanics was "beautifully consistent." But he could not accept the ontological ramifications of the theory because, "I simply do not branch." Everett's annotations to DeWitt's letter are presented here as lettered footnotes.*

In his reply to DeWitt, Everett referred to the dilemma faced by science when Copernicus asserted that the Earth moved around the sun, not the reverse. Making an analogy to his branching observer states, Everett asked DeWitt if he could "feel the motion of the earth?" He went on to explain his notion of how model theories connect to reality—an idea which he had developed in Appendix II of his long thesis and which had been cut from the short thesis that DeWitt read (pg. 168). Two decades later, in a referee report, DeWitt explained that Everett's letter convinced him that he was correct. "His reference to the anti-Copernicans left me with nothing to say but 'Touché'."[hx]

[hx] See DeWitt (1988).

DeWitt to Wheeler, May 7, 1957

THE UNIVERSITY OF NORTH CAROLINA
CHAPEL HILL

DEPARTMENT OF PHYSICS   May 7, 1957

Professor John A. Wheeler[hy]
Palmer Physical Laboratory
Princeton University
Princeton, New Jersey

Dear John,

You had asked for comments on Everett's paper "On the Foundations of Quantum Mechanics," and, although preoccupation with other matters has kept me from meeting the May 1st deadline, I thought you might nevertheless be interested to receive an evaluation from Chapel Hill. I have studied the paper and your assessment of it rather carefully, and I have several remarks to make:

In the first place, it seems to me that the professional philosopher will have a greater appreciation of Everett's work than will the average physicist, at least for the present.[hz] I say this as a matter of perspective, for it seems extremely unlikely that current physics (including quantum gravidynamics!) will be much affected by the new point of view. On the other hand, since the days of Boltzmann and Poincaré it has become increasingly clear that physicists themselves are obliged to be their own epistemologists, since no other persons have the necessary competence. Therefore Everett's effort is to be praised.

Of the total lack of experimental motivation facing such an effort Everett is doubtless keenly aware. Even Einstein, in developing the special theory of relativity, was motivated by experiment as well as by a general world view. Consequently, his theory had an immediate impact upon experimental physics as well as upon philosophy. It is hard for me to believe that Everett's ideas, on the other hand, will appreciably affect experimental physics (even on a cosmical level) although this, of course, remains to be seen.

In any event it is not at this point that a criticism of Everett's work should be aimed. Even if his ideas should have no experimental consequences whatever, his work would still be valuable. It is important to clarify the

---

[hy] Wheeler writes at the top of the page: "MBP. Pls send photocopy to Everett & return to me. JAW 13 May."

[hz] DeWitt himself is well informed concerning the philosophy of science, physics, and mathematics. He refers to canonical philosophical sources, many relatively recent, throughout the letter.

world views underlying the structure of theoretical physics and to point out inconsistencies, wherever they occur. This is largely the motivation of the whole general attempt to quantize the gravitational field, even at a much more pedestrian level—the lack of experimental data is identical.

As far as it goes, a parallel can validly be drawn between Everett's ideas and the theory of relativity. Your use of the term "relative state," in fact, seems to be a conscious effort to draw such a parallel. The role of the observer relative to the rest of the universe is emphasized in both theories.[ia]

Before coming to the real criticisms which I have of Everett's ideas, I should like to analyze this parallelism somewhat further by pointing out two of its aspects which seem to me rather suggestive, and at which you yourself have hinted, without going into detail. The first is an aspect of historical parallelism: The conventional interpretation of the formalism of quantum mechanics in terms of an "external" observer seems to me similar to Lorentz's original version (and interpretation) of relativity theory, in which the Lorentz–Fitzgerald contraction was introduced *ad hoc*. Everett's removal of the "external" observer may be viewed as analogous to Einstein's denial of the existence of any privileged inertial frame. The analogy fails, however, when one compares the short space of time which elapsed between Lorentz and Einstein with the three decades during which quantum mechanics, with its conventional interpretation, has managed to get along quite well. In answer to this, one may say that it was the pressure of experimental physics which in the case of relativity theory forced the application of "Occam's razor" and allowed the introduction of Minkowski's formalism, which led in such a naturally suggestive way to so many beautiful notions (including general relativity), whereas in Everett's case no such pressure exists. But *does* Everett's thesis constitute an entirely valid application of Occam's razor? Of this I am not so sure.

The second aspect of parallelism which struck me lies somewhat deeper: Consider the general theory of relativity. This is a physical theory based on Riemannian geometry, which in turn is a mathematical theory of continuous transformations of sets of continuous variables and of the idealized spaces of idealized points defined by such sets of variables. The use of such a geometry to describe the physical world has its basis in "the rough data of our senses" and in our crudest experiments. (Poincaré, "Science and Hypothesis."[ib]) The mathematical continuum from which the elements of this geometry are taken is, as has been pointed out many times, by no means identical with the physical continuum with which we start. However, we are obliged to introduce it in order to avoid the intolerable contradiction

---

[ia] In his reply, Everett reports that DeWitt's note here of the relationship between relativity and the relative state formulation of quantum mechanics is "correct and penetrating" (pg. 252).

[ib] Poincaré (1905).

$a = b$, $b = c$, $a < c$ inherent in the description of any sequence of "events" a, b, c of the physical continuum which are sufficiently closely spaced so that a is experimentally indistinguishable from b, and b from c, although a is distinguishable from c. By removing the contradiction the mathematical continuum is able to serve as a good model of the physical world; it permits the results of experiments to be expressed in mathematical terms, and once this is possible, we may be tempted to suppose, the metaphysical problem ends.

However, there is more to the matter than this. Not only crude observations but also our most refined experiments are describable within the framework of a mathematical continuum. Moreover, this continuum is found to be Euclidean to a very high degree of accuracy, when one follows the natural procedure of using solid bodies in space and reproducible regularities in time to define the metric. It is owing to the existence of the symmetries and regularities implied thereby that one is able to make what may be called a "good" observation. (I do not insist on the approximately Euclidean nature of space-time; other types of symmetry might do as well.)

At this point Einstein enters the picture and, with the general theory of relativity, replaces the solidity of the Euclidean framework by the shifting sands of general coordinate transformations. *The mathematical continuum which originally sprang from attempts to describe primitive sensory data is, however, still retained*—this, in spite of the fact that the level of description at which the general theory of relativity aims is vastly more sophisticated. On the other hand, the sensation of chaos produced by the introduction of general coordinate transformations makes the mathematical continuum seem a much less real thing than before, and tempts one to wonder if it could not be dispensed with entirely. Solid ground is only regained when the "observer" is given the dominant role in the theory. If you ask what the result of a given experiment will be, general relativity will (or at least should) always give you a definite answer—although a simple scheme may not always be ready at hand for classifying the answer. (Incidentally, Misner's search for a "superpotential" and Bergmann's—and my own—search for "true" observables, are nothing but attempts to discover such a scheme for purely gravitational effects.)

In this way you will find that general relativity, at the classical level, contains its own theory of measurement. *It will, in fact, give you back again the rough data of the senses from which its mathematical formalism sprang.* This, I think, is abundantly clear already from the well known linear approximations of the theory, and should persist at an even more advanced level. But it is a result which was certainly not obvious at the outset. It is not merely a tautologous consequence of the original mathematical removal of the contradictions inherent in the continuum of physical experience. The fact that there exist covariant nonlinear equations which guarantee the self-consistency of the theory is striking indeed.

The parallel with Everett's theory is now plain. Although quantum mechanics is founded on experimental results which are somewhat remote from primitive experience, the inferences to be drawn from them are clear and unambiguous: Among the many statistical aspects which Nature displays, certain ones are of a fundamental, irreducible, "built-in" variety. We describe this situation by introducing a wave function ψ which has a certain probability interpretation. The results of experiments may be expressed—now on a statistical basis—in terms of this wave function, provided it be assumed to satisfy a certain linear differential equation, the Schrödinger equation, the existence of which is just as striking as that of Einstein's equation. It is not initially supposed that ψ expresses an independent physical reality any more than do the numbers which designate a point of the mathematical continuum. In fact, ψ suffers discontinuous changes depending on the amount of our knowledge (or that of an observing apparatus) about a system. However, as Everett has shown, when the mathematics of Hilbert space (of which ψ is a "point") is combined with the Schrödinger equation, the whole scheme is found to possess more features of independent reality than were initially apparent and to mirror the physical world with a previously unanticipated fidelity. In short, the scheme is found to contain its own theory of the measurement process, by giving back again—*but at a new level*—the same elements of statistical interpretation which were put into it at the beginning, just as the combination of the mathematics of a Riemannian space with Einstein's equation gives back again (in first approximation) the laws of motion from which we can construct the rigid Euclidean codification of distances and times, which in turn contains, incidentally, the physical interpretation of the mathematical continuum, i.e., everything that is pertinent for a description of the physical continuum. The parallel occurrence of this phenomenon of self-consistency, or "self-containment", in two quite different branches of physics is rather remarkable and deserves further study.

Now for the criticisms: The first is a major one and cuts, I think, to the heart of any controversy which is likely to arise over Everett's work.[ic] It concerns the question of what is meant or can be meant by the word "correspondence"—a better word would be "isomorphism," but you seem to have avoided using it—particularly when applied to the ensemble of Everett's relative state vectors $\psi^O[A, B, C\ldots]$ as compared with the experience of a real physical observer.[id] I think the history of

[ic] Everett puts a check mark next to this paragraph.

[id] Everett himself used the term *isomorphism* in the long thesis, then he refined his description by using the term *homomorphism*. Everett's suggestion in the second appendix to the long thesis is that the relative record states of an observer are homomorphic with the observer's experience (pgs. 168–69). See also pg. 253. That DeWitt wanted an *isomorphism* between the model and the real world may explain why he ultimately ended up with a many-worlds interpretation of Everett. See the discussion in the conceptual introduction (pgs. 51–54).

the development of knowledge during the past century has so thoroughly conditioned the modern physicist that he will be quite willing to go along with you when you say that "terminology has to adjust itself in accordance with the kind of physics that goes on." However, there are limits.

Certain terms, such as "isomorphism," would seem likely to be useful under almost any conceivable circumstances and yet retain meanings which are held within fairly rigid bounds. I am afraid that it is at precisely the most crucial point in Everett's argument where many people, including myself, will be unable to swallow your implication that the word "isomorphism" applies.[ie]

It has been many years now since the logicians first began defining the natural numbers as "classes of all classes having so-and-so many members." I don't object when, advancing farther in the same direction, a philosopher like Russell ("The Analysis of Matter") defines a "point" (in space-time) as a certain class of events, or an "event" as a certain group of observations (both "real" and "virtual"). I understand what an isomorphism is, and am prepared to agree that for many practical purposes the word "isomorphism" may be replaced by the word "identity," and, indeed, that if there are any differences be[tween[if] isomorphic entities, the differences may often be purely semantic, even in the domain of the physical world.

What I am *not* prepared to accept relative to the subject at hand, however, is that the temporal behavior of the *superposition* of relative observer states $\psi^O[\alpha_i^1, \alpha_i^2, \ldots]$ is isomorphic to the "trajectory" of the memory configuration of a real physical observer, whether human] or inanimate. As Everett quite explicitly says: "With each succeeding observation ... the observer state 'branches' into a number of different states." The trajectory of the memory configuration of a real physical observer, on the other hand, does *not* branch. I can testify to this from personal introspection, as can you. I simply do *not* branch.[ig]

I do agree that the scheme which Everett sets up is beautifully consistent; that any single one of the states $\psi^O[\alpha_i^1, \alpha_i^2 \ldots]$, when separated from the superposition which makes up the total or "universal" state vector $\psi^{S_1+S_2+\cdots+O}$, gives an excellent representation of a typical memory configuration, with no causal or logical contradictions, and with "built-in"

[ie] If one took the physical world to be fully isomorphic to pure wave mechanics, one might take this to require that there be a one-to-one onto map between the entities one quantifies over in the theory and entities in the physical world. DeWitt objects to this since he believes that this would require many copies of each observer—one copy for each observer relative state. He later endorses just such an interpretation.

[if] Wheeler's secretary miscopied the letter, so the following bracketed text was likely never seen by Everett. The reinserted text here comes from Wheeler's copy.

[ig] Everett writes "!" in the margin. Everett responds to this objection directly in his letter to DeWitt by asking, "Do you feel the motion of the earth?" Everett then draws an analogy between the historical reception of Copernican astronomy and how it explains our experience and the present situation involving pure wave mechanics (pg. 254).

statistical features. The whole state vector $\psi^{S_1+S_2+\cdots+O}$, however, is simply too rich in content, by vast orders of magnitude, to serve as a representation of the physical world. It contains all possible branches in it at the same time. In the real physical world we must be content with just one branch. Everett's world and the real physical world are therefore not isomorphic.[ih]

The central difficulty in interpreting quantum mechanics lies, of course, in the notion of "probability," that most elusive of all concepts with which mathematicians and physicists have to concern themselves.[ii] For Poincaré, who did so much to elucidate the delicate ideas underlying probability theory, "chance" was strictly "the measure of our ignorance." He writes ("Science and Method"):[ij]

"What is chance? The ancients distinguished between phenomena which seemed to obey harmonious laws, established once for all, and those that they attributed to chance, which were those that could not be predicted because they were not subject to any law. In each domain the precise laws did not decide everything, they only marked the limits within which chance was allowed to move. In this conception, the word chance had a precise, objective meaning; what was chance for one was also chance for the other and even for the gods.

"But this conception is not ours. We have become complete determinists. . . ."

One would give anything to know what Poincaré would have said if his life span had been shifted the few years necessary for it to have encompassed the advent of the theory of Heisenberg, Schrödinger and Bohr. I, for one,

---

[ih] Everett writes in the margin: "! reply to this". DeWitt's argument is that while the universal state gives a beautiful representation of a *typical memory sequence*, it cannot be taken to be isomorphic to the actual properties of any particular physical observer since it is simply too rich a structure. Everett replies that the relationship is one of homomorphism: we can be more charitable and allow all the elements to coexist since those that do not describe one's experience won't cause any trouble (pg. 255). See also the discussion of empirical faithfulness in the conceptual introduction (pgs. 51–54).

[ii] For Everett the central problem in interpreting quantum mechanics was not the elusive nature of the concept of probability in general; rather, it was the quantum measurement problem. That said, he did recognize the problem of understanding probability within pure wave mechanics since there are no random events nor are there epistemic limitations regarding the relative state of an observer and his object system immediately following an experiment. Hence, one might argue, neither of the traditional approaches to understanding probability (as a measure of objective chance or as a measure of ignorance) are available in any straightforward way. Everett's strategy was to deny that there were probabilities, objective or subjective, in pure wave mechanics, then to recover the statistical predictions of quantum mechanics by arguing that a typical memory sequence would exhibit the quantum statistics. See for example pg. 77. One should expect the standard quantum statistics to be exhibited in one's own experience precisely insofar as one expects one's relative memories to be *typical* in Everett's sense.

[ij] Jules Henri Poincaré (29 April 1854–17 July 1912) was a French mathematician, theoretical physicist, and philosopher of science and mathematics. Here DeWitt refers to Poincaré (1921, pg. 395).

do not think he would have felt obliged to return to the *absolute* chance of the ancients. I think he would have resisted it, just as Einstein did.[ik]

The point at issue is quintessentially summed up in a single phrase of Everett's—in the section headed "Quantitative Interpretation, Measure.... etc."—:

"Probability theory is equivalent to measure theory mathematically...."

Yes, but *not* epistemologically. There is a vast difference between the two. The trouble with quantum mechanics is that it drives us, at least formally, perilously close to the absolute chance of the ancients. I say perilously, because if you accept the concept of absolute chance, and attempt to describe it *mathematically*, you have no choice but to do what Everett has done, namely, to introduce a mathematical world which branches. But the real world does not branch, and therein lies the flaw in Everett's scheme.[il] Of the seductive nature of the path which Everett has taken there can be no doubt. It is the natural extension of the route which has led us to the triumphs of quantum mechanics, and which has its foundation in the absoluteness of the indeterminacy principle. But at the end of the trail one comes suddenly upon a vast contradiction.[im]

Two other paths remain open for avoiding the contradiction. One is the traditional path espoused by Bohr and the Copenhagen school. According to Bohr the wave function ψ has an interpretation only with respect to an "external" classically describable observer. This observer is not a member of an ensemble, nor is the observation he makes one of an ensemble of observations; both stand alone. Therefore it is meaningless to apply the mathematics of probability to the observer or his observation, although it is quite validly applied to the underlying world observed. Although Bohr's observer can never predict with certainty the outcome of his observation, he nevertheless always obtains only one result.

The existence of a classical level of description, although not always explicitly stated, is absolutely essential to the Copenhagen view. This fact is most forcefully expressed by Dave Bohm at the end of his book on quantum

---

[ik] Insofar as pure wave mechanics is deterministic, Everett might be taken as showing one way to resist objective chance in quantum mechanics. Everett also knew that one could avoid objective chance in quantum mechanics by way of a hidden variable theory like Bohmian mechanics (pgs. 153–55).

[il] Everett again objects to this conclusion with a marginal "!". His objection, in his reply, is on at least two grounds: (1) the proper issue here concerns what one would experience according to the theory and about the *real* world and (2) he takes himself to have already shown why one would never notice any "branching" process since typically no branching events are represented in the sequences of memory records of an observer treated within the theory (pgs. 254–55).

[im] DeWitt is primarily worried here that the theory contradicts the empirical fact that we experience only a single determinate measurement outcome, which seems to be governed by objective probabilities.

theory:

"Without an appeal to a classical level, quantum theory would have no meaning... Quantum theory *presupposes* the classical level and the general correctness of classical concepts in describing this level. *It does not deduce classical concepts as limiting cases of quantum concepts.*"[in]

Such a statement, coming a few pages after his lucid analysis of the process of measurement (the results of which form the point of departure of Everett's work!), shows that Bohm saw the situation very clearly. It was a situation which bothered him, however, just as it has bothered Everett; and, as we all know, it eventually led him to abandon the Copenhagen view in favor of his theory of "hidden variables", which is the *other* path that may be taken in order to avoid the previously mentioned contradiction.

I think Heisenberg is wrong, at least as far as Bohm is concerned, in saying of the opponents of the Copenhagen interpretation, (in his article in "Niels Bohr and the Development of Physics") "It would, in their view, be desirable to return to the reality concept of classical physics or, more generally expressed, to the ontology of materialism; that is, to the idea of an objective real world, whose smallest parts exist objectively in the same way as stones and trees, independently of whether or not we observe them."[io] I think Bohm, like Everett, has reacted rather to the hybrid character of the Copenhagen view, which uses the classical theory as a "crutch" and which allows the continuity of the mathematical elements of the quantum theory to suffer damage as a result of this crutch. Heisenberg does have a cogent objection when he points out that theories like Bohm's destroy the symmetry properties of quantum mechanics which are expressed in the principle of complementarity. On the other hand, Bohm's theory succeeds in maintaining continuity while at the same time describing "the transition from the possible to the actual" (Heisenberg's words)[ip] which Everett fails to do.[iq]

It is quite possible that Poincaré might have been attracted by Bohm's theory. As I have said, Poincaré regarded chance solely as a measure of our ignorance. According to him ("Science and Method"[ir]) one is able to speak of the "laws" of chance only because small causes are able to produce large effects. If one is ignorant of the small causes one may then ascribe a random behavior to the large effects, *provided* the distribution of small causes is describable by an analytic function. In Bohm's theory this analyticity is insured by the underlying wave function. In accepting Bohm's theory, Poincaré would actually be able to loosen somewhat his rigidly

[in] Bohm (1951, pg. 625).

[io] Heisenberg (1955, pg. 17).

[ip] Heisenberg (1955, pg. 22).

[iq] Everett's response to this is that there is no such transition in his theory. See pgs. 254, 234, and 189.

[ir] Poincaré (1921).

deterministic pre-quantum views. For the small causes would now remain forever unknowable to any and all observers, despite their best endeavors, even if the whole universe were considered at once.

Let me sum up again quoting Heisenberg (loc.cit.): "[The] probability concept [of quantum mechanics] is closely related to the concept of possibility, the 'potentia' of the natural philosophy of the ancients such as Aristotle; it is, to a certain extent, a transformation of the old 'potentia' concept from a qualitative to a quantitative idea. On the other hand, the single quantum jump of Bohr, Kramers and Slater is "factual" in nature; it "happens" in the same manner as an event in everyday life... "[is] . That is to say, the real world does not branch. It is constantly in the process of passing from the possible to the actual—not *many* actuals, but *one* actual.[it]

It is obvious that what we have here is a first class dilemma—or perhaps I should now say "trilemma," in view of what Everett has done. On the one hand we have the Copenhagen view with its disturbing invocation of discontinuities and a classical level. On the other hand we have theories like Bohm's which, although simultaneously self-contained and isomorphic with the real world, are repugnant because of the mathematical superfluousness of the "hidden variables." Although he never expressed himself in these terms I believe that Einstein felt keenly the existence of such an issue, and that this was basically the motivation for his famous controversy with Bohr. The merit of Everett's work is that the issue is now presented so clearly.

As far as your own espousal of Everett's views is concerned, while I sympathize with your remark that "terminology has to adjust itself in accordance with the kind of physics that goes on," if this means that I am not, for example, to be allowed to use such terms as "*many* actuals" and "<u>one</u> actual," (see above) or that it is meaningless to make a distinction between the two, I am afraid I can't go along with you. For me, Everett's world is not a faithful representation of the real world.[iu]

My other criticisms of Everett's work are relatively minor. I feel, for example, that some note should have been taken of Bohm's solution of the metaphysical problem, and of his general work on the theory of

---

[is] Heisenberg (1955, pg. 13). It is interesting that Heisenberg explicitly took the collapse to be a real event for Bohr–insofar as this is fair, it brings Bohr's position closer to the standard von Neumann–Dirac collapse theory.

[it] For Everett's reply see pg. 255 and his note added in proof (pg. 189).

[iu] Everett does not require pure wave mechanics to provide a faithful representation of the real world—he just requires it to provide a faithful representation of experience within a substructure of the model. Indeed, Everett insisted that one could do no more. Curiously, with his many-worlds interpretation, DeWitt later held that Everett's theory was isomorphic to the real world in a sense that was stronger than Everett himself believed one could ever know. See for example pgs. 253 and 168–72.

measurement. Certainly no one working in this field has failed to read the closing chapters of Bohm's book.[iv]

Also I should like to have seen a careful, abstract, mathematical demonstration that "good" observations can actually be made; that is, I should like to see a generalization, applicable to arbitrary operators, of the von Neumann example, or of the old standby, the Stern-Gerlach experiment which everybody, including Bohm, uses. This would have allowed Everett to give a more careful definition of what a "good" observation is than he has done. I think that when Everett uses the term "good observation" he often means "*perfect* observation,"[iw] and that when he says "all averages of functions over any memory sequence.... can be computed from the probabilities $a_I^* a_I$, except for a set of memory sequences of measure zero," he really means "...except for a set of memory sequences having a measure *which converges extremely rapidly to zero as the observations making up the memory sequence tend toward perfection.*" His attempt to obtain zero measure solely by taking the limit as the number of observations tends to infinity is not quite right I think. It is certainly incorrect if the observations are only "fair" to "poor."[ix]

Finally, it seems to me an unnecessary, although convenient, restriction to insist that the system state, if it is an eigenstate of the quantity being measured, shall remain unchanged after the measurement process is over. It is not necessary that an observation be repeatable to be significant, provided one can take the lack of repeatability quantitatively into account after a given period of time. The removal of this restriction would allow one to discuss the types of experimental observations and the types of memory sequences which are actually encountered in practice.

I hope that these remarks will serve as an indication to you that Everett's ideas are extremely stimulating. Let's have more of the same. Kindest regards.

Yours sincerely,

Bryce DeWitt

BDeW:cwj

[iv] In his textbook on the subject Bohm had provided one of the clearest accounts of quantum measurement at the time (Bohm, 1951).

[iw] Everett means perfect correlation when he says good observation, but he also treats approximate measurements in his long thesis (pgs. 145–48). DeWitt did not know this since he had only read a version of the short thesis at this point.

[ix] Although DeWitt has an interesting worry here, this point is again related to how Everett treats approximate measurement. On careful analysis, one finds that whatever imperfections there are in measurement also show up in the limiting state as described by terms associated with an appropriate corresponding measure (pgs. 145–48).

EVERETT TO DEWITT, MAY 31, 1957

Hugh Everett, III
Arlington Towers, T-438
Arlington, Virginia
May 31, 1957

Dr. Bryce DeWitt
Department of Physics
University of North Carolina
Chapel Hill, North Carolina

Dear Dr. DeWitt:

Professor Wheeler has sent me a copy of your letter concerning my paper, "On the Foundations of Quantum Mechanics."[iy] I found your comments quite interesting and well expressed—so well that I am taking the liberty of sending parts of your letter to others with whom I am corresponding.[iz]

I find your analysis correct and penetrating on most essential points, particularly with respect to the parallel between my theory and the theory of Relativity.[ja] I must take issue, however, with what you call the major flaw in the theory, the question of the transition from "possible to actual."

[iy] The letter from DeWitt to Wheeler was dated May 7, 1957 (pg. 242). Much of what Everett argues in this letter concerning the proper status of a physical theory generally and pure wave mechanics in particular is contained in the second appendix to Everett's long thesis, which DeWitt had not read (pgs. 168–72).

[iz] A note at the bottom of Everett's copy of DeWitt's letter indicates that copies were sent to Wheeler and to Wiener, as Everett promised (pg. 234).

[ja] DeWitt notes at least two parallels between relativity and the relative state formulation of quantum mechanics and spends much of his letter discussing them. One parallel is historical, the other conceptual. The *historical parallel* is between Everett's relative state formulation of quantum mechanics and Einstein's special relativity. Just as the collapse dynamics was introduced in an ad hoc way in the standard external formulation of quantum mechanics, the Lorentz–Fitzgerald contraction was introduced in an ad hoc way in Lorentz's original version of relativity (see Lorentz, 1892). And just as Everett denies the necessity of an external observer, Einstein denies that there is a preferred inertial frame. The *conceptual parallel* is between the relative-state formulation of quantum mechanics and general relativity. DeWitt holds that each theory, while at first thought to be inconsistent with our ordinary experience, recaptures our ordinary experience to first approximation by treating the observer as a physical system within the theory. There is also a closely related parallel between Everett's relative-state formulation and special relativity that concerns a similarity in *how* the relative-state formulation and special relativity recover ordinary experience. In both special relativity and in the relative-state interpretation the observer plays a role in individuating descriptions of other physical systems for the purpose of explaining ordinary experience. More specifically, just as a system's apparent properties are relative to the observer's inertial frame in special relativity, the apparent state of a physical system depends on the relative state of the observer in Everett's formulation of quantum mechanics. In any case, Everett clearly approves of DeWitt's analogies between relativity and the relative-state formulation of quantum mechanics and quickly proceeds to consider his criticism of the theory.

First, I must say a few words to clarify my conception of the nature and purpose of physical theories in general.[jb] To me, any physical theory is a logical construct (model), consisting of symbols and rules for their manipulation, *some* of whose elements are associated with elements of the perceived world. If this association is an isomorphism (or at least a homomorphism) we can speak of the theory as correct, or as faithful.[jc] The fundamental requirements of any theory are logical consistency and correctness in this sense.

However, there is no reason why there cannot be any number of different theories satisfying these requirements, and further (somewhat arbitrary) criteria such as usefulness, simplicity, comprehensiveness, pictorability, etc., must be resorted to in such cases.[jd] There can be no question of which theory is "true" or "real"—the best that one can do is reject those theories which are *not* isomorphic to sense experience.[je]

When one is using a theory, one naturally pretends that the constructs of the theory are "real" or "exist." If the theory is highly successful (i.e. correctly predicts the sense perceptions of the user of the theory) then the confidence in the theory is built up and its constructs tend to be identified with "elements of the real physical world." This is, however, a purely psychological matter. No mental constructs (and this goes for everyday, prescientific conceptions about the nature of things, objects, etc., as well as elements of formal theories) should ever be regarded as more "real" than any others. We simply have more *confidence* in some than others.

A crucial point in deciding on a theory is that one does *not* accept or reject the theory on the basis of whether the basic world picture it presents is compatible with everyday experience. Rather, one accepts or rejects on

---

[jb] See the second appendix to the long thesis for an extended discussion of Everett's understanding of the nature and purpose of physical theories (pgs. 168–72).

[jc] Everett thinks of an empirically faithful physical theory as being fully characterized by a formal set of symbols, a set of syntactic rules for manipulating the symbols, and a homomorphism between *some* of the syntactic structures in the theory and experience. Empirical faithfulness, then, requires only that one be able to find a significant portion of one's experience somehow represented in the syntactic structures of the theory.

[jd] If a theory is both consistent and empirically faithful, then it can be taken as a serious contender, but it has been long recognized that these two conditions are not enough to specify a single physical theory. This is the problem of empirical underdetermination. See Quine (1951) for an example of a contemporary discussion of this problem. Because of underdetermination, serious contenders must be judged on other grounds. That Everett characterized these other grounds as "somewhat arbitrary" suggests that he did not take any of them to be essential attributes of a theory. Nevertheless, Everett held that pure wave mechanics did better than the other empirically faithful formulations of quantum mechanics on the grounds of simplicity, comprehensiveness, and pictorability. See for example his discussion of Bohmian mechanics (pgs. 153–55).

[je] This anti-metaphysical view of theories is echoed in his extended discussion of the nature of theories in the long thesis (pgs. 168–72).

the basis of whether or not the *experience which is predicted by the theory* is in accord with actual experience.

Let me clarify this point. One of the basic criticisms leveled against the Copernican theory was that "the mobility of the earth as a real physical fact is incompatible with the common sense interpretation of nature." In other words, as any fool can plainly see the earth doesn't *really* move because we don't experience any motion. However, a theory which involves the motion of the earth is not difficult to swallow if it is a complete enough theory that one can also deduce that no motion will be felt by the earth's inhabitants (as was possible with Newtonian physics). Thus, in order to decide whether or not a theory contradicts our experience, it is necessary to see what the theory itself predicts our experience will be.[jf]

Now in your letter you say, "the trajectory of the memory configuration of a real physical observer, on the other hand, does *not* branch. I can testify to this from personal introspection, as can you. I simply do *not* branch." I can't resist asking: Do you feel the motion of the earth?

In another place: "...[Everett's theory] contains all possible branches in it at the same time. In the real physical world we must be content with just one branch. Everett's world and the real physical world are therefore not isomorphic." Yet another: "But the real world does not branch, and therein lies the flaw in Everett's scheme."

I must confess that I do not see this "branching process" as the "vast contradiction" that you do. The theory is in full accord with our experience (at least insofar as ordinary quantum mechanics is). It is in full accord just because it *is* possible to show that no observer would ever be aware of any "branching," which is alien to our experience as you point out.

The whole issue of the "transition from the possible to the actual" is taken care of in a very simple way—there is no such transition, *nor is such a transition necessary for the theory to be in accord with our experience.*[jg]

From the viewpoint of the theory, all elements of a superposition (all "branches") are "actual," none any more "real" than another. It is completely unnecessary to suppose that after an observation somehow one element of the final superposition is selected to be awarded with a mysterious quality called "reality" and the others condemned to oblivion.

[jf] In light of DeWitt's comments, Everett added a note in proof to his *Reviews of Modern Physics* paper describing this analogy (pg. 189). It is curious that DeWitt did not think that Everett's theory was compatible with ordinary experience when DeWitt seems to have taken one of the analogies between Everett and Einstein to be that they each showed how to recapture ordinary experience in a counterintuitive context. For his part, Everett is clear that if he cannot somehow recapture the ordinary and determinate quantum experiences of observers modeled within the theory, then pure wave mechanics is not a contender; but, if he can, then pure wave mechanics must be taken seriously.

[jg] This is Everett's reply to DeWitt's main criticism (pg. 250). See also the discussion of the relationship between physical theories and experience in the conceptual introduction (pgs. 51–54).

We can be more charitable and allow the others to coexist—they won't cause any trouble anyway because all the separate elements of the superposition ("branches") individually obey the wave equation with complete indifference to the presence or absence ("actuality" or not) of any other elements.[jh]

This is only to say that the theory manages to avoid the difficulty of "the transition from possible to actual"—and I consider this to be not a weakness, but rather a great strength of the theory. The theory *is* isomorphic with experience when one takes the trouble to see what the theory itself says our experience will be. Little more can be asked of it without exposing a naked philosophic prejudice of one kind or another.

Of course, I do not hold that this theory is the only possible acceptable interpretation of quantum mechanics. I believe that any number of theories can be constructed which will adequately portray our experience, so that selection among them must be largely a matter of taste.

I do believe, however, that at this time the present theory is the simplest adequate interpretation.[ji] The hidden variable theories are, to me, more cumbersome and artificial—while the Copenhagen interpretation is hopelessly incomplete because of its a priori reliance on classical physics (excluding *in principle* any deduction of classical physics from quantum theory, or any adequate investigation of the measuring process), as well as a philosophic monstrosity with a "reality" concept for the macroscopic world and denial of the same for the microcosm.[jj]

I would like to point out that from my point of view there is no preference for deterministic or indeterministic theories. That my theory is

---

[jh] In a note on his copy of the letter from Wiener, Everett explains that one does not say "Case A is actually realized," except relative to some other state. See fn. hg on pg. 233, his explanation to Wiener (pg. 234), and the note added in proof inspired by these two exchanges (pg. 189). In short, Everett held that there was no transition from possible to actual, but rather that all possibilities are actually realized as relative states. In the passage here, Everett makes at least two significant points. The first is that the linear dynamics allow one to think of each element of a superposition as evolving on its own. This can be understood as meaning that the elements do not interact with each other except by way of interference effects, which are also required by the linearity of the dynamics. The second point concerns precisely *how* the theory can be taken as empirically faithful. When Everett says that we can "allow the others to exist" he is agreeing with DeWitt that at least some of the branches are *not* compatible with our particular experience. The theory is empirically faithful then because there exists at least one branch among the many branches that *is* isomorphic to our experience. This, it seems, is the homomorphism between the theory and experience that Everett referred to earlier in the letter (pg. 253). See also pg. 169 and the discussion of empirical faithfulness in the conceptual introduction (pgs. 51–53).

[ji] Everett also took pure wave mechanics to be both the simplest and the most comprehensive formulation of quantum mechanics. See the discussion following pg. 168.

[jj] This description of the problems faced by hidden variable theories like Bohmian mechanics (pgs. 153–55) and the Copenhagen interpretation (pgs. 152–53) reflects Everett's discussion in the long thesis.

fundamentally deterministic is not due to any deep conviction on my part that determinism holds any sacred position. It is quite conceivable that an adequate *stochastic* interpretation could be developed (perhaps along the lines of Bopp's theories) where the fundamental processes of nature are pictured as stochastic processes *whether or not* they are undergoing observation.[ik] I only object to mixed systems where the character changes with mystical acts of observation.

With respect to your "minor" criticisms, most of them are explicitly dealt with in the original work from which the article was condensed.[il] I hope, sometime soon, to revise it and make it available, as it contains a much fuller discussion of the various points, as well as a discussion of the present alternative formulations of quantum mechanics. It is just impossible to do full justice to the subject in so brief an article as the one you read.

Sincerely yours,

Hugh Everett, III

Note: Address after August 1:

607 Pelham Street
Alexandria, Virginia

HE:nge

Copies sent to
Wiener
Wheeler

[ik] See Bopp (1947, 1953a and 1953b). See also Everett's discussion of Bopp's theory in the long thesis (pgs. 155–56).

[il] DeWitt also for example wanted an account of approximate measurement. Everett gives this in the long thesis that DeWitt had not yet read. See the discussion following pg. 145.

CHAPTER 17

Correspondence: Everett and Frank (1957)

In Spring 1957, Everett corresponded with several supporters of his theory. Yale University professor Henry Margenau, for example, a physicist with a philosophical bent, wrote to Everett in April, after reading his preprint, commenting "The problem with which you deal has irritated many minds. I, for one, find your disposal quite acceptable."

In his Reviews of Modern Physics *article on Everett's theory, Wheeler referenced the philosophical writings of Philip Frank, a philosopher of science at Harvard University. In May 1957, Everett sent Frank the* Reviews of Modern Physics *preprints, along with a letter in which he identified himself with Frank's broadly empiricist world view. Frank had also written about the historical difficulty of gaining acceptance of new scientific theories, such as the sun-centered solar system proposed by Copernicus. It is probably not a coincidence that Everett referred to Copernicus in his May 31, 1957, correspondence with DeWitt. Writing to Frank on the same day, he referenced the treatment of Copernicus by the 16th century religious and scientific establishment as a way of explaining the rejection of his relative-state formulation by Bohr and his followers.[jm]*

Everett to Frank, May 31, 1957

Hugh Everett, III
Arlington Towers, T-438
Arlington, Virginia
May 31, 1957

Professor Philipp G. Frank
Department of Physics

[jm] Everett's identification of his understanding of the proper cognitive status of physical theories with Philip Frank's provides a helpful connection for interpreting Everett. Frank's late empiricism does mesh well with Everett's discussion of the status of his theory in the second appendix to the long thesis (pgs. 168–72). See also the discussion of Everett's notion of empirical faithfulness in the conceptual introduction (pgs. 50–54) in this volume.

Harvard University
Cambridge, Massachusetts

Dear Professor Frank:

As a result of membership in the "Library of Science" book club, I have received several of your works on the philosophy of science,[jn] I have found them extremely stimulating and valuable. I find that you have expressed a viewpoint which is very nearly identical with the one which I have developed independently in the last few years, concerning the nature of physical theory.[jo]

For this reason I am enclosing a copy of a paper I have recently written entitled, "On the Foundations of Quantum Mechanics," as well as an evaluation of it by Professor J. A. Wheeler of Princeton, both to be published in the July issue of *Reviews of Modern Physics*.[jp]

I think that you may be interested in it, as I am, as an example of a certain type of theory—a completely abstract mathematical model which is ultimately put into correspondence with experience.[jq] It has the interesting feature, however, that this correspondence can be made only by invoking the theory itself to predict our experience—the world picture presented by the basic mathematical theory being entirely alien to our usual conception of "reality."[jr] The treatment of observation itself in the theory is absolutely necessary. If one will only swallow the world picture implied by the theory, one has, I believe, the simplest, most complete framework for the interpretation of quantum mechanics available today.[js]

I think that it will be quite interesting to see the various reasons for accepting or rejecting it that will be advanced by various physicists. Already

---

[jn] Everett may have read the following works in part or whole: *Foundations of Physics* (Frank, 1946), *Relativity: A Richer Truth* (Frank, 1950), *The Validation of Scientific Theories* (Frank, 1954), and *Modern Science and Its Philosophy* (Frank, 1955).

[jo] Whereas Philipp Frank was a logical positivist and an original member of the Vienna Circle, by the time he was at Harvard his position might be best understood as a form of empirical pragmatism or operationalism. In any case, Frank's views at this time did indeed fit well with what Everett said concerning the nature of physical theories in the second appendix to the long thesis (pgs. 168–72) and in his letter to DeWitt (pg. 253). See Barrett (2011b) for a discussion of both Frank's and Everett's views on the proper cognitive status of physical theories.

[jp] This was Everett's short thesis and Wheeler's companion paper as included in the present volume.

[jq] The correspondence is the homomorphism that Everett discusses in the second appendix to the long thesis (pg. 169) and in the letter to DeWitt (pgs. 253–55).

[jr] For Everett one invokes pure wave mechanics to predict experience by modeling physical observers themselves within the theory then asking what their sequence of memory records would be. It is significant that pure wave mechanics itself explains why one would not detect the splitting of the quantum mechanical state.

[js] Everett took the simplicity, comprehensiveness, and pictorability of pure wave mechanics to be among its greatest virtues.

the comments that have been received show an astonishing variety of viewpoints. Because of its particular nature this theory is especially suited to illustrate a number of points which can arise in any discussion of the nature of theories in general.

If you have any comments or criticisms on the theory proper, I would be grateful to receive them. If you have any interest in the matter with respect to the more general questions of the philosophic, psychological or sociological reasons which influence the attitudes toward physical theories, I will be happy to send you copies of the comments and criticisms that I receive.

Sincerely yours,

Hugh Everett, III

Note: Address after August 1:

607 Pelham Street
Alexandria, Virginia

HE:nge

FRANK TO EVERETT, AUGUST 3, 1957

Cambridge, August 3, 1957

Dear Mr. Everett:

Please excuse my long delay in answering your kind letter of May 31. I have been all the time on a trip abroad and did not receive your letter until my return to Cambridge four days ago.[jt]

Therefore, it may take some time before I have read your paper as carefully as it deserves. To be written with the first glance, your purpose seems rather attractive to me. I have always disliked the traditional treatment of "measurement" in Quantum Theory according to which it seems as if "measurement" would be a type of fact which is essentially different from all other physical facts. However, I shall reserve my judgment until I actually have studied your paper.

Anyway, I would like to receive from your paper and comments concerned with the general philosophical and sociological problems which are connected with physical Theory.[ju]

[jt] Frank's handwritten note is difficult to read and his language is somewhat nonstandard. A scan of the original text is available online at the UCIspace archive.

[ju] It is unclear whether Everett sent him anything further.

After Sept. 15 I shall be in New York, where I shall teach at Columbia
Philosophy of Science during the academic year 1957/8.
Very nicely yours,
　Philipp Frank
1558 Massachusetts St
Cambridge 38, Mass
　After Sept. 15
Department of Philosophy Columbia University
　New York City

CHAPTER 18

Correspondence: Everett and Jaynes (1957)

In May 1957, the first of a pair of related papers by E. T. Jaynes appeared in the journal Physical Review. *In these papers Jaynes sought to apply Shannon information theory to statistical mechanics with the aim, in part, of providing an account of thermodynamic probabilities.*

Everett saw a direct connection between Jayne's project of providing probabilities for classical statistical mechanics and his own project of providing probabilities for pure wave mechanics, but he disagreed with Jaynes on the proper way to carry out such a project. Everett argued that Jaynes' subjectivist approach did not avoid the problem of having to postulate special a priori *probabilities. In particular, Everett pointed out that Jayne's proposed measure of information was only one of a class of suitable information measures—a fact that Everett explicitly acknowledged in his own work. For Everett, the class of potential information measures was, in part, determined by the specific character of the system being considered; from which Everett concluded that Jaynes' principle of maximum entropy cannot be construed as providing an* a priori *or canonical standard for statistical inference. Concerning his own work, while Everett held that the norm-squared probability measure was special in pure wave mechanics, he also knew that this measure represented one option among many.[iv]*

Jaynes wrote a lengthy reply to Everett's letter, dated 17 June 1957. He thanked Everett for his comments, but firmly defended his position. In particular, Jaynes argued that the special character of an object system may always be accounted for in terms of learned information which causes one to update their subjective probabilities from earlier prior probabilities. Hence, Jaynes concluded, "You claim that my theory is only a special case of your theory, with one particular information measure. I can, with equal justice claim that your theory is a special case of mine.*"*

It seems that Everett did not respond to Jaynes' letter nor to a subsequent letter Jaynes wrote after reading the short version of

---

[iv] See, for example, Everett's presentation of his prior typicality measure for pure wave mechanics in the long thesis (pgs. 123–30). However, the proper status of probabilities in pure wave mechanics remains an open question.

Everett's thesis when it was published in the Review of Modern Physics. *Copies of these letters can be found in the archive of Jaynes' papers at Washington University in St. Louis.*

EVERETT TO JAYNES, JUNE 11, 1957

Hugh Everett, III
Institute for Defense Analyses
The Pentagon
Washington, D.C.
June 11, 1957

Dr. E.T. Jaynes
Department of Physics
Stanford University
Stanford, California

Dear Dr. Jaynes,

I am writing with respect to your article, "Information Theory and Statistical Mechanics" in the May 15 Physical Review.

While I fully sympathize with the "subjectivist" view of statistical mechanics that you express, I must point out a rather fundamental and inescapable difficulty with the principle of maximum entropy as you have stated it. It occurred to me also several years ago that one might take such an approach—that one might be able to circumvent the reliance on dynamical laws by basing deductions on a minimum information principle, with the subjectivist interpretation of probabilities as the justification. This is indeed an appealing idea.

The difficulty is that one is seduced by this method into believing, as you apparently do and I did, that one has circumvented the problem of assigning a priori probabilities when one has in fact *done no such thing*, but has tacitly admitted a particular a priori distribution.

Briefly, the trouble lies in using the expression $\sum p_i \ln p_i$ to measure information. I shall demonstrate shortly that this choice is highly prejudicial and is equivalent to merely assuming *equal* a priori probabilities for each state.

It has occurred independently to several people, myself included, that the proper definition of information is a *relative* one, the information of a probability distribution *relative* to some underlying (basic) measure (or probability distribution) already given.

Thus for a discrete set of states $\{S_i\}$, with basic measure $\{\mu_i\}$ (simply a set of weights in the discrete case), the relative information of a distribution

$\{p_i\}$ over the states is defined as:

$$I = \sum_i p_i \ln \frac{p_i}{\mu_i}$$

The reasons for this more general definition will become clear shortly. First, it enables one to define information for *any* probability distribution, i.e., for arbitrary probability measures over completely arbitrary sets of unrestricted cardinality. (Judging by your footnote, you are aware of the difficulties of even the continuous case for the ordinary definition.) This general definition comes about as follows: Consider an arbitrary set X, with probability measure p and underlying (I call it an information measure) measure μ. Now consider a *finite* partition $\mathcal{P}$ of X into subsets X_i. We then have a finite distribution over these sets and an information (in the relative sense) defined:

$$I^{\mathcal{P}} = \sum_i p(X_i) \ln \frac{p(X_i)}{\mu(X_i)}$$

Now it is an easily proved theorem that any *refinement* of $\mathcal{P}$ will never decrease the information (i.e., $\mathcal{P}'$ is a refinement of $\mathcal{P}$ implies that $I^{\mathcal{P}'} \geq I^{\mathcal{P}}$). Hence $I^{\mathcal{P}}$ is a *monotone function* on the *directed set* of all finite partitions, and *always* has a limit, which we define as the information of P relative to μ. Thus this relative definition generalizes, while the usual one doesn't, as you know.

A second advantage lies in the application to *stochastic processes*. If one defines the information of a distribution *relative to a stationary distribution of the process*, one can prove the theorem that *information never increases with time* (entropy never decreases).

Example: Two state Markov process with transition probabilities $T_{i \rightarrow j}$:

$$\begin{array}{c} \\ 1 \\ 2 \end{array} \begin{array}{cc} 1 & 2 \\ \left(\begin{array}{cc} 1/3 & 2/3 \\ 1/3 & 2/3 \end{array} \right. & \left. \begin{array}{c} \\ \\ \end{array} \right) \end{array}$$

has the stationary distribution $p_1^* = 1/3$, $p_2^* = 2/3$. *For this process* the information of a distribution p_1, p_2 should be:

$$I = \sum P_i \ln \frac{p_i}{p_i^*} = p_1 \ln \frac{p_1}{1/3} + p_2 \ln \frac{p_2}{2/3}$$

It is only for *doubly-stochastic* processes (where $\sum_i T_{ij} = 1$ *as well as* $\sum_j T_{ij} = 1$) that the stationary measure is *uniform*, and one can get away with the old definition $\sum P_i \ln p_i$.

But notice that to determine the *stationary measure*, one *must know the dynamics* of the system. If you try to make predictions about this example using a minimum $\sum p_i \ln p_i$ you will make worse predictions about this example than I, who use $\sum p_i \ln p_i / p_i^*$, since I take into account the known fact that this system is *not* equally likely to be in any of its states.

Similarly in the case of statistical mechanics of gases. *Only after one has established* that the measure one is using is stationary (Liouville's Theorem) is one justified in using it. The central problem is, as always, *discovering the basic measure* (or a priori probabilities, if you will). It is just fortuitous 1) that Lebesgue measure is the proper one for phase space so that $\int p \ln p \, dt$, which is *really* information relative to Lebesgue measure, is correct, and 2) that for doubly-stochastic processes, the type almost always encountered in physics, the *uniform* measure is stationary and hence $\sum p_i \ln p_i$ correct. These two circumstances are, I believe, what cause people to be seduced into believing a special case can be regarded as a general principle.

I really have a lot more to say on the subject, but time doesn't permit. I hope this rather hastily written letter conveys adequately the nature of my objection that you have not really sidestepped the fundamental problem of assigning a priori probabilities—which *does* depend on dynamical laws, ergodic properties, etc.—but have only camouflaged it.

Nevertheless, I sympathize with your viewpoint, and believe a lot *can* be done in this line. I hope you will continue, and that my remarks on the more general definition of information may be of help to you.

Sincerely yours,

Hugh Everett, III

P.S.   Re the axioms which "uniquely" determine the definition of information (or entropy):

1. This general definition satisfies continuity in p_i's.
2. It satisfies composition law, when basic measure composed also same way as probabilities.
3. It satisfies condition that $\min(I)$ is monotone decreasing with the number of allowed states (increasing uncertainty). This is analogous, and more natural, that the requirement $I(1/n, \ldots, 1/n)$ is monotone decreasing.

It is, then, on quite as firm ground as the more restricted form, and your statement, "Therefore one expects that deductions made from any other information measure, if carried far enough, will eventually lead to contradictions," is a bit too strong.

HE:ne

PART IV

Post-thesis Correspondence and Notes

CHAPTER 19

Transcript: Conference at Xavier University (1959)

In 1959, Everett and his young family traveled to Copenhagen and spent several weeks vacationing. Mixing business with pleasure, Everett met with Bohr several times to discuss his theory. But the two men found little to agree on concerning quantum mechanics. Bohr smoked his briar pipe and mumbled obscurely in their meetings, as was his trademark behavior in old age. Everett spent a good deal of time bar-hopping. And in the bar of the Hotel Østerport, after imbibing a few beers, he invented and wrote down on hotel stationery the generalized Lagrange multiplier method, sending it off to his WSEG colleagues, who were astonished at its usefulness for operations research.

In early 1959, Everett had discussed his theory with Xavier University physics professor Boris Podolsky in New York City. Podolsky asked Everett for a copy of his long thesis. Everett said he would send it to him after he returned from "arguing about it with Bohr for a month or so" (Hugh Everett, 1959).

In 1962, Everett was invited to explain his interpretation of quantum mechanics to a panel of physicists convened at Xavier University in Cincinnati, Ohio, to discuss the quantum measurement problem. The conference was chaired by Podolsky. A transcript of the five-day proceeding was made public forty years later (in 2002). It documents an intimate, sometimes angry, sometimes mirthful colloquy among several leading physicists of the day, including Eugene Wigner, Paul Dirac, Yakir Aharonov, Wendell Furry, Boris Podolsky, and Nathan Rosen. They explored several possible solutions to the measurement problem, ranging from Bohm's hidden variable theory to non-physical explanations of collapse that had been proposed by Wigner. Everett despised public speaking, but he could not have asked for a smarter, better informed group of physicists to address. It was the first of only two known occasions in which he explained his theory of pure wave mechanics to a public gathering of his peers.

CONFERENCE ON THE FOUNDATIONS OF
QUANTUM MECHANICS

Xavier University
Cincinnati 7, Ohio
October 1–5, 1962

Chairman
Boris Podolsky

Conference Organizer
and
Department Chairman
John Hart

Conference Reporter
Frederick G. Werner

Sponsored jointly by the National Aeronautics and Space
Administration, the Office of Naval Research, and the Judge Robert
S. Marx Foundation.

Copyright ©, 1962, by
Xavier University, Ohio

This is a limited edition of the conference manuscript to be used for editing
by the conferees.

Roster of Limited-Attendance Portion of Conference[iw]

<u>Main Participants</u>

Professor Y. Aharonov	Professor Boris Podolsky
Yeshiva University	Xavier University
Professor P. A. M. Dirac	Professor Nathan Rosen
Cambridge University (England)	Technion, Haifa, Israel
Professor Wendell H. Furry	Professor Eugene P. Wigner
Harvard University	Princeton University

[iw] Boris Podolsky (1896–1966) was a co-author with Nathan Rosen (1909–1995) and Einstein on the famous 1935 EPR paper (Einstein et al., 1935) (See also Einstein and Rosen, 1935). Podolsky served as chairman of the conference, and took an active role throughout in leading the conversation. Yakir Aharonov (b1932) was a young theoretical physicist who

Limited-Attendance Group

Dr. William Band
Washington State University

Dr. Eugen Merzbacher
University of North Carolina

Dr. Dieter R. Brill
Yale University

Dr. Jack Rivers
University of Missouri

Dr. Gideon Carmi
Yeshiva University

Dr. O. von Roos
Jet Propulsion Laboratory
California Institute of Technology

Dr. Harold Glaser
Office of Naval Research

Dr. Solomon L. Schwebel
University of Cincinnati

Dr. Eugene Guth
Oak Ridge National Laboratory

Dr. Abner Shimony
M. I. T.

Dr. Arno Jaeger
University of Cincinnati

Dr. Jack A. Soules
Office of Naval Research

Dr. Kaiser S. Kunz
New Mexico State University

Dr. Frederick G. Werner
Xavier University

Dr. Michael M. Yanase, S. J.
Institute for Advanced Studies,
Princeton, N.J.

Director of audio-visual recording: Mr. Thomas Fischer

Assisted by: Robert Podolsky
Austin Towle

had been a student of David Bohm (see Aharonov and Bohm 1959). He was interested in the foundations of quantum mechanics. Paul Dirac (1902–1984) was largely responsible for the original statement of the standard formulation of quantum mechanics (Dirac, 1930). Wendell Furry (1907–1984) was a professor of physics at Harvard University who made contributions to foundations of quantum field theory. Eugene Wigner (1902–1995) was a professor of physics at Princeton University. He favored an extension of the von Neumann–Dirac collapse theory where it is the intervention of the conscious mind of an observer that causes the collapse of the quantum mechanical state (Wigner, 1961). In addition to the main participants, several prominent physicists and philosophers took part in the meeting. Abner Shimony (b1928) was a philosopher and physicist. He received his Ph.D. in philosophy from Yale (1953) and a Ph.D. in physics in (1962) from Princeton. Gideon Carmi, another of David Bohm's students, was a theoretical physicist with an interest in the foundations of quantum mechanics. He later collaborated with both Aharonov and Bohm. Kaiser Kunz was a theoretical physicist who did significant work in electrodynamics.

CONFERENCE ON THE FOUNDATIONS OF
QUANTUM MECHANICS[ix]
October 1–5, 1962

Public Relations Policy

In order that each participant may feel free to express himself sponta-neously in the spirit of the *limited attendance portion* of this conference, the chairman has adopted the following policy regarding references. It is understood that each person present, before referring in publication to remarks made by a participant during these sessions is expected to check such material with the participant or participants concerned. Reports of general conclusions are to be checked with the chairman. This policy applies as well to the published report of the proceedings, which is to be edited by the chairman.

Dr. Podolsky: In the heat of an argument I can make a statement, and I probably will, which any freshman with a pencil and paper and five minutes can prove to be nonsense. Perhaps a few minutes later I might regret having made the statement. Now, we don't want such statements to get out. The principal reason for this is to make sure the participants won't stop to worry about whether or not what they're saying is really so, or whether it is nonsense. We want the participants to feel free to express themselves spontaneously, and afterwards, in more sober discussions, withdraw these statements without things getting out in the newspapers.

EXCERPT FROM THE MONDAY MORNING SESSION

A short discussion followed about the possibility of inviting Everett to discuss his point of view about the reduction of the wave packet. Podolsky asks Rosen if there is something he could say about Everett's ideas. He

---

[ix] The original typescript of this document contains numerous handwritten amendments to Everett's statements. Since there is every reason to believe that these changes were made by Everett himself, we have simply included most of these changes without comment in the text in order to make the text as easy to read as possible. The conference reporter F. G. Werner likely made an audio recording of the conference proceedings and then transcribed the recording. Werner then likely sent his original typescript to the participants and subsequently contacted the participants in accordance with the conference's public relations policy. Werner then noted the changes requested by the participants to their statements. These revisions were made in red pen. The handwriting appears to be the same for revisions to different participants' statements throughout the document, suggesting that Werner compiled the suggested changes from the participants, then made the changes to the original document in his own hand. Everett, at least, may have provided his changes by phone. At one point in the typescript Liouville's theorem is written in Werner's hand as, "Lidovilles Theorem." While it is unlikely that Everett would have gotten the name of the theorem wrong, it is possible that Werner may have misheard it over the phone.

explains that according to Everett, it is not necessary to worry about the problem of the reduction of the wave packet, because all the different possibilities after measurement are on an equal footing. The various possible results of a measurement correspond to a kind of branching so that if you get one result it means that you are just on one of the branches. But since all of the other branches exist on the same footing, one describes all the possible measurements as one huge tree. Each time after a given result is found, one simply goes along one of the branches and from this branch one continues into further branching by making another measurement, and so on. We all seem to feel that the measurement does something decisive. For example, when we cast a die and get a given result, we have in some way singled out this result out of all the possible results.

Podolsky: Oh yes, I remember now what it is about—it's a picture about parallel times, parallel universes, and each time one gets a given result he chooses which one of the universes he belongs to, but the other universes continue to exist.[jy]

Aharonov: Perhaps Professor Rosen will be willing to introduce the idea a little bit more fully with perhaps a little bit more on the mathematical side.

Rosen: I just have some recollection of the paper. It's not a question of mathematics, it seems to me, but rather a question of interpretation. The mathematics involved is very simple: you expand a wave function as a linear combination of eigenfunctions of the observed quantity. In other words, if you have two systems interacting, one of them being the measured system and the other the measurer, you can use what Professor Furry will talk more about in the afternoon, namely, correlation between the measurer and the measured system. Then you get an expansion involving eigenfunctions of the system, multiplied by the eigenfunctions of the measuring instrument. The usual belief is that when the measurement is over, one of these terms is singled out and the others are thrown away. That is what is referred to as the reduction of the wave packet. The other point of view, that of Everett, is to keep all the terms.[jz]

Aharonov: There seems to be a problem here. It raises the question is time reversible? If you look on the process of branching you see that it has a definitely preferred direction of time.[ka] You never experience any collection of past branching connected together with one observer in the present. So the

---

[jy] Podolsky had discussed Everett's formulation of quantum mechanics with Everett in early 1959. Rosen's recollections are likely based on Everett's *Reviews of Modern Physics* article. See the discussion in the conceptual introduction (pg. 40).

[jz] Rosen's summary closely follows what Everett actually said in the short thesis. It is unsurprising that the group considers him to be the best informed on Everett's views. At this point in the typescript the following is crossed out in red pen: "in spite of the fact that all one gets out of the measurement is experience."

[ka] Everett, however, held that the branching process occurs in *both* temporal directions and that the irreversibility of measurement or of branching is only apparent. More specifically, since the linear dynamics is time reversal symmetric, irreversibility, for Everett was a feature of the subjective experience of observers who were themselves treated within the theory, not a

observer described by this method is always going in one direction of time, namely, more and more branching toward the future and not vice-versa. In other words, it seems that the idea of the unique direction of time is basic for this theory, and one should therefore explain why a reversible equation for a closed system somehow irreversibly measures in this idea of branching.

EXCERPT FROM THE TUESDAY MORNING SESSION

Rosen: *(continues)* Well, in talking about measurements and the reduction of the wave packet we come upon this relevant point. Just what does happen in the measurement? The fact is that at some stage we have to think of the measurement as making a decision among a number of different possibilities, singling out one result from a number of potential results. That is the essential feature in the final stage of the measurement. Simply calling one thing an object and the other an instrument in itself does not insure this, because one could treat both of them quantum-mechanically. As Professor Wigner pointed out, you have the same problem about carrying out the measurement on the instrument that you had in carrying out a measurement on the object by using the instrument so that, according to this line of thought, you can have an endless chain. Yet somehow or other, we are able to cut this chain and say that there are certain instruments that we call classical ones which have the property that they make a decision and give us one answer out of many possibilities. When we get that answer, we have singled out one term in the expansion which Professor Furry wrote down, so that we get one term instead of the whole series. This is what we call the reduction of the wave packet. Now at this point I think it is appropriate to mention Dr. Everett's point of view, in which he does not accept the idea of the reduction of the wave packet. I hope he will correct me if I say this incorrectly and I hope he will add something to what I say. As I understand it, he considers this whole series as continuing to exist even after the measurement has been carried out. He does not want to distinguish between the actual result as obtained in a given case and the other possible results which might have been obtained, so that even after the measurement he has the series of terms, instead of one term. He thinks of the wave function as changing only in accordance with the Schroedinger equation, in a continuous way, without the possibility of this sudden change in the wave function, which we call the reduction of the wave packet. My own feeling is that such a point of view is tenable and consistent, but should be interpreted as referring not to what one observer finds but what many observers carrying out the same sort of measurement on the same sort of system

feature of the dynamics. See pgs. 143–44, 224, and 240 for discussions. See also Everett's note on his copy of the Bell paper (fn. kz on pg. 287).

would find. If Dr. Everett does not agree with me, I hope he will present his point of view himself. Would you care to say something at this stage of our discussion?

Hugh Everett: I think you said it essentially correctly. My position is simply that I think you can make a tenable theory out of allowing the superpositions to continue forever, even for a single observer.

Shimony: It seems to me that if this is the case, there are two possibilities. The two possibilities involve awareness. One possibility is that ordinary human awareness is associated with one of these branches and not with the others. Then the question becomes, how does your formalism permit this solution? The other possibility is that awareness is associated with each branch.

Rosen: Wait just a moment. I think perhaps it would help the group if you (Everett) could give us a little bit of background on this. I threw you into the middle of the discussion, and I am not sure that everybody in the audience is familiar with your theory. Would you mind saying a few words?

Everett: Well, the picture I have is something like this: Imagine an observer making a sequence of observations on a number of, let's say, originally identical object systems. At the end of this sequence there is a large superposition of states, each element of which contains the observer as having recorded a particular definite sequence of results of observation. I identify a single element as what we think of as an experience, but still hold that it is tenable to assert that all of the elements simultaneously coexist. In any single element of the final superposition after all these measurements, you have a state which describes the observer as having observed a quite definite and apparently random sequence of events. Of course, it's a different sequence of events in each element of the superposition. In fact, if one takes a very large series of experiments, in a certain sense one can assert that for almost all of the elements of the final superposition the frequencies of the results of measurements will be in accord with what one predicts from the ordinary picture of quantum mechanics.[kb] That is very briefly it.

Podolsky: Perhaps it might be a little clearer to most people if you put it in a different way. Somehow or other we have here the parallel times or parallel worlds that science fiction likes to talk about so much.[kc] Every time a decision is made, the observer proceeds along one particular time while the other possibilities still exist and have physical reality.

[kb] As Everett later explains, the "certain sense" here is that a typical element and in the limit almost all of the elements *in the normed-squared coefficient measure* exhibit the standard quantum statistics. In this precise sense, the experience of a typical *relative* observer is what is predicted by the standard collapse theory for this sort of experiment. See pgs. 187–89 and 122–23.

[kc] Although Everett has used the language of his thesis in describing his theory, the discussion changes significantly when Podolsky suggests that they use the parallel-worlds language of science fiction to make the theory "a little clearer to most people". Everett does not seem particularly worried about the change in language. For examples of discussions concerning the best language to use to describe his model see pgs. 68–69, 206, 210, and 223. See also the discussion of this exchange in the conceptual introduction (pgs. 40–41).

Everett: Yes, it's a consequence of the superposition principle that each separate element of the superposition will obey the same laws independent of the presence or absence of one another. Hence, why insist on having a certain selection of one of the elements as being real and all of the others somehow mysteriously vanishing?

Furry: Actually, wouldn't you prefer to say that no decisions were made, but to the observer looking back it looks in retrospect as if the decisions were made? The observer also exists in all the other states, and in each of them as he looks back, it looks as if the appropriate decisions were made. This means that each of us, you see, exists on a great many sheets or versions, and it's only on this one right here that you have any particular remembrance of the past. In some other ones we perhaps didn't come to Cincinnati.

Everett: The picture that it leads to I do think is tenable, and I think it's the simplest one that can arise. We simply do away with the reduction of the wave packet.[kd]

Podolsky: It's certainly consistent as far as we have heard of it. The question arises as to what happens if we have a large number of observers and how these worlds of individual observers fit in together.

Everett: Well, again, all of the consistency of ordinary physics is preserved by the correlation structure of this state. You'll always find that an observer who repeats the same measurement will always get the same answer, and even more will always agree when interacting with another observer measuring the same system. This consistency can be deduced from the structure of wave mechanics.

Podolsky: It looks like we would have a non-denumerable infinity of worlds.

Everett: Yes.

Podolsky: Each proceeding with its own set of choices that have been made.

Furry: To me, the hard thing about it is that one must picture the world, oneself, and everybody else as consisting not in just a countable number of copies but somehow or another in an undenumerable number of copies, and at this my imagination balks. I can think of various alternative Furrys doing different things, but I cannot think of a non-denumerable number of alternative Furrys.

(Podolsky chuckles)

Everett: I'd like to make one final remark here. Imagine a very large series of experiments made by an observer. With each observation, the state of the observer splits into a number of states, one for each possible outcome, and correlated to the outcome.[ke] Thus the state of the observer is a constantly branching tree, each element of which describes a particular history of observations. Now, I would like to assert that, for a "typical" branch, the frequency of results will be precisely what is predicted by ordinary quantum mechanics.[kf] Even more

---

[kd] Everett explains here both what pure wave mechanics is (the standard theory without collapse) and the resulting virtue (simplicity).

[ke] Here Everett returns to the language of splitting states.

[kf] See Everett's discussions of typicality of branches pgs. 187–89 and 122–23.

strongly, I would like to assert that, as the number of observations goes to infinity, almost all branches will contain frequencies of results in accord with ordinary quantum theory predictions. To be able to make a statement like this requires that there be some sort of a measure on the superposition of states. What I need, therefore, is a measure that I can put on a sum of orthogonal states. There is one consistency criteria which would be required for such a thing. Since my states are constantly branching, I must insist that the measure on a state originally is equal to the sum of the measures on the separate branches after a branching process.[kg] Now this consistency criterion can be shown to lead directly to the squared amplitude of the coefficient, as the unique measure which satisfies this. With this unique measure, deduced only from a consistency condition, I then can assert: indeed, for almost all (in the measure theoretical sense) elements of a very large superposition, the predictions of ordinary quantum mechanics hold.[kh] Now I could draw a parallel here to statistical mechanics, where the same sort of thing takes place.[ki] Here we like to make statements for almost all trajectories. They are ergodic and things like that. Here also you can only make such a statement if you have some underlying measure that you regard as fundamental, since any such statements would be false if I take a measure that had only non-zero measure on the exceptional trajectories. In statistical mechanics it turns out there is uniquely one measure of the phase space which you can use, the Lebesgue measure. This is because it is preserved under the transformation of phase space (by Liouville's[kj] Theorem). being essentially the only measure giving the conservation of probability. It is precisely this analogue that I use on the branching of the state function and I can therefore assert that the probabilistic interpretation of quantum mechanics can be deduced quite as rigorously from pure wave mechanics as the deductions of statistical mechanics.

Carmi: (Some discussion of questions raised by Dr. Gideon Carmi were incompletely recorded at this point in the session.)

Podolsky: *(to Shimony)* Do you wish to comment on this?

Shimony: You eliminate one of the two alternatives I had in mind. You do associate awareness with each one of these.

[kg] This is a diachronic form of Everett's additivity condition. See the discussion following pg. 123 for the presentation of this condition in the long thesis.

[kh] It is not that the measure of typicality is deduced from pure wave mechanics alone; rather, it is deduced from pure wave mechanics together with the additivity, or constancy, condition and a small handful of other conditions. Everett also needs to assume that the standard of typicality is a function of the coefficents associated with the branches alone, rather than something like a count of the branches where a particular relative fact obtains, and that the phase information represented by the coefficients is irrelevant to the measure. See the discussion following pg. 123.

[ki] Everett does just this in the long thesis. See for example pgs. 191–92.

[kj] Original text reads in red pen: "Lidovilles Theorem."

Everett: Each individual branch looks like a perfectly respectable world where definite things have happened.[kk]

Shimony: Then the question that I have about the alternatives that you have chosen is: what, from the standpoint of any one of these branches, is the difference within a branch, between your picture of the world and one in which there are stochastic elements?

Everett: None whatever. The whole point of this view-point is that a deduction from it is that the standard interpretation will hold for all observers. In addition, however, one can, within this viewpoint, get some hold on approximate measures and this type of thing.

Podolsky: Thank you, Dr. Everett.

Rosen: I am reminded of a story by O. Henry that I once read, about a man who walked along a road and came to a fork with several roads leading from it. He decided to follow one of them and certain things happened. Then the story went back to the same point and he decided to go along another road from the fork and something else happened to him, and so on for three or four versions, according to which road he chose.[kl]

Aharonov: I think we should be happy because other parts of us are perhaps doing much nicer life because they have chosen different branches.

Excerpt from the Panel Discussion

Kaiser Kunz: I remember in my elementary work having to work out certain problems involving, let us say, a quadratic equation. I get two solutions. Then the question is, are they both good or not? We substitute back and find that one of them is an extraneous solution. It seems to me that there is a certain parallel case here of a more sophisticated kind. We're simply saying that quantum mechanics will give us a right or correct solution. But it gives us too many and the collapse of the wave function, so to speak, is that which actually occurs. Whether it occurs during cognition, or whether somehow or another we blame it on the process of measurement that occurs, seems to be the debate. The basic thing seems to be pretty clear. It is that quantum mechanics gives us multiple values, so to speak, and our problem philosophically is, when do we pick the solution. We make it. We simply force it to agree with what we have observed. So this observation is taken as the correct solution.

[kk] Here Everett refers to each branch as looking like an ordinary world. This is the only place where Everett himself directly ties the notion of a branch to that of a world in the documents we have. See the discussion of this exchange in the conceptual introduction (pgs. 40–41).

[kl] In O. Henry's short story "Roads of Destiny" (Henry, 1909) the main character David chooses between the "left branch," the "right branch," and the "main road." Insofar as all branches are equally actual for Everett there is no choice involved. All branches are actual and associated with relative states for the component systems.

Furry: If you're positivistic minded enough, there is no problem, there is no trouble. The logical positivists love this.[km]

Podolsky: The question is, really, what is it you do observe and how do we observe it?

Kunz: I think this morning we got even another viewpoint, which is that even the observation doesn't determine which one we really have. Regardless of whether we get the multiple valuedness, it continues on indefinitely. It seems to me that we all agree that there is multiple valuedness and that quantum mechanics gives it to us. But where we disagree is, if we can select out one of these values, and when the solution is made.

Wigner: It depends also on whether we select out.

Kunz: Yes.

Wigner: Yes it does. Now that is the point of view of Dr. Everett.

Rosen: Would you like to comment, Dr. Everett?

Everett: Yes. Well, what he said pretty much covers it.[kn]

At this point there is a missing page in the original typescript. After the typescript resumes, Everett no longer speaks.

EXCERPT FROM THE FRIDAY MORNING SESSION

Aharonov: You will say a few words about what Everett makes of the reduction of the wave packet?

Shimony: Everett simply doesn't. Everett makes the reduction of the wave packet not an ultimate thing. That is, ultimately the universe has one state, and its propagation is governed by the Schrödinger equation. What seems like reduction is really only appearance versus reality. Namely, at one of the crucial junctures where reduction seems to occur, or appear, one has a branching of the relative state, that is, the state having left out part of the universe, in various directions. Now as to that, there are various questions which one can ask. One is, is awareness associated with only one of these, but not with all of them? That's certainly a possibility. Everett's answer was no, so maybe we shouldn't even consider that. He says, no, if there is awareness, it is equally associated with every possibility.

[km] Kunz suggests that the problem of understanding pure wave mechanics is one of having a theory that makes *too many* predictions. As Furry notes, however, this would not necessarily be a problem for a logical positivist. The thought may be that it is enough for the positivist that one has the right empirical predictions for the typical experience of an observer. This is reminiscent of the argument of the second appendix to the long thesis (pgs. 168–72). See also DeWitt's similar worry that Everett's theory is "too rich in content" (pg. 247) and Everett's reply that the surplus content will not cause any trouble (pg. 255).

[kn] Presumably Everett is referring here to Kunz's statement immediately above.

Aharonov: In that case then, each possibility doesn't know about the others, each possibility has no way to know the others.[ko]

Shimony: That's right, and if this is the case, well, it seems to me that the thing to ask is how is a situation as visualized in one of the branches to be distinguished operationally, or by any other way, from a situation in which you don't suppose that the other branches are real, but only suppose that there is one branch where a stochastic jump has occurred. In other words, what are the differences, if any, between one part of it which is enclosed in one branch and one part of it which is enclosed in another branch? What is the difference from that standpoint between his theory of multiple branching, and the theory which has only one branch, but has chance events? And his answer is that there is no difference observationally, there is only a difference logically, and his claim is that the theory he is proposing is more logical. Well, I don't know what this means. I think that if you have two statistical theories each equally consistent, you can't claim one is more logical than the other. It seems to me that in some sense these are equivalent ways of talking about the same thing. One way is more elaborate in its terminology than the other. I think one should invoke Occam's razor: Occam said that entities ought not to be multiplied beyond necessity. And my feeling is that among the entities which aren't to be multiplied unnecessarily are histories of the universe. One history is quite enough.[kp]

Aharonov: I don't see that you point to any inconsistencies. The question is, are there any inconsistencies?

Shimony: I think that my answer is that either there is not a very apparent equivalence between his way of talking and a way of talking which is much more intuitive. You know, the equivalences are often buried when one makes transformations. That is, if you analyze very carefully the meanings of all terms, there is such an equivalence. The other possibility is one that I mentioned, that they are somehow different, but this is one place where it's certainly reasonable to invoke Occam's razor.

Podolsky: Dr. Band.

Band: : If your friend gives you information spontaneously without being asked, does this cause jumping from one branch to another?

Shimony: No, there seems to be a possibility that when this branching occurs most of them are dead and one is alive, but he doesn't want to say this; he wants to say that in the other branch he made a foolish decision or in the other branch he made a wise decision, whatever comfort that would be to you. And in the other branch you were aware of your faulty decision.

Podolsky: There seems to me to be a possibility that when you have two observers simultaneously observing an instrument, that both of them produce reduction of

---

[ko] Except by way of something like a Wigner's Friend experiment. See Everett's discussions (pgs. 176 and 73–75). David Albert later investigated how experiments that detect other Everett branches might be performed as self-measurements (Albert, 1986).

[kp] Crossed out here is the following: "This is simple. That's essentially my analysis of Everett, but it's a very different one than one gets from looking at all the possible answers."

a wave packet, but not the same wave packet. In other words the wave function may be a sufficiently subjective sort of a thing, so each observer produces a wave packet for his own consideration.

Shimony: Fine, that's one of the possibilities, but then I ask what is it in the nature of things that allows intersubjective agreement? Is it what Leibniz has called, 'pre-established harmony?' Well, that's a desperate and quite ad hoc answer. Ordinarily we believe in an agreement between us when we make an observation, that certain physical conditions for the observation are the same for us. That is, there is something there that we are both observing, and there is similarity enough to describe it.

Podolsky: I see the difficulty.

A Conference Report by F. G. Werner[kq]

The lively spirit of the extended discussions on the problems of quantum mechanics, so evident in the panel discussion, carried through to the limited-attendance sessions. Hugh Everett flew to Cincinnati from Washington to present his relative-state formalism.

[kq] The following is an excerpt from a report by Werner on the Xavier conference entitled *The Foundations of Quantum Mechanics* that was published in the January 1964 *Physics Today*.

CHAPTER 20

Notes: Everett on DeWitt (1970)

In July 1970, DeWitt lectured on "The Many-Universes Interpretation of Quantum Mechanics" at the Enrico Fermi International School of Physics, Villa Monastero, Varenna, Italy. DeWitt sent Everett a preprint of the paper and Everett penciled two remarks on it. The first remark "Goddamit, you don't see it" was in response to DeWitt's assertion in a footnote that Everett's derivation of the Born rule was "not entirely satisfying." The second comment, a slashing "yes" related to DeWitt's remark that while branching worlds do not all have to conform to the laws of physics as we know them, one would never observe worlds that are too wild to allow for the evolution of sentient life. DeWitt's article was reprinted three years later in The Many Worlds Interpretation of Quantum Mechanics *(1973).*

31

$$\leq \frac{1}{\varepsilon} \langle \Psi | \Psi \rangle \sum_{s_1 \cdots s_N} \delta(s_1 \cdots s_N) w_{s_1} \cdots w_{s_n}$$

$$= \frac{1}{N\varepsilon} \langle \Psi | \Psi \rangle \sum_s w_s (1 - w_s) \leq \frac{1}{N\varepsilon} \langle \Psi | \Psi \rangle. \tag{4.16}$$

From this it follows that no matter how small we choose ε we can always find an N big enough so that the norm of $|\chi_N^\varepsilon\rangle$ becomes smaller than any positive number. This means that

$$\lim_{N \to \infty} |\Psi_N^\varepsilon\rangle = |\Psi\rangle. \tag{4.17}$$

It will be noted that, because of the orthogonality of the basis vectors $|s_1\rangle|s_2\rangle\cdots$, this result holds regardless of the quality of the measurements, i.e., independently of whether or not the condition

$$\langle \Phi[s_1 \cdots s_N] | \Phi[s_1' \cdots s_N'] \rangle = \langle \Phi | \Phi \rangle \prod_{n=1}^{N} \delta_{s_n s_n'} \tag{4.18}$$

for good measurements is satisfied or not.

A similar result is obtained if $|\Psi_N^\varepsilon\rangle$ is redefined by excluding, in addition, elements of the superposition (4.8) whose memory sequences fail to meet any finite combination of the infinity of other requirements for a random sequence. Moreover, no other choice for the w's but (2.4) will work. The conventional statistical interpretation of quantum mechanics thus emerges from the formalism itself. Nonrandom memory sequences in the superposition (4.8) are of measure zero in the Hilbert space, in the limit $N \to \infty$.* Each automaton (that is, apparatus cum

---

*Everett's original derivation of this result[1] invokes the formal equivalence of measure theory and probability theory, and is rather too brief to be entirely satisfying. The present derivation is essentially due to R. N. Graham[7] (see also DeWitt[8]). A more rigorous treatment of the statistical interpretation question, which deals carefully with the problem of defining the Hilbert space in the limit $N \to \infty$, has been given by Hartle.[9]

Figure 20.1. Everett's notes on DeWitt's "Many-Universes Interpretation" paper.

memory sequence) in the superposition sees the world obey the familiar statistical quantum laws. This conclusion obviously admits of immediate extension to the world of cosmology. Its state vector is like a tree with an enormous number of branches. Each branch corresponds to a possible universe-as-we-actually-see-it.

The alert student may now object that the above argument contains an element of circularity. In order to derive the <u>physical</u> probability interpretation of the numbers w_s, based on sequences of observations, we have introduced a nonphysical probability concept, namely that of the measure of a subspace in Hilbert space. The latter concept is alien to experimental physics because it involves many elements of the superposition at once, and hence many simultaneous worlds, which are supposed to be unaware of one another.

The problem which this objection raises is like many which have arisen in the long history of probability theory. It should be stressed that no element of the superposition is, in the end, excluded. All the worlds are there, even those in which everything goes wrong and all the statistical laws break down. The situation is similar to that which we face in ordinary statistical mechanics. If the initial conditions were right the universe-as-we-see-it <u>could</u> be a place in which heat sometimes flows from cold bodies to hot. We can perhaps argue that in those branches in which the universe makes a habit of misbehaving in this way, life fails to evolve, so no intelligent automata are around to be amazed by it.[*]

---

[*]It may also happen that the arrow of time is reversed for some of the branches. This would be the case if the state vector of the universe were invariant under time reversal.

Figure 20.1. Continued.

CHAPTER 21

Notes: Everett on Bell (1971)

Although he did no more work on quantum mechanics after his dissertation was published in 1957 (as far as we know), Everett kept tabs on reactions to his theory as it gained fame. He paid particular attention to a critique written by John Stewart Bell, a staff physicist at CERN, the particle accelerator complex in Geneva, Switzerland.

Bell was an experimentalist whose passion was exploring the foundations of quantum mechanics. Attracted to Bohm's hidden variables theory, he published a famous paper in 1964 (Bell, 1964) in which he proved that quantum mechanics ruled out the existence of the locally deterministic hidden variables of the sort that Einstein, Poldolsky, and Rosen had wanted (Einstein et al., 1935), but it did not rule out nonlocal hidden variables like Bohm's theory.

Partly motivated by DeWitt's Physics Today *and Varenna papers, Bell was led to investigate the work of Everett's formulation of pure wave mechanics, which shared with Bohmian mechanics a denial that the wave function ever collapses but differs from Bohm's theory in not adding so-called hidden variables to the standard quantum mechanical state.*

Among Everett's effects is a preprint of a paper written in 1971 by Bell, "On the Hypothesis That the Schrödinger Equation Is Exact."

Whereas Bell took Everett's theory to be as worthy of thoughtful criticism, Bohr's apologist, Leon Rosenfeld, wrote to Bell in November 1971 regarding "On the Hypothesis That the Schrödinger Equation Is Exact":

My dear Bell,

Many thanks for the preprint of your last paper which I did read because you are one of the very few heretics from whom I always expect to learn something, and, indeed, I found this new paper of yours exceedingly instructive. To begin with, it is no mean achievement to have given Everett's damned nonsense an air of respectability by presenting it as a refurbishing of the idea of preestablished harmony. . . . Is it not complacent of you to think that you can contemplate the world from the point of view of God? (Byrne, 2010, pg. 316)

Ref.TH.1424-CERN

On the Hypothesis that the Schroedinger Equation is Exact

J. S. Bell
CERN—Geneva

Contribution to the international colloquium on issues in contemporary physics and philosophy of science, and their relevance for our society, Penn State University, September 1971.

Ref.TH.1424-CERN
27 October 1971

5.[kr] EVERETT (?)[ks]

The Everett (?) theory of this section will simply be the pilot-wave theory without trajectories. Thus instantaneous classical configurations x are supposed to exist, and to be distributed in the comparison class of possible worlds with probability $|\psi|^2$. But no pairing of configurations at different times, as would be effected by the existence of trajectories, is supposed. And it is pointed out that no such continuity between present and past configurations is required by experience.

I would really prefer to leave the formulation at that, and proceed to elucidate the last sentence. But some additional remarks must be made for readers of Everett and DeWitt, who may not immediately recognize the formulation just made.

A) First there is the "many-universe" concept given prominence by Everett and DeWitt. In the usual theory it is supposed that only one of the possible results of a measurement is actually realized on a given occasion, and the wave function is "reduced" accordingly. But Everett introduced the idea that *all* possible outcomes are realized every time, each in a different edition of the universe, which is therefore continually multiplying to accommodate all possible outcomes of every measurement. The psycho-physical parallelism is supposed such that our representatives in a given "branch" universe are aware only of what is going on in that branch. Now it seems to me that this multiplication of universes is extravagant, and serves no real purpose in the theory, and can simply be dropped without repercussions. So I see no reason to insist on this particular difference between the Everett theory and the pilot-wave theory—where, although the *wave* is never reduced, only *one* set of values of the variables x is

[kr] Everett makes no comments in the first four sections of Bell's paper, aside from a question mark in the margin of the introduction. There, Bell says, "[T]here are ... theories in which the linear Schroedinger equation is supposed to be exact, and in which no wave-packet reduction occurs. Of course, the wave function is then no longer held to be a complete description of physical reality. It is still necessary to recognize facts, and since these are not now incorporated into the wave function by reduction, some additional variables have to be invoked to represent them. This paper is an analysis of two theories of this kind, mainly due, respectively, to de Broglie 2) and to Everett 3),4) It seems to me that the close relationship of the Everett theory to that of de Broglie has not been appreciated...." Everett puts his question mark next to Bell's claim that the wave function is no longer held to be a complete description of physical reality. The remainder of Everett's comments appear in the fifth section of Bell's paper. These marginal comments are reproduced here, along with the fifth section of Bell's paper. The original manuscript is available online (Bell, 1971); a revised version was published as Bell (1987).

[ks] Everett writes in the margin: "? Why?". Bell clearly referred to the theory as "Everett (?)" since he did not know if he had correctly understood Everett. For his part, it is possible that Everett did not fully understand the implications of Bell's Everett (?) theory. See fn. 1b on pg. 288.

realized at any instant.[kt] Except that the wave is in configuration space, rather than ordinary three-space, the situation is the same as in Maxwell-Lorentz electron theory.[1] Nobody ever felt any discomfort because the field was supposed to exist and propagate even at points where there was no particle. To have multiplied universes, to realize all possible configurations of particles, would have seemed grotesque.

B) Then it could be said that the classical variables x do not appear in Everett and DeWitt. However, it is taken for granted there that meaningful reference can be made to experiments having yielded one result rather than another. So instrument readings, or the numbers on computer output, and things like that, are the classical variables of the theory. We have argued already against the appearance of such vague quantities at a fundamental level. There is always some ambiguity about an instrument reading; the pointer has some thickness and is subject to Brownian motion. The ink can smudge in computer output, and it is a matter of practical human judgement that one figure has been printed rather than another. These distinctions are unimportant in practice, but surely the theory should be more precise. It was for that reason that the hypothesis was made of fundamental variables x, from which instrument readings and so on can be constructed, so that only at the stage of this construction, of identifying what is of direct interest to gross creatures, does an inevitable and unimportant vagueness intrude. I suspect that Everett and DeWitt wrote as if instrument readings were fundamental only in order to be intelligible to specialists in quantum measurement theory.

C) Then there is the surprising contention of Everett and DeWitt that the theory "yields its own interpretation". The hard core of this seems to be the assertion that the probability interpretation emerges without being assumed. In so far as this is true it is true also in the pilot-wave theory. In that theory our unique world is supposed to evolve in deterministic fashion from some definite initial state.[ku] However, to identify which features are details critically dependent on the initial conditions (like whether the first scattering is up or down in an α particle track) and which features are more general (like the distribution of scattering angles over the track as a whole) it seems necessary to envisage a comparison class.[kv] This class we took to

---

[1] But the following difference of detail is notable. In the Maxwell-Lorentz electron theory particles and field interacted in a reciprocal way. In the pilot-wave theory the wave influences the particles but is not influenced by them. Finding this peculiar, de Broglie always regarded the pilot-wave theory as just a stepping-stone on the way toward a more serious theory which would be in appropriate circumstances experimentally distinct from ordinary quantum mechanics. [See Bell (1971) for his references.]

[kt] Everett writes in the margin: "Not consistent."
[ku] Everett writes in the margin: "Unique measure."
[kv] Everett writes in the margin: "Why x2."

be a hypothetical ensemble of initial configurations with distribution $|\psi|^2$. In the same way Everett has to attach weights to the different branches of his multiple universe, and in the same way does so in proportion to the norms of the relevant parts of the wave function.[kw] Everett and DeWitt seem to regard this choice as inevitable.[kx] I am unable to see why, although of course it is a perfectly reasonable choice with several nice properties.[ky]

D) Finally there is the question of trajectories, or of the association of a particular present with a particular past. Both Everett and DeWitt do indeed refer to the structure of the wave function as a "tree", and a given branch of a tree can be traced down in a unique way to the trunk. In such a picture the future of a given branch would be uncertain, or multiple, but the past would not.[kz] But, if I understand correctly, this tree-like structure is only meant to refer to a temporary and rough way of looking at things, during the period that the initially unfilled locations in a memory are progressively filled, labeling the different branches of the tree only by the macroscopic-type variables describing the contents of the locations. When a more fundamental description is adopted there is no reason to believe that the theory is more asymmetric in time than classical statistical mechanics. There also apparent irreversibility can arise (e.g., the increase of entropy) when coarse-grained variables are used. Moreover, DeWitt says "...every quantum transition taking place on every star, in every galaxy, in every remote corner of the universe is splitting our local world in myriads of copies of itself". Thus DeWitt seems to share our idea that the fundamental concepts of the theory should be meaningful on a microscopic level, and not only on some ill-defined macroscopic level. But at the microscopic level there is no such asymmetry in time as would be indicated by the existence of branching and non-existence of debranching. Thus the structure of the wave function is not fundamentally tree-like. It does not associate a particular branch at the present time with any particular branch in the past any more than with any particular branch in the future.[la] Moreover, it even

---

[kw] Everett writes in the margin: "But why."

[kx] Everett writes in the margin: "as also classical stat. mech!" See Everett's analogy with classical statistical mechanics (pgs. 125–26).

[ky] Everett writes in the margin: "re-read Proof!" Everett's proof of the uniqueness of the typicality measure, however, depends on a handful of background assumptions. It is the necessary acceptance of these background assumptions that Bell questions. See Everett's proofs (pgs. 189 and 123).

[kz] Everett writes in the margin: "Tree both ways!" This fits with Everett's view that pure wave mechanics is only *apparently* irreversible to observers treated within the theory. See for example pgs. 143–44 and 240.

[la] Everett writes in the margin: "Yes, either way." Everett also writes at the bottom of the page: "branching only relative to choice of basis—can make either way!" That branching is relative to choice of basis fits with the absence of any notion of a physically preferred basis in Everett's characterization of relative states. See for example pg. 180. See also the above note (fn. kz) for Everett on irreversibility.

seems reasonable to regard the coalescence of previously different branches, and the resulting interference phenomena, as the characteristic feature of quantum mechanics. In this respect an accurate picture, which does not have any tree-like character, is the "sum over all possible paths" of Feynman.

Thus in our interpretation of the Everett theory there is no association of the particular present with any particular past.[lb] And the essential claim is that this does not matter at all. For we have no access to the past. We have only our "memories" and "records".[lc] But these memories and records are in fact *present* phenomena. The instantaneous configuration of the x's can include clusters which are markings in notebooks, or in computer memories, or in human memories. These memories can be of the initial conditions in experiments, among other things, and of the results of those experiments. The theory should account for the present correlations between these present phenomena. And in this respect we have seen it to agree with ordinary quantum mechanics, in so far as the latter is unambiguous.

The question of making a Lorentz invariant theory on these lines raises intriguing questions.[2] Leaving these aside it would seem that the Everett theory provides a resting place for those who do not like the pilot-wave trajectories but who would regard the Schroedinger equation as exact. But a heavy price has to be paid. We would live in a present which had no particular past, nor indeed any particular (even if unpredictable) future. If such a theory were taken seriously it would hardly be to take anything else seriously.[ld] So much for the social implications.[le]

In conclusion it is perhaps interesting to recall another occasion when the presumed accuracy of a theory required that the existence of present historical records should not be taken to imply that any past had indeed occurred. The theory was that of the creation of the world in 4004 B.C. During the 18th century growing knowledge of the structure of the earth seemed to indicate a more lengthy evolution. But it was pointed out that God in 4004 B.C. would quite naturally have created a going concern. The trees would be created with annular rings, although the corresponding number of years had not elapsed. Adam and Eve would be fully grown,

---

[2] Would it be necessary to restrict attention to the here as well as the now? Point-sized reminiscers?

[lb] Everett writes in the margin: "still unique measure." Everett may not have understood how radical Bell's Everett(?) theory was. On Bell's view, observers would fail to have histories. See the discussion of Bell's Everett(?) theory in Barrett (1999).

[lc] Everett writes in the margin: "correct." He also makes a checkmark next to the word "records".

[ld] Everett writes: "(what difference from probabilistic?—also no unique past!)". The second part of this comment fits with Everett's view that branching works both toward the future and the past. See fn. kz on pg. 287.

[le] Everett writes in the margin: "ha".

with fully grown teeth and hair.[3] The rocks would be typical rocks, some occurring in strata and bearing fossils—of creatures that had never lived. Anything else would not have been reasonable:[4]

> Si le monde n'eut été à la fois jeune et vieux, le grand, le sérieux, le moral, disparaissaient de la nature, car ces sentiments tiennent par essence aux choses antiques... L'homme-roi naquit lui-même à trente années, afin de s'accorder par sa majesté avec les antiques grandeurs de son nouvel empire, de même que sa compagne compta sans doute seize prin-temps, qu'elle n'avait pourtant point vécu, pour être en harmonie avec les fleurs, les oiseaux, l'innocence, les amours, et toute la jeune partie de l'univers.[lf]

[3] They would also have navels, although they had not been born. See, for example, P.H. Gosse, Omphalas (1857).

[4] F. de Chateaubriand, "Genie du Christianisme" (1802).

[lf] Chateaubriand (1802, Vol I, pg. 162).

REFERENCES

. . .

9) In particular it is not clear to me that Everett and DeWitt conceive in the same way the division of the wave function into "branches." For DeWitt this division seems to be rather definite, involving a specific (although not very clearly specified) choice of variables (instrument readings) to have definite values in each branch. This choice is in no way dictated by the wave function itself (and it is only after it is made that the wave function becomes a complete description of DeWitt's physical reality). Everett on the other hand (at least in some passages) seems to insist on the significance of assigning an arbitrarily chosen state to an arbitrarily chosen subsystem and evaluating the "relative state" of the remainder.[lg] It is when arbitrary mathematical possibilities are given equal status in this way that it becomes obscure to me that any physical interpretation has either emerged from, or been imposed on, the mathematics.

[lg] Everett puts a check mark in the margin. The suggestion is that he agrees with the distinction that Bell draws here between Everett and DeWitt. That Everett does not require a preferred basis fits well with, for example, his description of the fundamental relativity of states (pg. 180) and with Everett's earlier marginal note (fn. la on pg. 287).

CHAPTER 22

Correspondence: Jammer, Wheeler, and Everett (1972)

In 1972–1973, while Max Jammer was researching his book The Philosophy of Quantum Mechanics *(1974), he corresponded with Everett, who revealed much about his motivation in writing the thesis and his current assessment of the theory.*

JAMMER TO WHEELER, JANUARY 11, 1972

Professor M. Jammer Ph. D.
  January 11, 1972

Professor John A. Wheeler
Department of Physics
Princeton University
Princeton, New Jersey

Dear Professor Wheeler:

I would be grateful to you if you could send me the address of Dr. Hugh Everett III who in March 1957 wrote his Ph. D. thesis at your institute.

Being interested in certain historical details (in connection with my forthcoming book on the historical development of interpretations of quantum mechanics) concerning the "Everett-Wheeler-etc" interpretation, but being unable to find Dr. Everett's address in any of the available directories, I thought to approach you in this matter.

May I also renew my request, on this occasion, to be on your mailing list for reprints of papers written by you or members of your Institute. In particular, I would appreciate receiving—in connection with the above—reprints of your paper[lh] in the TRANS. OF THE NEW YORK ACADEMY OF SCIENCES December 1971 ("From Mendeleev's Atom to the Collapsing Star")[li] and, of course, if available, of any other papers (more technical) on this subject.

With my best wishes for the New Year

Sincerely yours,
Max Jammer

[lh] In the original letter, "in THE SCIENCES (June-July 1971) and" appears here, but is crossed out in pen.

[li] Wheeler (1971).

WHEELER TO JAMMER, MARCH 19, 1972

John A. Wheeler
Department of Physics: Joseph Henry Laboratories
Jadwin Hall
Post Office Box 708
Princeton, New Jersey 08540

19 March 1972

Professor Max Jammer
Physics Department
Bar-Ilan University
Ramat-Gan, Israel

Dear Professor Jammer:

It was a pleasure to hear from you. I am a devotee of your books and would feel myself so fortunate if someday we could spend walking up and down and talking.—I have been very busy with the book of Misner, Thorne and myself, *Gravitation*, Freeman & Co., S. Francisco, 1973, hence my delay in replying. The latest address I have for Hugh Everett III is 607 Pelham St., Alexandria Virginia.[lj] —I am myself now, for myself, writing to University Microfilms, Ann Arbor, Michigan, for a "photo copy book" of Hugh Everett's thesis (long compared to printed article).[lk] —I encouraged him to spend, and he did spend, some weeks in Copenhagen later on, but Bohr did not take to his way of describing q. mechanics, as he also earlier had not accepted Feynman's way of describing q. mechanics.

I am asking Mrs. Witt to send to you[ll]. the Mendeleev paper and some other papers; I imagine you have the *Battelle Rencontres* or its German equivalent, *Einstein's Vision* (Springer) and Harrison, Thorne, Wakano, Wheeler, *Grav. Theory and Grav. Collapse* (U. of Chicago Press).

I send you every good wish!

Sincerely,

John Wheeler

[lj] Wheeler writes in the margin: "If this doesn't work, let me know and I'll try via phone, long distance."

[lk] Here Wheeler is misremembering the long version to have been Everett's proper thesis.

[ll] Wheeler writes in the margin: "separately."

JAMMER TO EVERETT, AUGUST 28, 1973

BAR-ILAN UNIVERSITY
Ramat-Gan, Israel
Department of Physics
Professor M. Jammer, Ph. D.
August 28, 1973

Dr. Hugh Everett III
8114 Touchstone Terrace
McLean, Virginia 22101
U. S. A.

Dear Dr. Hugh Everett,

Having finally obtained your address from Prof. Bryce DeWitt, I allow myself to approach you most respectfully with the following request.

I have just completed my historical study of the interpretations of quantum mechanics, a sequel volume of my book on THE CONCEPTUAL DEVELOPMENT OF QUANTUM MECHANICS (McGraw-Hill, New York, 1966, 1972).[lm] The only item on which I was so far unable to obtain any historical data concerns the so-called MANY-WORLDS INTERPRETATION, associated with your name. I have not yet seen the book edited by B. DeWitt and N. Graham which probably contains in its introduction some historical notes—but I am not sure at all whether it really does.[ln] In any case I would prefer your own statements, if I may, on the following questions:

1. Where did you study until your thesis under J. A. Wheeler (in 1957)? And by whom, in addition to Wheeler, were you intellectually influenced?
2. Did you ever (before 1957) study philosophy or psychology? If so, what subjects and with whom?
3. What were the first reactions to your 1957 paper (Relative state formulation)? Was it criticized at that time, by whom?
4. Was there any specific motive or reason that induced you to propose your interpretation and measurement theory?
5. Are there any other historical data of relevance to these questions?

I hope I am not too inquisitive or intrusive. Needless to say, I shall greatly appreciate your cooperation.

Sincerely yours,

Max Jammer

P. S. Would it be possible to get a preprint of your paper "The Theory of the Universal Wave Function"?

[lm] Jammer (1966) and Jammer (1974).
[ln] DeWitt and Graham (1973).

DBS
CORPORATION

September 19, 1973

Professor M. Jammer
Bar-Ilan University
Ramat-Gan, Israel

Dear Professor Jammer:

Thank you very much for your kind inquiry concerning my doctoral thesis. I shall attempt to answer your questions in the order that you presented them. I did my undergraduate study at the Catholic University of America in the field of chemical engineering. My graduate studies were in the field of mathematical physics at Princeton University and my thesis was done under John Wheeler.

With respect to my thesis, the principal intellectual influences were two other residents of the Princeton Graduate College at that time, Charles Misner (currently a professor at the University of Maryland specializing in gravitation theory) and Aage Petersen who was at that time Neils Bohr's assistant and spending a year at Princeton. The basic ideas for my interpretation arose out of discussions with these two fellow students which I subsequently presented to Professor Wheeler, who encouraged me to pursue the matter further as a thesis. During the course of this pursuit I would say that perhaps the primary influences were von Neumann's book and the later chapters of Bohm's Introduction to Quantum Mechanics.[lo]

In answer to question 2, the only formal course in philosophy or psychology that I had was an introduction to epistemology at Catholic University.

With respect to question 3 on the reactions to the 1957 paper (which was a brief condensation of the ideas contained in the full document), as far as I can recollect there was very little attention given to the paper at that time. The only strong criticism of that era that I recollect was contained in a private correspondence to Professor Wheeler, from an author

---

[lo] The first part of Everett's answer agrees with what he reported to Misner in conversation (pg. 300). Everett also clearly relied heavily on John von Neumann's *Mathematical Foundations of Quantum Mechanics* (von Neumann, 1955), which had just been translated and published by Princeton University Press. The influence of David Bohm's book *Quantum Theory* (Bohm, 1951) is more subtle, but also unmistakable. The discussions in Bohm's book and the Wigner Friend thought experiment arguably formed the foundation for how Everett understood the measurement problem. See pgs. 176 and 73–75 for Everett's presentation of the problem.

who shall remain unnamed, to the effect that the implied splitting of the state of an observer by my theory could not possibly be true because I (the critic) am unaware of any such splitting effect.[lp] This criticism led to the insertion of a footnote in my published paper drawing the parallel with the manifest absurdity of the Copernican system of planetary motions because I, as an observer, do not feel any such motion.[lq] The point being, of course, that if the theory itself predicts one's sensory perceptions to be what they are in fact, the weight of this criticism is somewhat diminished. The unwillingness of most physicists to accept this theory, I believe, is therefore due to the psychological distaste which the theory engenders overwhelming the inherent simplicity of the theory as a way of resolving the apparent paradoxes of quantum mechanics as conventionally conceived. Thus, the theory was not so much criticized, as far as I am aware, but simply dismissed.

Subsequent to the publication of the paper, I had informal discussions with a number of physicists concerning the subject (including Bohr and Rosenfeld in Copenhagen, in 1959, Podolsky and Wigner and a number of others active in the field at a conference at Xavier University several years later).[lr] I was somewhat surprised, and a little amused, that none of these physicists had grasped one of what I considered to be the major accomplishment of the theory—the "rigorous" deduction of the probability interpretation of quantum mechanics from wave mechanics alone.[ls] This deduction is just as "rigorous" as any of the deductions of classical statistical mechanics, since in both areas the deductions can be shown to depend upon an 'a priori' choice of a measure on the space.[lt] In classical statistical mechanics this measure is standard Lebesgue measure on the phase space whereas in quantum mechanics this measure is the square of the amplitude of the coefficients of an orthonormal expansion of a wave function.

What is unique about the choice of measure and why it is forced upon one is that in both cases it is the only measure that satisfies a law of conservation of probability through the equations of motion. Thus, logically in both classical statistical mechanics and in quantum mechanics, the only possible

---

[lp] The author in question was DeWitt. See the discussion surrounding pg. 250.

[lq] Here Everett credits his exchange with DeWitt (pg. 254) for the note in proof (pg. 189).

[lr] See pgs. 268–69.

[ls] Indicating its importance to him, Everett tried twice to make this point at the Xavier conference. See pgs. 273–75.

[lt] For Everett's analogy with statistical mechanics see pgs. 125–26. Note that Everett takes the deduction of the measure of typicality to constitute the rigorous deduction of the probability interpretation of quantum mechanics. Specifically, he does not claim that he has deduced a metaphysical interpretation of pure wave mechanics from the theory alone, as Bell seems to have thought (pgs. 286 and 290). That said, Everett also required background assumptions for the deduction of the uniqueness of his measure of typicality and hence did not get even this from pure wave mechanics alone. See pgs. 187–89 and 122–23.

statistical statements depend upon the existence of a unique measure which obeys this conservation principle.

One of my major points was, therefore, that the probability interpretation of quantum mechanics—that somehow the measuring process was "magic" and subject to a separate axiom governing the collapse of the wave function—did not have the status of an independent postulate at all, but was in fact a deduction from pure wave mechanics alone. That this point was essentially completely overlooked at that time I can now only ascribe to my failure in writing the paper. This point occurred rather far into the paper at a point where I now realize most readers would have stopped reading.

To question 4, were there any specific motives or reasons that induced me to propose my interpretation and measurement theory—I must answer in all candor the primary motive was, of course, to obtain a thesis. However, I must also admit to a strong secondary motive to resolve what appeared to me to be inherent inconsistencies in the conventional interpretation. I was of course struck, as many before and also many since, by the apparent paradoxes raised by the unique role assumed by the measurement process in quantum mechanics as it was conventionally espoused. It seemed to me unnatural that there should be a "magic" process in which something quite drastic occurred (collapse of the wave function), while in all other times systems were assumed to obey perfectly natural continuous laws.

I thought at that time that perhaps the pursuit of this apparent difficulty would lead to a new and different theory which, while resolving the apparent paradoxes, would also lead to entirely new predictions. Unfortunately, as it turned out, the theory which I constructed resolved all the paradoxes and at the same time showed the complete equivalence with respect to any possible experimental test of my theory and that of the conventional quantum mechanics.[lu] The net result of my theory therefore is simply to give a complete and self-consistent picture (without any particular "magic" associated with measurement) that in all practical predictions will of course be identical to the predictions of the conventional formulation.

To me, therefore, the real usefulness of this picture or theory of quantum mechanics is simply as an alternative which could be acceptable to those who sense the paradoxes in the conventional formulation, and therefore save much time and effort by those who are also disturbed by the apparent inconsistencies of the conventional model. As you know, there have been a large number of attempts to construct different forms of quantum mechanics to overcome these same apparent paradoxes. To me, these other attempts appear highly tortured and unnatural. I believe that my theory

---

[lu] This is not quite true. The collapse dynamics and pure wave mechanics make different empirical predictions precisely in the context of the Wigner's Friend experiments that Everett used to motivate the measurement problem. In particular the collapse theory would not give the interference predictions that Everett expected from the pure state of the friend. See pgs. 176 and 73–75.

is by far the simplest way out of the dilemma, since it results from what is inherently a simplification of the conventional picture, which arises by dropping one of the basic postulates—the postulate of the discontinuous probabilistic jump in state during the process of measurement—from the remaining very simple theory, only to recover again this very same picture as a deduction of what will *appear* to be the case to observers.[lv] I therefore believe that my formulation is by far the simplest from an axiomatic point of view. The acceptability, however, clearly is a matter of personal taste.[lw]

With respect to question 5, I cannot think of any other historical data of relevance at this time. My own history subsequent to leaving Princeton University in the summer of 1956, for what it may be worth, is as follows:

I entered the field of operations research and was employed by the Weapons Systems Evaluation Group in the Pentagon studying problems of strategic nuclear warfare until the summer of 1964. At that time, I and several of my colleagues formed a company—Lambda Corporation—to apply operations research and computer-modeling techniques to problems facing the United States government and also commercial enterprises. Most recently, in July of this year, I and another colleague left Lambda Corporation to form yet another enterprise, DBS Corporation—devoted to supplying data processing services to managers and policy makers in the federal government and industry.[lx] My colleague in this enterprise, Donald Reisler, is also an ex-physicist, and indeed one who worked in the foundations of quantum mechanics. Only after working together at Lambda Corporation for some time did we discover that we had both done our theses in the field of foundations of quantum mechanics. His thesis is entitled "Einstein, Podolsky, Rosen Paradox: 'Can the Quantum Mechanical Description of Physical Reality be Considered Complete?'" Since our mutual discovery, however, realizing the pitfalls that might divert our energies, we have entered into a mutual pact not to read one another's thesis for ten years. Accordingly, we have placed a copy of each of our theses in the DBS Corporate files, to be exhumed ten years hence and read, when presumably we could afford the luxury of such a diversion.

With respect to your postscript on the possibility of getting a preprint of my paper (presumably the full document rather than the published

---

[lv] Note that the characterization of pure wave mechanics is simply the standard collapse theory without the collapse postulate.

[lw] The virtue that Everett constantly returns to is simplicity. The claim is that he gets all the empirical predictions of the standard theory with a simpler, self-consistent theory. Acceptability of pure wave mechanics is largely a matter of taste because criteria that go beyond logical consistency and empirical adequacy are at most pragmatic. See pgs. 168–72 for Everett's extended discussion of theoretical virtues.

[lx] Hence the DBS logo at the head of the letter.

paper) I am sorry to answer that the only copy in my possession is the aforementioned, on file with DBS Corporation. The only other copy that I am aware of is the one I sent to Professor Bryce DeWitt. I would suggest, therefore, that you inquire whether he has perhaps copied that document.

Sincerely,

Hugh Everett III
Chairman
DBS Corporation

CHAPTER 23

Transcript: Everett and Misner (1977)

As the many-worlds theory gained in popularity, Wheeler gradually disassociated himself from it, eventually disavowing it. In 1977, Wheeler and DeWitt invited Everett to give a talk on his theory at the University of Texas in Austin. Dressed in his customary black suit, he chain smoked while pacing and speaking in a rapid fire staccato. According to the physicist David Deutsch, who was there, Everett spoke of the many worlds as if they were physically, ontologically real, while dismissing the notion of a preferred basis problem (Byrne, 2010, pgs. 321–333).

A few weeks after the Austin talk, Charles and Suzanne Misner spent an evening at the Everett's house celebrating the Misner's 18th wedding anniversary. Suzanne encouraged the men to make a tape recording about how Everett's theory came about during their Princeton years. It is the only known recording of Everett.

MAY, 1977, CHARLES MISNER INTERVIEW OF HUGH EVERETT, III (46 YRS.)

Charles Misner: Well, it's been a great evening. I enjoyed spending our 18th anniversary with you, Hugh. And why don't you lead on after your drink by telling us how you got started with John Wheeler.[ly]

Hugh Everett: Well, I got started—,

Charles: *(interrupting Hugh)* What ever made you decide to pick him for a thesis advisor?

Hugh: Well, actually, I didn't. Somebody assigned him to me, I think, and I needed one to get my fellowship renewed.

Charles: Right, so it was all bureaucracy....

Hugh: That's correct. My first year was actually spent in the Mathematics Department.

Charles: When did you get on to, ah, weird Quantum Mechanics?

[ly] Both Everett's and Misner's voices are notably slurred in the recording, presumably by dint of the festivities.

Hugh: Oh, it was because of you and Aage Petersen, one night at the Graduate College after a slosh or two of sherry, as you might recall.[lz] You and Aage were starting to say some ridiculous things about the implications of Quantum Mechanics and I was having a little fun joshing you and telling you some of the outrageous implications of what you said, and as we had a little more sherry and got a little farther in the conversation... Don't you remember, Charlie? You were there!

Charles: I don't remember that evening actually, but I do remember that Aage Petersen was around—that's entirely possible.

Hugh: You had too much sherry, that's the problem.

Charles: Was that after I was starting on the Rainich Theory?[ma] Which we didn't know was the Rainich Theory at the time because ...

Hugh: Oh, I remember that... Yes, that was that same ...

Charles: You read his book and told us, "Look, this was a great idea—why did he stop there instead of going on to finish the job?"

Hugh: I do remember that, yes. Whatever happened to that, I wonder?

Charles: Well, we eventually decided to go on and finish the job.

Hugh: How'd it come out?

Charles: Well, only many years later we discovered—well, not so many years later, but actually Peter Bergmann wrote us a letter and said, "Look, it's not in his book, but actually in some other publications, he did all this stuff that you were hoping to have for your thesis, Misner."[mb]

Hugh: *(laughs)* Why didn't he publish it?

Charles: He did, but in some obscure place.

Hugh: I see.

Charles: So I had to do something else for my thesis instead of Rainich Theory.

Hugh: I never knew that.

Charles: Yeah, I spent all those years on that and eventually had to go back and turn out something on Feynman quantization.

Hugh: I had to study for generals—that's what happened to me—I had to study for generals, so I couldn't follow any of that up.

Charles: Hmm, didn't you take generals when I did?

Hugh: No, I took them a year later.

Charles: You did; I see.

Hugh: I think, let's see. Yeah, yeah, right. It was the Army–McCarthy hearings as I recollect... I did not that year. I remember, I was freely watching them and all, but you had to occasionally study and couldn't watch all of the hearings...

[lz] Aage Petersen was Bohr's assistant from 1952 to 1962 at the University of Copenhagen. He had visited Princeton with Bohr in 1954 and had befriended both Everett and Misner.

[ma] This was work that turned out to be closely tied to the earlier work of the mathematical physicist George Yuri Rainich (March 25, 1886–October 10, 1968).

[mb] The story is that the physicist Peter Gabriel Bergmann (1915–2002) told Misner about Rainich's earlier work that predated Misner's thesis.

Charles: Oh don't think they ever had any in the morning when I went to class. *(pauses)* I must have gone at least once a week to Bergmann's class.[mc]
(Hugh and Charles laugh)

Hugh: Once a week! All right, yeah. Oh, goodness.

Charles: Now that just shows you that Notre Dame is so much better than Catholic U. I had all I needed to pass generals from Notre Dame, and you still had to study.

Hugh: That's true, I had to I guess, or at least I thought I had to.

(Hugh and Charles laugh)

Suzanne: You haven't talked about Wheeler in here yet.

Hugh: Yeah, how did Wheeler get into this? How did you pick Wheeler?

Charles: How did I pick Wheeler? Well, I picked Wheeler because after spending the Army–McCarthy hearings studying for generals, or the other way around, I decided I should—because they told me I should—find a thesis advisor. And I had been working with Wightman.[md]

Hugh: Mhmm.

Charles: So I first went to Wightman and he unfortunately was an honest man and said "I'd be delighted to have you for a thesis student; I have a lot of interesting things to do, but you probably should take notice that of my last three students, none of them has got their Ph.D. in less than seven years."

Hugh: You know, Charlie, much the same happened to me. Now I forget who gave me advice like that but somebody said something much the same as that—"You can get out quicker with Wheeler." *(laughs)* Let's see, who was that, I wonder, hmm.

Charles: So anyway, when I went around to talk to John Wheeler he had all kinds of wonderful ideas, all of them nebulous, going off into the mysteries of the universe, and anything I wanted to do was just so exciting that he couldn't possibly restrain himself. And it seemed that everything was possible and I might as well do it quickly and get finished and solve the problems of the universe easily at the same time!

Hugh: What year was Aage there?

Charles: I don't remember when Aage was there.

Hugh: Was that our first year? I believe it was.

Charles: My first year was '53–'54.

Hugh: Yeah, remember we were both there in the same year.

Charles: That's right. We were roommates together.

Hugh: That's right.

Charles: We had for some reason . . .

Hugh: Well, that was later, I think. Let's see, at first . . .

[mc] Peter Bergmann served as Einstein's research assistant at the Institute for Advanced Study.
[md] Arthur Wightman (b1922) was a mathematical physicist. John Wheeler served as his thesis advisor some years before Everett began his work at Princeton.

Charles: We had this luxurious room, remember, a whole suite, two bedrooms, a bathroom and a little sitting room in the Graduate College—I had never lived in such luxury before in my life. We were down on the first floor in the Graduate College someplace.

Hugh: That was you and Harvey.[me]

Charles: Oh, was that right?

Hugh: Yeah.

Charles: Oh, that's right.

Hugh: It was me and Hale.[mf]

(Hugh and Charles laugh)

Charles: So the four of us all got mixed together...

Hugh: Right, that was in the third year.[mg]

Charles: So then we had...Okay, yeah you were...

Hugh: I'm trying to pin down when Aage was there, because that's ...Oh boy. Hmm...

Charles: Well, there's all these...trying to bring back ancient days.

Hugh: I know, I know, but we could probably pin it down. Well, anyway, the whole business started with those discussions, and my impression is, I went to Wheeler then later and said "Hey, how about this? This is the thing to do."

Charles: You already had some game theory going, is that right? I think you had some publications coming up...

Hugh: Oh, yeah. I had to do that to get my NSF renewed. I had no hope in the Physics Dept. in my first year because nobody knew me, so I did something in mathematics...well, I got it renewed.

Charles: Yeah.

Hugh: Well, hmm. Gosh!

Charles: I don't know how much I did with Wheeler the first year. I can't remember that at all, but I must have seen him occasionally—I don't know whether I even took the Relativity course...I know the year before I came he taught a Relativity course and he took all his students, including Art Komar—

Hugh: *(interrupts Charles)* Yeah, I heard about that.

Charles: —to this famous visit to Einstein and I missed that and I always thought that was one of the great catastrophes of my life. But I suppose Wheeler must have taught Relativity again. Did you have Relativity from him the first year you were there?

Hugh: Um...

Charles: He taught it that year; I suppose he taught it the next year, but I'm not sure.

[me] In her rough typescript of the taped conversation Nancy Everett reports this is Harvey Arnold.

[mf] Hale Trotter.

[mg] That is 1955–1956.

Hugh: I think it was all about geons or something like that.[mh]

Charles: Oh, of course it was, yeah. We certainly heard about seminars about geons and whatnot.

Hugh: And wormholes, we certainly heard about wormholes.[mi]

Charles: Yes, and I was learning all kinds of mathematics from Hale Trotter.[mj] 'Cause we always went to Mathematics Tea—well, there only was one...

Hugh: There only was one Tea! *(laughs)*

Charles: There only was one Tea, everyone went there, and that's where I learned all kinds of fancy mathematics. Of course I had a big start at Notre Dame from Arnold Ross who was the mathematics chairman there. He had me through half of Bourbaki and things like that while I was an undergraduate.[mk] But I remember Hale and various mathematicians were helping me with algebraic topology and this, that, and the other thing.

Hugh: Well, I remember the day when we thought that elementary particles would be obviously the way different knots would be knotted in multiply connected space and we went over there and said all we've got to know is the classification of knots and we'll have the answer—they weren't able to help us. Remember that?

Charles: I don't remember that, but I do remember lots of discussions about knots...

Hugh: Because Hale got into that.

Charles: Wasn't Milnor solving some knot problems as an undergraduate shortly before or after that—various people were full of knot theory.[ml]

Hugh: Yes, whatever happened to knot theory?

Charles: I think all the problems got solved.

Hugh: Oh. I thought they were undecidable or something.

Charles: Oh well, that's a solution; you should know with your...

Hugh: Oh yes, right.

Charles: The word problem got solved while I was there.

Hugh: My problem was solved?

Charles: Yes, that's right.

Hugh: There was also this Russian who...I think it was parallel solutions or something like that. Goodness. I am trying to remember the exact history of that and I can't. All I knew is that it really stemmed from that night with the really fun discussions with Aage—flushing out the paradox.

[mh] Geons were introduced by Wheeler as spatially confined gravitational waves.

[mi] Wormholes are postulated connections between otherwise distant space-time regions. Such structures are permitted by the field equations of general relativity, and were a particular interest of Wheeler.

[mj] Hale Trotter is emeritus professor of mathematics at Princeton University.

[mk] Nicolas Bourbaki is the pseudonym of a group of mathematicians who aimed to produce a fully rigorous development of mathematics from first principles. Their project has extended to many volumes.

[ml] John Milnor (born 1931) won the Fields Medal for his work in differential geometry.

Charles: I don't really know which year he was there.

Hugh: I believe it was our first year. Though I'm not sure of that—could have been the second.

Charles: When did Einstein give his seminar; that was about our third year there? No....

Hugh: I don't recollect.

Charles: Because that was certainly an occasion—remember that—when Wheeler arranged that Einstein give a seminar at the University and I don't think he'd given one before...

Hugh: I don't think I attended.

Charles: Sure you did. I remember you were there.

Hugh: Really, what did he say? I don't remember it.

Charles: Well, maybe that's true—maybe I had to take notes and try to tell you—no, I would have thought you were there.

Hugh: I sure don't remember it, and I think I would have. *(laughs)*

(Change of reel)

Hugh: I know nothing about this whatsoever! I'm outraged! *(laughs)* Where was I? Was it on a weekend?

Charles: I can't remember. I doubt it. I think it was a Tuesday night or something like that.

Hugh: Maybe I was in Trenton. I don't know. Whatever that one was, I don't know—I don't remember it.

Charles: I doubt I have any notes of it. I do remember it... maybe I was actually trying to remember a few of the things he said to carry them on to you because he made some remarks about interpretations of quantum mechanics—he wasn't quite as determined about classical interpretations as he had been in the famous debates and things like that. You know, he admitted that quantum mechanics was certainly correct in all its predictions...

Hugh: But somehow unclean.

Charles: Yeah, but he was shaken from the kind of uncleanness he worried about in his debates with Bohr and he felt that he had a basic solidity; anything that he did to it would have to recognize that some of the crazy things it said were verifiable.

Hugh: Hmm. Right. Somewhat restricts your possibilities of getting out. *(laughs)* Well, I don't know.

Charles: Which year was it that I went to Leiden with Wheeler? Was that our second year there? It probably was. I would think that's right.

Hugh: I don't remember—oh, I do remember you going to Leiden.

Charles: It was a spring term and Joe Weber, Peter Putnam, and I all tagged along with John Wheeler for a semester when he was a visiting professor—[Lorentz] Professor at Leiden.

Hugh: Well, hmmm. What was I doing during that time?

Charles: Maybe you were studying for generals.

Hugh: Maybe he wasn't even my professor then.

Charles: That's possible.

Hugh:: Who was? I had somebody else in my first year. For the life of me I can't remember how the hell I got connected with Wheeler. Somebody advised me that I ought to get connected to him!

Charles: You probably already had these quantum mechanical ideas and just needed someone to talk to about them and he was obviously the kind of person who...

Hugh: I have that slight impression, but I can't be sure.

Charles: We may have started—no—in Leiden I was already working on these Rainich ideas which you had proposed. You found Rainich's book, and he said that algebraically the Maxwell tensor can be expressed in terms of the....

Hugh: Okay, that was indeed in our second year. I think—I have that feeling.

Charles: That's probably true.

Hugh: I remember that Christmas vacation that I was home I played around with that and then I had to drop it I felt because I really had to work on the generals for that spring. So that was the second year.

Charles: You talked to me about it and so I began taking it up with Wheeler and we followed it out at Leiden, which would have been my second year, spring '54–55.

Hugh: Okay, okay, so maybe it was that I wasn't doing anything...oh, all right, so maybe this whole business started in the second year, hmm.

Charles: So that would be...

Hugh: Because I think that pre-dated me—quantum mechanics—uh, I think.

Charles: Fall of '55, maybe, when I was back...uhm...

Hugh: Uh, noooo, no, no no—because I had the whole thing written by the fall of '55. That whole document was done...

Charles: Quantum Mechanics?

Hugh: Yeah.

Charles: At the end of your second year?

Hugh: Yeah. As a matter of fact, it was somewhat embarrassing because I had most of it written...the real embarrassment was that Wheeler at one point was threatening to get me my Ph.D. before the third year had run out, and as you remember, the draft was still in force in those days and the last thing in the world I wanted to do was do that. No, that thing was written in the winter of '55 or at least the first draft of it was. No, it was the third year, '55–'56 was the third year. He was threatening....

End of one side tape.

Hugh: Yeah, anyway, he was threatening to get me out in mid-term of the third year, and that did not coincide at all with my plans, especially given the draft situation at that time. So that was in the third year...

Charles: So that was...your involvement in the spring of '56.

Hugh: Uh, somewhere in the second year I got involved with him; I had no involvement whatsoever in my first year.

Charles: And half of the second year he was in Leiden. It was the beginning part of the year we must have been talking with him because...

Hugh: Maybe it was... it very well... in fact it probably was because I did nothing but study for generals and things like that in the second half of that year—

Charles: ...First half of that year you were probably talking to Wheeler.

Hugh: Well, maybe so.

Charles: Because I certainly was talking to him a bit about the Rainich stuff, although I—

Hugh: O.K., so when did the quantum thing... because I must have talked to him some, and, I certainly, you know, that summer is when I wrote really— was writing that first thing and Nance was typing it and everything else and the whole thing was ready by that autumn and that winter, which is what caused the threat!

Charles: Did we spend only one year in the Graduate College, then?

Hugh: We spent two years in the Graduate College. It was in our third year that we were on Linden Lane.

Charles: But then, I have to get that straight, because it was certainly that year in Linden Lane that I started taking Dutch lessons with Renée Franken, the Dutch girl, remember.

Hugh: Yes, so maybe you didn't go to Leiden til then!

Charles: That's right. It must not have been.

Hugh: Aha! That makes a lot more sense.

Charles: Okay. So that's how it was.

Hugh: All right, so it happened in the second year and the only question is where and what. That must have been when Aage was there.

Charles: Did we room together in the second year. Is that possible?

Hugh: Ahh, no. The first year I roomed with an Englishman and...

Charles: I was with Harvey Arnold, down on that first floor by the passageway.

Hugh: Maybe you were in the second year too. In the second year I roomed with Hale. And in the third year all four of us roomed together in Linden Lane. *(pauses)* All right, so what exactly happened in that second year? I guess we conclude that that's when Aage was there.

Charles: Yeah, must've been. One of those summers, probably Wigner had arranged for me to get a job at Bell Labs.

Hugh: Yeah.

Charles: One summer I had a job at Bell Labs. Whether that was the first summer or the second, I'm not sure.

Hugh: All right, the Rainich business was the second year.

Charles: Probably got straight the second year.

Hugh: I remember working one Christmas at home, actually calculating Christ-awful symbols [i.e., Christoffel symbols]. *(laughs)*

Charles: I would allow that the second year for me to have had a course from John on relativity.

Hugh: Okay. And I don't think I had that course, that year or...

Charles: —or maybe I never took it actually, because I had a course in relativity at Notre Dame, so I may have thought I didn't have to take one at Maryland. I might have just joined his lunch group, you know, and talked with people in his office. He used to have games with people in his office for brown bag lunches or for discussions or something. I might have joined that group without partaking, taking part in the...

Hugh: I was very much out of that group, I remember...that's probably why I didn't get invited to the Einstein thing. *(laughs)* Well, it went something like this: We'd had those discussions with Aage and all that, and somehow, and I don't yet know, I got connected with Wheeler or whatnot. The question is of the thesis topic, and you know, what about looking into this mess here, you know, this obvious inconsistency in the theory or whatever I thought of it then...

Charles: It's strange that he would be so interested in it—all in all, because it certainly went against the normal tenets of his great master, Bohr.

Hugh: Well, he still feels that way a little bit, even as recently as last month in Austin he was a little bit feeling that way...

Charles: How'd he work that out? Because he certainly had great respect for Bohr.

Hugh: Oh yes, indeed.

Charles: But he was always intrigued by an imaginative idea—he tried to get you and Bohr to agree.

Hugh: Yes, I spent six weeks—as you know, you were there—and that was a hell of a...doomed from the beginning...

Charles: You sat trying to talk to Bohr about...

Hugh: ...well he refought the battles of 1925 or something.

Charles: You didn't get the chance to say something, and then he would relight his pipe seventeen times—that was marvelous that you could even make that attempt—I don't know how old he was at the time—sixty or something?

Hugh: Well, you know what really came out of the trip to Copenhagen was the invention of the LaGrange multiplier.

Charles: For what—

Hugh: No reason—it's not relevant to physics why it was on that record...It was the truly great accomplishment of that Copenhagen trip, though; it made Lambda Corporation and several other things.

Charles: Which year was that Copenhagen trip?

Hugh: That was 1959, six weeks.

Charles: That was '59; that was the year we married.

Hugh: Because we went over on one of the earliest Pan Am jets. *(pauses)* Oh, my goodness.

Charles: Well, Wheeler certainly kept at it once he got into it.

Hugh: Yeah.

Charles: I'm sure he would have been happy if you had converted his professors to your viewpoint, or at least not have two people he's usually in favor of disagreeing...

Hugh: Yes, well you should have been in Austin a month ago. This was on human consciousness—whether computers are conscious and so on. [...] work is quite relevant to this—Wheeler actually read it before the assembled multitudes and all that. Then at your place last week, he confessed, he actually now believes it, except on Tuesdays, once a month—he said he feels he really has to reserve one day a week to disbelieve in it, and so he...

Charles: I see, well, that's a good perception. Most things that he's in doubt about he has a Monday, Wednesday, Friday ....

Hugh: Ohhhhhhhhh! Okay. Yes, right. Yes, he believes in being very conservative, I think, for somebody with such far-out ideas as Wheeler has, to call himself a conservative.

Charles: Oh, well, at first, Dynamic Conservativism—wasn't that his battle cry when he was trying to sell geons originally? He said, "The only thing revolutionary about this is that I want to believe the equations and see what their solutions are."

Hugh: Yes, that's right...

Charles: "What are you going to do about it?"

Hugh: That's right. Yes, marvelous concept. Well, that's what I did in Quantum Mechanics, too—perhaps that's where it came from, that same idea. Let's just believe the basic equations—what's this extra jazz for? So you do get a weird and funny picture!

Charles: That's certainly what he was doing with his—well, I recall—Kugelblitzes the year before I came there, maybe my first year there.[mm]

Hugh: Oh, yes.

Charles: He gave a presidential address to the Physical Society about Kugelblitzes, but then they eventually became geons, and that even got threatened shortly thereafter when he got a letter from General Electric Company saying that the "Geon" was their trademark name for Freon.

(both laugh)

Charles: And if he wished to use the word "geon" would he please acknowledge their copyright.

Hugh: Back to Kugelblitzes.

Charles: I guess that was somehow resolved without extensive payments from the Friends of Elementary Particle Physics. Did you ever benefit from the Friends of Elementary Particle Physics?

Hugh: Not knowingly. What....

Charles: Well, they were a Princeton benefactor...

[mm] See Wheeler (1955) for an early account of Wheeler's Kugelblitz (ball-lightning) idea. Misner reports that Wheeler later renamed these geons.

Hugh: A subversive Mafia-like organization?

Charles: I don't know, various people who always seemed to be students of Wheeler's. I must have gotten sent to a Physical Society meeting from Leiden to Utrecht, or I don't know what...

Hugh: I never went anywhere!

Charles: Tullio Regge, I'm pretty sure, came up from Italy to visit Wheeler in Leiden on their generosity, and I've heard various other people who benefited from the Friends of Elementary Particle Physics, which seemed to be an account at Princeton that Wheeler... [mn]

Hugh: Now you tell me! I never got any of this largesse, that I know of.

Charles: Maybe he figured your father was too rich.

Hugh: My father was an Army colonel, not rich!

Charles: But the suspicion gradually dawned on most graduate students that the Friends consisted primarily of John.

Hugh: Oh, really?

Charles: Yeah.

Hugh: Ah, huh! Gee whiz. Well, I didn't know any about that.

Charles: He was preaching this idea that you ought to just look at the equations and obey the fundamentals of physics while you follow their conclusions and give them a serious hearing. He was doing that on these solutions of Einstein's equations like wormholes and geons.

Hugh: I've got to admit that; that is right, and might very well have been totally instrumental in what happened.

Charles: ...encouraging you to follow up your own ideas on this...

Hugh: obviously ridiculous! *(laughs)*

Charles: Obviously ridiculous argument that was only intended to shut up obstreperous friends like me and old Bohr and Aage Petersen.

Hugh: Yeah, right.

Charles: I can't think I ever had very many serious thoughts about basic quantum mechanics, but of course Petersen...

Hugh: Yeah, none of us really took it seriously at the time, *(laughs)* except for Aage, of course.

Charles: Oh, actually I went through a very strange experience. I don't know whether you went through it, but I certainly did, as an undergraduate getting taught by people who had learned quantum mechanics in the thirties. And to them, quantum mechanics was really a big philosophical change, and they were shocked by the whole idea and so forth. And somehow we were...

Hugh: ...didn't seem all that funny!

Charles: ...and felt that well, you know, every new course in physics you get some new kind of nonsense which seems to make sense a little bit later, so quantum

---

[mn] Tullio Regge (born 1931) is an Italian theoretical physicist.

mechanics is no worse than electromagnetic fields, or $F = ma$, or whatever it might be.

Hugh: Yeah, electromagnetism is still mysterious. *(both laugh)* Well, goodness, I don't know.[mo]

Charles: Well, how did John pick you up? I mean, you had these ideas that stirred up in general conversation; and you began to work out more of the detailed mathematics of it. Where did John begin to play a role in the whole affair? Did you have a whole thesis written before you talked to him?

Hugh: No, no. [That's the part I can't quite bring back. I mean, we did have a conversation or two, and uh...][mp] Yeah, I guess...no, uh, he did indeed play a role; it's coming back now. You know, I said, "There's obviously something wrong here." I showed the paradoxes and whatnot, that something should be done to change it. He did keep saying "Why?" you know, his ultimate conservatism, as he put it, kept coming through, you know, "Why, why?"

Charles: You mean he kept saying, "Well, maybe it's right?"

Hugh: Well, he kept doing funny little experiments with balls running down inclined slopes or something, I never quite got the gist of what they were, but... I can't put my finger on it. On where or when...

Charles: Now for me his main influence—at least a very important influence—was to always point out possible opposite solutions that were so far fetched that you felt you had to put Johnny on the straight and narrow track and sort of... *(Hugh laughs)* catch...and show how physics was really done!

Hugh: Yes.

Charles: Because he was always off in the wild blue yonder...that was stretching your imagination and not allowing you to ever do any mundane problems. But, you evidently might have had a chance of being one ahead of him on this kind of stuff. Did he have the same attitude towards you? I mean, did he try to make you think more exotically, or was he helping you focus down towards the real test of the questions?

Hugh: Well, I don't know what to say about that. In any interaction both sides focus and do...very, very hard [to untangle.
We're going to run out...][mq]

[mo] From this point forward drumming can be heard in the background of the recording. At the end of Nancy's typescript she remarks that "Mark's drum in the background must have been distracting."

[mp] This sentence is in Nancy Everett's rough transcription but is no longer a part of the audio record.

[mq] This bracketed section appears in Nancy Everett's transcription but is not a part of the current audio record.

CHAPTER 24

Correspondence: Everett and Lévy-Leblond (1977)

In 1976, a conference called "Fifty Years of Quantum Mechanics" was held at the University of Louis Pasteur, Strasbourg. It was attended by more than a dozen physicists, including a young French scientist, Jean-Marc Lévy-Leblond. Wheeler's presentation was called, "Include the Observer in the Wave Function?" Wheeler had a history of vacillating on Everett's theory, but now, 20 years after he had endorsed it in Reviews of Modern Physics, *he rejected Everett's inclusion of the observer in the wave function, essentially upholding the Copenhagen interpretation.*

In 1977, Jean-Marc Lévy-Leblond wrote a letter to Everett, including a copy of a paper he had presented at the conference that compared DeWitt's and Everett's views. Everett responded by agreeing with Lévy-Leblond's analysis of the differences between the two views. Everett especially liked the short spoof of foundational writing in physics (on "zipperdynamics") that Lévy-LeBlond had tacked onto the end of his paper.

August 17, 1978[mr]

H. Everett III
8114 Touchstone Terrace
McLean, Virginia 22101
USA

Dear Dr. Everett,

I obtained your address through the kindness of Prof. Wheeler, who suggested that I directly ask your opinion on what I believe to be a crucial question concerning the "Everett and no-longer-Wheeler" (if I understood correctly!) interpretation of Qu. Mech.[ms]

[mr] The date on Lévy-Leblond's letter is August 17, 1978, but the correct date, as confirmed by Lévy-Leblond, is August 17, 1977.

[ms] As suggested by Lévy-Leblond's comment, Wheeler no longer endorsed Everett's theory.

The question is one of terminology: to my opinion there is but a single (quantum) world, with its universal wave function. There are not "many worlds," no "branching," etc., except as an artefact due to insisting once more on a *classical* picture of the world. This idea is developed in greater details in pp. 184–185 of the enclosed preprint.[mt] Your comments on this point, and other ideas in this paper, would be much appreciated, as well as any recent work of yours on these questions.

Sincerely yours.

Jean-Marc Lévy-Leblond

[mt] The preprint paper was Lévy-Leblond (1976). It concerns the conceptual structure of the quantum measurement problem, the Copenhagen interpretation, and Everett's pure wave mechanics. Lévy-Leblond takes Everett's position that the central problem with the Copenhagen interpretation is that it relies too much on classical intuition and thus sacrifices a quantum understanding of the measurement process. He argues that DeWitt's splitting-worlds interpretation of Everett distorts Everett's views by insisting with Bohr on a classical world. This section of Lévy-Leblond's paper follows:

> Once more, under a question of terminology lies a deep conceptual problem. [Everett's] interpretation in effect has been called by several people, especially DeWitt, one of his main propagandists, the "many-worlds (or many-universes) interpretation of quantum theory." The rejection of the postulate projection [*sic*] leaves us with the "universal" state vector. Since, with each successive measurement, this state-vector "splits" into a superposition of several "branches", it is said to describe "many universes," one for each of these branches. Where the Copenhagen interpretation would arbitrarily choose "one world" by cutting off all "branches" of the state-vector except one (presumably the one we think we sit upon), one should accept the simultaneous existence of the "many worlds" corresponding to all possible outcomes of the measurement. Now, my criticism here is exactly symmetrical of the one I directed against the orthodox position: the "many worlds" idea again is a left-over of classical conceptions. The coexisting branches here, as the unique surviving one in the Copenhagen point of view, can only be related to "worlds" described by classical physics. The difference is that, instead of interpreting the quantum "plus" as a classical "or", DeWitt and al. interpret it as a classical "and". To me, the deep meaning of Everett's ideas is not the coexistence of many worlds, but on the contrary, the existence of a single quantum one. The main drawback of the "many-worlds" terminology is that it leads one to ask the question of "what branch we are on," since it certainly looks as if our consciousness definitely belonged to only one world at a time: But this question only makes sense from a classical point of view, once more. (Lévy-Leblond, 1976, pgs. 184–85)

EVERETT TO LÉVY-LEBLOND, NOVEMBER 15, 1977

November 15, 1977

Prof. J. M. Lévy-Leblond
Laboratoire de Physique Theorique et Hautes Energies
Tour 33—1er etage
Universite Paris VII
2, Place Jussieu
75221 Paris Cedex 05
France

Dear Professor Lévy-Leblond:

The reason for the delay in acknowledging receipt of your pre-print, "Towards a Proper Quantum Theory," is that it is one of the more meaningful papers I have seen on the subject, and therefore deserving of a reply. This is always a mistake for me to make, as I very rarely complete a thorough review of papers, despite all good intentions. In this case, your observations seem entirely accurate (as far as I have read.)[mu] I especially enjoyed your inspired conclusion, pgs. 192–4.[mv] Line 4 under *Conclusion*— shouldn't 'urging' be 'urgent'?

I have not done further work in this area since the original paper in 1955. (not published in its entirety until 1973, as the "Many-Worlds Interpretation etc."). This, of course, was not my title as I was pleased to have the paper published in any form anyone chose to do it in! I, in effect, had washed my hands of the whole affair in 1956.[mw]

[mu] As Lévy-Leblond explained in his introductory note, the point of sending the paper was to ask whether he was right in interpreting Everett's position as involving one quantum world, not many classical worlds.

[mv] This refers to Lévy-Leblond's spoof that illustrated the nature of Copenhagen explanation. This was published as an appendix to LeBlond's paper (Lévy-Leblond, 1976).

[mw] In Everett's original draft of this letter, this second paragraph reads as follows:

> I have not done further work in this area since the original paper in 1955 (not published in its entirety until 1973, as the "Many-Worlds Interpretation etc."). This, of course, is not my title as I was pleased to have the paper published in any form anyone chose to do it in! I, in effect, had washed my hands of the whole affair in 1955. Far be it for me to look a gift Boswellian writer in the mouth! But your observations are entirely accurate (as far as I have read).

In both versions of the letter, Everett agrees that his formulation of quantum mechanics is best understood in terms of a single quantum world. In the draft version, Everett also acknowledges his debt to DeWitt's Boswellian efforts. James Boswell (1740–1795) chronicled the life and views of his friend Samuel Johnson in such a compelling way that he virtually

Fearing that I am unable to top your "Theoretical zipperdynamics,"[mx] I will let well enough alone. Thank you for sending me your most interesting paper. I hope I can comment in further detail at some future date.

Very truly yours,

Hugh Everett, III
8114 Touchstone Terrace
McLean, VA 22101

HE/ne

cc: Prof. J. A. Wheeler

---

guaranteed Johnson's enduring fame. His biography of Johnson is also famous for taking liberties with the details of Johnson's life for the purpose of engaging narrative. Concerning how far Everett had read, he had clearly read the section involving his own formulation of quantum mechanics and DeWitt's interpretation of it as a many-worlds theory as he explicitly responded to the material in this section. Everett's notes on his copy of the paper and what he says in his letter indicate that he also closely read at least the conclusion and the zipperdynamics sections. The earlier date in the draft letter is Everett's.

[mx] This is the title of Lévy-Leblond's spoof of the Copenhagen explanation (Lévy-Leblond, 1976).

CHAPTER 25

Correspondence: Everett and Raub (1980)

After the Austin meeting, Wheeler and DeWitt encouraged Everett to return to academia and do more work on quantum theory. But Everett was not interested. He was consumed by financial and personal troubles, including his debilitating addictions to alcohol, food, tobacco, and sex. However, in 1980 he responded to a letter from a physics enthusiast, David Raub, with several important remarks about how he viewed his theory.

EVERETT TO RAUB, APRIL 7, 1980

Hugh Everett, III
8114 Touchstone Terrace
Mclean, Va.   22102
(703-356-8931)

April 7, 1980

L. David Raub
2559 Brandon Road
Columbus, OH 43221

Dear Mr. Raub,

In answer to your letter of March 31, 1980, I offer the following comments. I certainly still support all of the conclusions of my thesis. I definitely agree with you that it still is the only completely coherent approach to explaining both the contents of quantum mechanics and the appearance of the world. To this day, all other attempts to come to grips with the apparent paradoxes of the measurement process seems to me to be far more artificial and unsatisfactory.[my]

Dr. Wheeler's position on these matters has never been completely clear to me (perhaps not to John either). He is, of course, heavily influenced by

---

[my] Here the claim is that his theory is both the most natural and the only completely coherent view. This is somewhat stronger than his earlier claims that it is the simplest, most comprehensive theory. See for example pgs. 157, 168–72, and 296.

Bohr's position (he was a student of Bohr) which essentially regards the entire formalism as merely a calculating device, and does not worry any further about "reality".[mz] It is equally clear that, at least sometimes, he wonders very much about that mysterious process, "the collapse of the wave function". The last time we discussed such subjects at a meeting in Austin several years ago he was even wondering if somehow human consciousness was a distinguished process and played some sort of critical role in the laws of physics.[na]

I, of course, do not believe any such special processes are necessary, and that my formulation is satisfactory in all respects. The difficulties in finding wider acceptance, I believe, are purely psychological. It is abhorrent to many individuals that there should not be a single unique state for them (the world view), even though my interpretation explains all subjective feelings quite adequately and is consistent with all observations.[nb]

I have encountered a number of other scientists who "subscribe" to my view, but have no list. By and large they seem to be the younger crop who did not start with preconceived ideas. Perhaps Wheeler has a list of such persons.

The only other work I did on this subject was a much fuller exposition, which predates (1955) the shorter paper, and which has recently been published by P.U. Press. I enclose a copy with my compliments.[nc]

Sincerely,

Hugh Everett, III

[mz] This passage suggests that Everett thought of his theory as more than just an algorithm for making predictions. More specifically, it was of central importance to him that his theory allowed him to model the process of observation and, most importantly, to represent the records of observers. Everett believed that it was only by dint of this that his theory could be put into correspondence with experience and hence be considered empirically faithful. See pg. 157 for a brief description of Everett's understanding of the virtues of pure wave mechanics.

[na] This was a position most directly associated with Wigner (Wigner, 1961). For his part, however, Wheeler did continue to think that measurement was somehow special. When later referring to his views from the 1970s on, Wheeler described his position as being that "the true essence of quantum mechanics... is *measurement*". He further explained: "Measurement, the act of turning potentiality into actuality, is an act of choice, choice among possible outcomes. After the measurement, there are roads not taken. Before the measurement all roads are possible—one can even say that all roads are being taken at once, (Wheeler and Ford, 1998, pg. 339–40).' Of course, having measurement play this sort of special role in determining what was actual was completely alien to Everett's project. For his part, Everett's views on the nature and cognitive status of physical theories seem to have been relatively stable over his career.

[nb] This is one of the two recorded instances where Everett mentions worlds. For the other mention see pg. 275. On the other hand, see Everett's agreement with Lévy-Leblond's one-quantum-world view (pg. 313). As Everett explains in the long thesis, there is typically no single post-measurement relative state for an observer (pgs. 121–22).

[nc] Everett sends a copy of "The Many-Worlds Interpretation of Quantum Mechanics" (1973), which includes the long thesis, which Raub had not read.

PART V

Appendixes

Everett's Notes on Possible Thesis Titles

The following handwritten documents were found in Everett's files. The first note was likely written before Everett began work on the thesis proper. It indicates what he thought of as the most important aspects of his project. As in the long and short theses, a main theme here is that pure wave mechanics allows one to consistently treat multiple, interacting observers *as quantum mechanical systems like any other. Other salient aspects of the project are the deterministic nature of the theory, the associated absence of postulates concerning probability, the interpretation of pure wave mechanics by way of correlations between subsystems, and the centrality of the universal wave function in characterizing these correlations.*

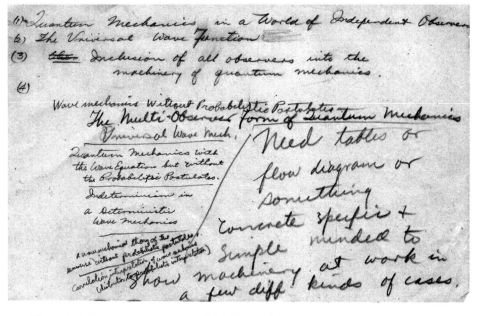

Figure A.1. Everett's notes on possible thesis titles.

Early Draft Outline for Long Thesis

This is perhaps the earliest extant outline of Everett's original long thesis. His thesis notes more generally indicate that he had originally intended to formulate the general concept of information in the context of its utility in hypothetical games. Also of interest here is Everett's early rejection of the Copenhagen interpretation's suggestion that the classical level be excluded from quantum mechanical description, the characterization of pure wave mechanics as a theory describing correlations by way of the universal wave function, and the mention of splitting and the amoeba analogy that he described further in the minipaper he wrote for Wheeler (pgs. 69–70). Insofar as Wheeler recommended against the inclusion of this analogy in the long thesis, advice that Everett followed, this outline may predate the associated minipaper of September 1955. Also of interest is Everett's preliminary conjecture regarding the possibility of an information-theoretic formulation of the uncertainty relation.

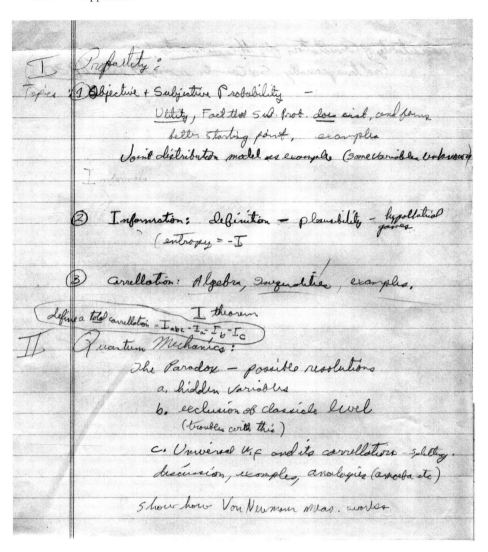

Figure B.1. Everett's early draft outline for the Long Thesis.

To be investigated: Uncertainty relation: is $I_x + I_\kappa \lessgtr \ln \frac{1}{\pi e}$
all wf's . ($\hbar \kappa = p$)

Modified Second law: direction of correlation entropies
example of class. stat mech: prove that _total_ entropy
invariant $\Rightarrow$ (second ent $\uparrow \Rightarrow$ Ecorrelation ent)

Correlations in field theory $\rightarrow$ expect C's to
decrease as function of distance
(maybe get a c _____)

Symmetry properties, _____

[Feynman method]

Transitivity of Correlation _for Measurements_ !
(not true generally, See Counter example around $\beta 30$)

Universal Wave Function Note

Although it is unclear when this small scrap was written, its importance is indicated by its inclusion in Everett's thesis notes file. The main point concerns Everett's belief that the theory of the universal wave function captured Bohr's talk of complementarity, which he took here to amount to rules of thumb for engineers, in a clear and consistent way that did not require one to deny the possibility of providing a quantum mechanical understanding of the classical behavior of an observer and her measurement apparatus. Also of interest is Everett's comment that one should not insist that everything in the correlation structure be measurable.

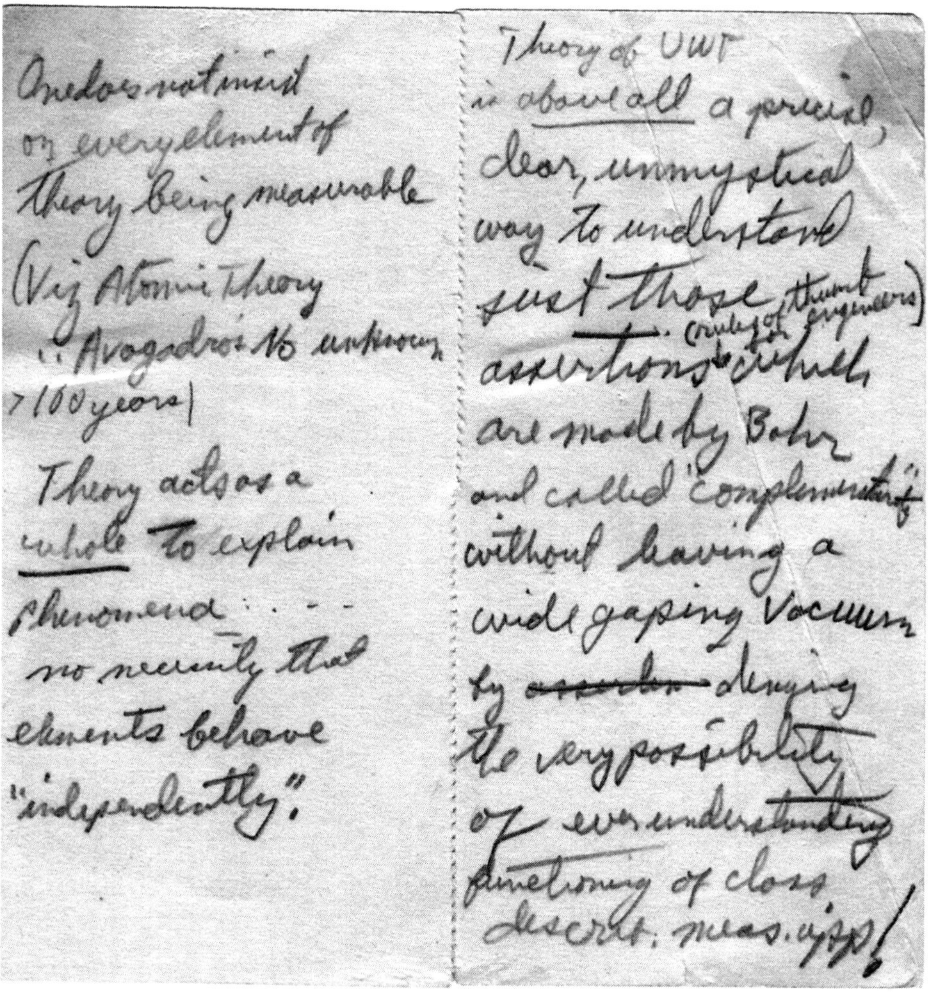

One does not insist
on every element of
theory being measurable
(Viz Atomic Theory
" Avogadro's N₀ unknown
7 100 years)

Theory acts as a
whole to explain
phenomena . . . .
no necessity that
elements behave
"independently".

Theory of UWF
is above all a precise,
clear, unmystical
way to understand
just those (crutch of engineers)
assertions which
are made by Bohr
and called "Complementarity"
without having a
wide gaping Vacuum
by denying
the very possibility
of ever understanding
functioning of closed
discrete, meas. ???

Figure C.1. Everett's note on the theory of the universal wave function.

Handwritten Draft Introduction to the Long Thesis

This is an early handwritten draft of the introduction to the long thesis. Of particular interest are Everett's revisions to the text and his notes on the backs of several of the pages. The central thesis of the project is to show "that the concept of 'universal wave function' and the associated correlation machinery provides a logically self consistent description of a universe in which several observers are at work." For Everett, everything follows from the universal wave function alone; that is, the correlation structure characterized by the universal wave function provides a complete model of a quantum universe representing multiple, interacting observers.

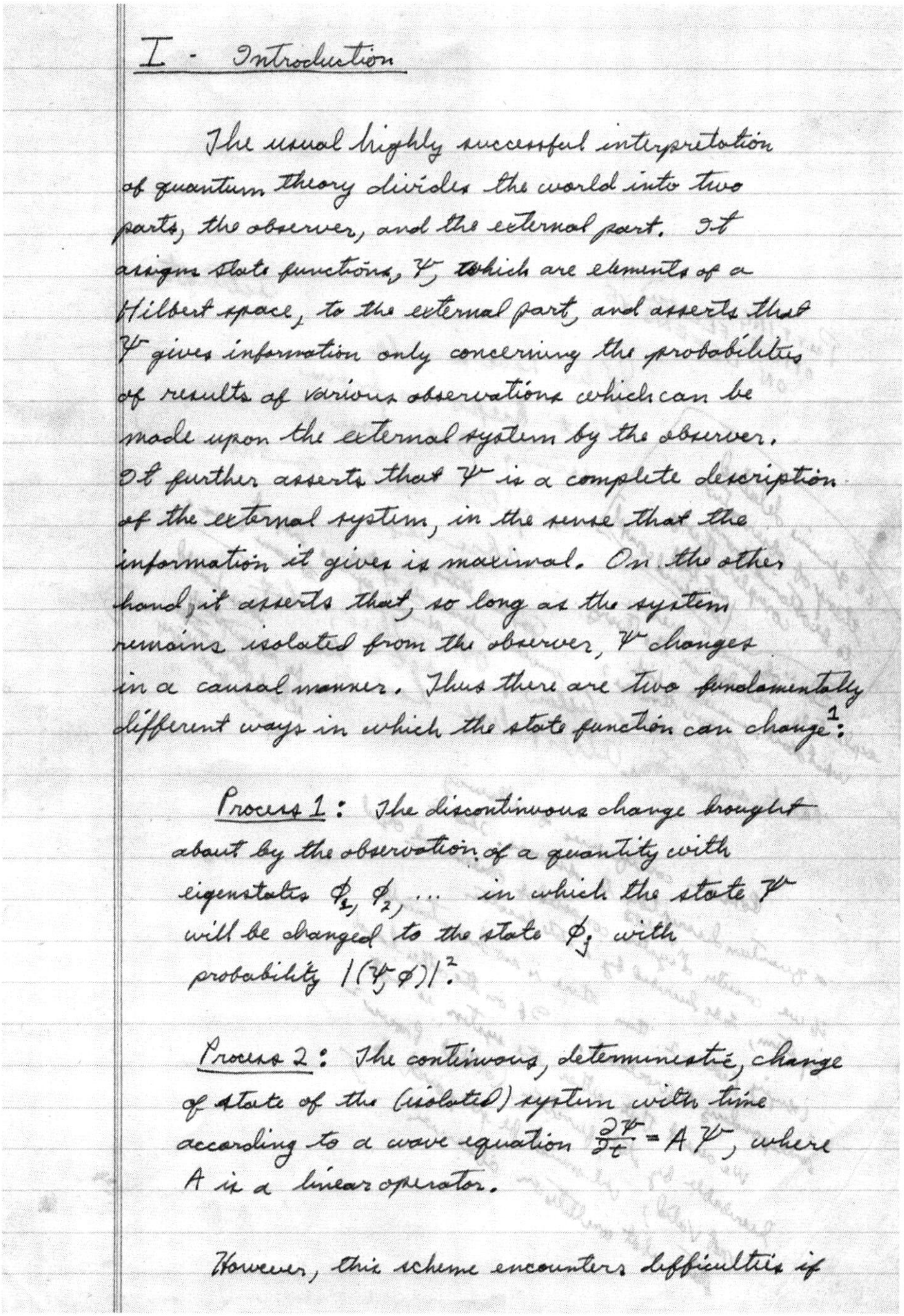

I · Introduction

The usual highly successful interpretation of quantum theory divides the world into two parts, the observer, and the external part. It assigns state functions, Ψ, which are elements of a Hilbert space, to the external part, and asserts that Ψ gives information only concerning the probabilities of results of various observations which can be made upon the external system by the observer. It further asserts that Ψ is a complete description of the external system, in the sense that the information it gives is maximal. On the other hand, it asserts that, so long as the system remains isolated from the observer, Ψ changes in a causal manner. Thus there are two fundamentally different ways in which the state function can change[1]:

Process 1: The discontinuous change brought about by the observation of a quantity with eigenstates $\phi_1, \phi_2, \ldots$ in which the state Ψ will be changed to the state ϕ_j with probability $|(\Psi, \phi_j)|^2$.

Process 2: The continuous, deterministic, change of state of the (isolated) system with time according to a wave equation $\frac{\partial \Psi}{\partial t} = A\Psi$, where A is a linear operator.

However, this scheme encounters difficulties if

Figure D.1. Everett's handwritten draft introduction to the Long Thesis.

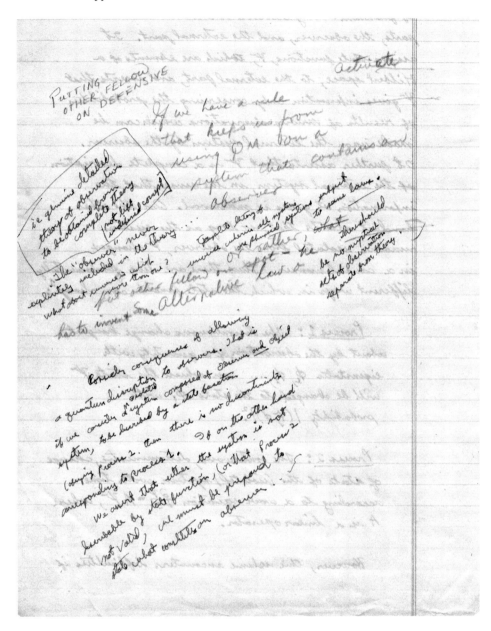

notebook (performs his observation). Having observed the notebook entry he turns to A and informs him in a patronizing manner that since his (B's) wave function just prior to his entry into the room, which he knows to have been a complete description of the room and its contents, had non-zero amplitude over other than the present result of the measurement, the result must have been decided only when B entered the room, so that A, his notebook entry, and his memory about what occurred one week ago had no independent objective existence until the intervention by B. In short, B implies that A owes his present objective existence to B's generous nature which compelled him to ~~condescend to~~ intervene *on his behalf*. However, to B's consternation, A does not react with anything like the respect and gratitude he should exhibit towards B, and ~~at~~ at the end of a somewhat heated reply, in which A conveys in a colorful manner his opinion of B and his beliefs, he rudely punctures B's ego by observing that if B's view is correct, then he has no reason to feel complacent, since the whole present situation may have no objective existence, but may depend upon the future actions of yet another observer.

At this point several alternatives appear ~~which~~ avoid the paradox:

<u>Alternative 1</u> :   To postulate the existence

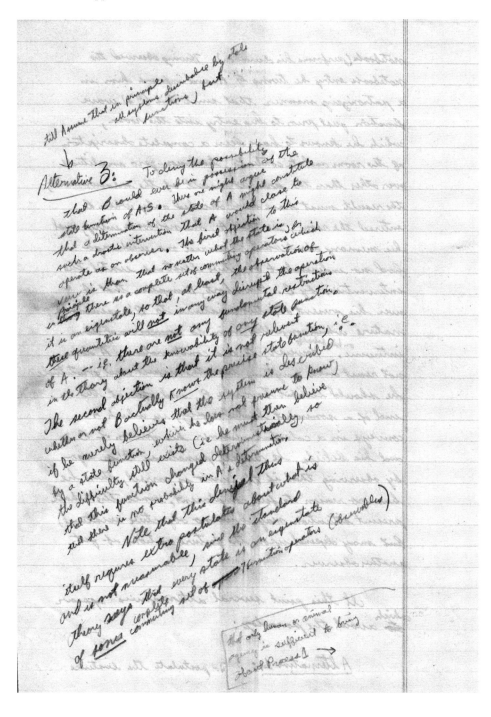

Alternative 3: To deny the possibility
that O could ever be in possession of the
state function of A+S. Here one might argue
that a determination of the state of A would constitute
such a drastic intervention that A would cease to
operate as an observer. The first objection to this
view is than that no matter what the state is, for
 there is a complete set of commuting operators of which
it is an eigenstate, so that, at least, the observation of
these quantities will not in any way disrupt the operation
of A. — i.e. there are not any fundamental restrictions
in the theory about the knowability of any state function
The second objection is that it is not relevant
whether or not O actually know the precise state function, i.e.
by a state function, which he does not presume to know)
the difficulty still exists (i.e. he must then believe
that this function changed deterministically, so
that there is no probability in A's determination.

Note that this denial this
itself requires extra postulates about what is
and is not measurable, since the standard
theory says that every state is an eigenstate
of some complete commuting set of Hermitian operators (observables)

that only human or animal
agency is sufficient to bring
about Process 1

it is contemplated that the universe contains more than one observer. Let us consider the case of one observer A, who is performing measurements upon a system S, the totality (A+S) in turn forming the object system for another observer, B. We wish to consider the consequences of B's use of a quantum mechanical description of A+S. If we are to deny this possibility for B, then we must be prepared to state what constitutes an "observer" or measuring apparatus in distinction to an ordinary physical system to which quantum mechanics is applicable, and we must then attribute to observers, or measuring apparatus a preferred position -- a kind of mystical aloofness from the natural laws.

On the other hand, if we do allow B to give a quantum description to A + S, by assigning a state function ψ_{A+S}, then so long as B does not interact with A + S its state changes causally according to process 2, even though A may be performing measurements upon S. From B's point of view nothing resembling process 1 can occur (there are no discontinuities), and the question of ~~the can~~ the validity of A's use of process 1 is raised. That is, ~~apparently~~ either A is incorrect in assuming process 1, with its probabilistic implications, to apply to his measurements, or else B's state function is an inadequate description of what is happening to A+S.

To better illustrate the situation we consider the following extremely hypothetical drama. Isolated somewhere out in space is a room containing an observer, A,

underline

Insert 1

underline

Insert 2

underline extremely hypothetical

who is about to perform a measurement upon a
system S. After performing his measurement he
will record the result in his notebook. We assume
that he knows the state function of S (perhaps as
a result of a previous measurement), and that it is
not an eigenstate of the measurement he is about to
perform. A being an orthodox quantum theorist then
believes that the outcome of his measurement is
undetermined and that the process is correctly described
by Process 1. ¶ In the meantime, however, there is

new Paragraph

another observer, B, outside the room, who is in
possession of the state function of the entire room,
including S, the measuring apparatus, and A, just
prior to the measurement. B is only interested in
what will be found in the notebook one week hence,
so he computes the state function ~~of~~ the room for
one week in the future according to Process 2.
One week passes, and we find B ~~still~~ in possession
of the state function of the room, which ~~he finds~~ ~~took~~
~~he is also an~~ *this equally* orthodox quantum theorist ~~of~~ believes
to be a complete description of the room. If ~~it is~~
~~now the case that~~ B's state function tells, exactly
going to be what is in the notebook, then A ~~was~~ incorrect
calculation before
in his belief about the indeterminacy of the
outcome of his measurement (~~because B could~~
~~predict this state function even before the measurement~~
~~took place~~). We therefore assume that B's state
function contains non-zero amplitudes over
several of the possible notebook entries. ¶ At this

Insert 3

new → Paragraph point B opens the door to the room and looks at the

of only one observer in the universe. This is the solipsist position, in which each of us must hold the view that we alone are the only valid observer, with the rest of the universe and its inhabitants obeying at all times Process 2 except when under our observation. This view is quite consistent, but one must feel uneasy when, for example, writing textbooks on quantum mechanics, describing process 1, for the consumption of other persons to whom it does not apply.

Alternative 2:   To limit the applicability of quantum mechanics by asserting that the quantum mechanical description fails when applied to observers, or to measuring apparatus, or more generally to systems approaching macroscopic size.

If we try to limit the applicability so as to exclude measuring apparatus, or in general systems of macroscopic size, we are faced with the difficulty of sharply defining the region of validity. For what n might a group of n particles be construed as forming a measuring device so that the quantum description fails? And to draw the line at human or animal observers, i.e. to assume that all mechanical apparatus obeys the usual laws, but that they are somehow not valid for living observers, is to do violence to the principle of psycho-physical parallelism[2], and constitutes a view to be avoided, if possible. To do justice to this principle we must insist that we be able to conceive of mechanical

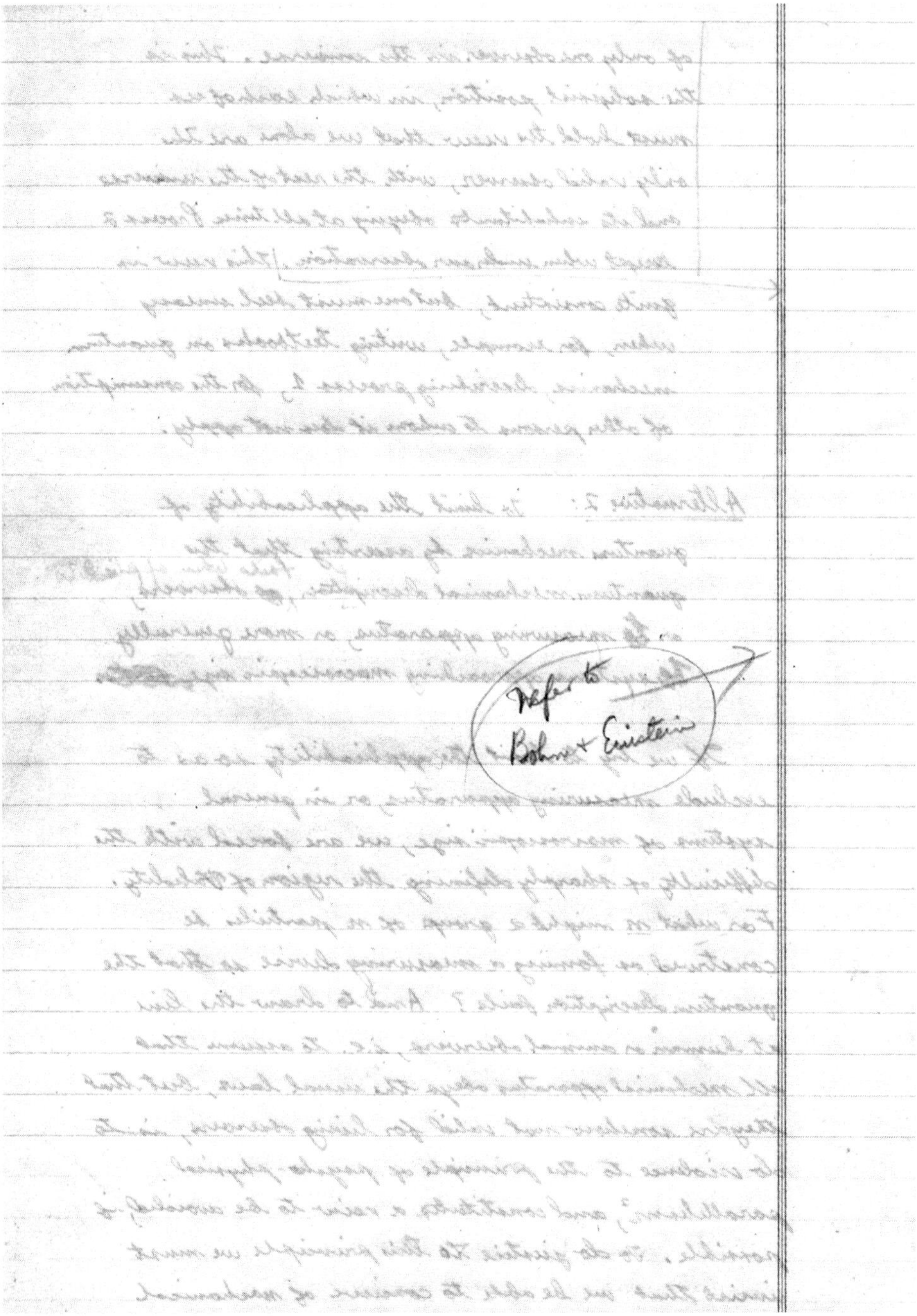

devices (such as servomechanisms), obeying natural laws, which we would be willing to call observers.

Alternative 4: To abandon the position that the state function is a <u>complete</u> description of a system. The state function is to be regarded not as a description of a single system, but of an ensemble of systems, so that the probabilistic assertions arise naturally from the incompleteness of the description.

It is assumed that the correct complete description, which would presumably involve further (hidden) parameters beyond the state function alone, would lead to a deterministic theory, from which the probabilistic aspects arise as a result of our ignorance of these extra parameters in the same manner as in classical statistical mechanics'. ~~This is the view held by Einstein, among others, and while quite appealing suffers from the fact that no one has yet presented a completely satisfactory theory along such lines.~~

Alternative 5: To assume the universal validity of the quantum description, and to treat all observation processes on equal footing with all other natural processes, by the abandonment of Process 1 (i.e. to assert the general validity of <u>pure</u> wave mechanics, without any statistical assertions, in which

Thesis: <u>We show that the concept
of "universal wave function"
and the associated correlation machinery
provides a logically self consistent
description of a universe in which
several observers are at work.</u>

Process 2 is at all times obeyed). An "observer" is always regarded simply as a subsystem of a composite system, which also contains the object-system of the observation in interaction with the observer, and whose state function always obeys Process 2.

Insert 5

ark

It is immediately evident that Alternative 5 is a theory of many advantages. It has the virtue of logical simplicity and it is complete in the sense that it is applicable to the entire universe. All processes are considered equally (there are no "measurement processes" which play any preferred role), and the principle of psycho-physical parallelism is fully maintained. Since the universal validity of the state function description is asserted, one can regard the state functions themselves as the fundamental entities, ~~and one can~~ and one can even ~~consider~~ the state function of the whole universe. In this sense this theory can be called the theory of the "universal wave function", since it can be based upon this all propositions alone. There remains, however, the question of whether or not such a theory can be put into correspondence with our experience.

Insert 6

Insert 7

It will be our purpose to show that this can, in fact, be done in a satisfactory manner. We shall ~~be able to~~ introduce into the theory ideal observers which represent observers which can be conceived as automatically functioning machines [servomechanisms] possessing recording devices [memory] and which are capable of responding to their environment into the theory. The behavior of which shall always

Insert 8

(cont.)

be treated entirely within the framework of wave mechanics. Furthermore, we shall ~~be able to~~ deduce the probabilistic assertions of Process 1 as subjective appearances to such ~~total~~ observers, thus placing the theory in correspondence with experience. We are then led to the novel situation in which the formal theory is objectively continuous and causal, while subjectively discontinuous and probabilistic. While this point of view thus shall ultimately justify our use of the statistical assertions of the orthodox view, it enables us to do so in a logically consistent manner, allowing for the existence of other observers. ~~and~~ At the same time it gives a deeper insight into the meaning of quantities systems, ~~and~~ the role played by quantum mechanical correlations.

~~To the indeterminism in a determ~~

Insert 8 (contd)

~~The key to the possibility of~~ such an interpretation the ~~lies in the proper~~ exploitation of the correlations between subsystems of a composite system which is described by a wave function.) A subsystem of such a composite system does not, in general, possess an independent state function. That is, in general a composite system can not be represented by a single pair of subsystem states, but can be represented only by a superposition of such pairs of subsystem states. For example, the Schrödinger wave function for a pair of particles, $\Psi(x_1, x_2)$, cannot generally be written in the form $\Psi(x_1, x_2) = \phi(x_1) \chi(x_2)$ but only in the form $\Psi = \sum a_{ij} \phi_i(x_1) \chi_j(x_2)$. In the latter case there is no single state for particle 1 alone or particle 2 alone, but only the superposition of such cases.

Insert 9

In fact, to any arbitrary choice of state for one subsystem there will correspond a {relative} state for the other subsystem, which will generally be dependent upon the choice of state for the first subsystem, so that the state of one subsystem is not independent, but correlated to the state the remaining subsystem. Such correlations between systems arise from interaction of the systems, and from our point of view all measurement and observation processes are to be regarded simply as ~~each~~ interactions between observer and object-system which produce strong correlations.

It is ~~then a consequence of regarding the~~ *inescapable* Let one observer as a subsystem of the composite system Observer + object-system. ~~That~~ *Then* after interaction there will not, in general, exist a single observer state, but a superposition, each element of which contains a definite observer state and a corresponding relative object~~tive~~-system state. Furthermore it is a general consequence of the superposition ~~principle~~ principle for linear wave equations that if the observation is performed upon an object system which is not an eigenstate of the measurement, that the result will be a superposition of states, each of which describes an observer which has obtained a different result of observation, and for which the relative *object* system state is nearly the eigenstate corresponding to the observed result, the whole superposition being combined with approximately the coefficients of the expansion of the original object-system state in terms of the eigenfunctions of the measurement.

Insert 10

Thus each element of the resulting superposition describes an observer who perceived a definite result, and to whom it appears that the object-system state has been transformed into the corresponding eigenstate. ~~Also In~~ this sense ~~and~~ the usual assertions of Process 1 appear to hold on a subjective level to each observer described by an element of the superposition. We shall see also that correlation plays an important role in preserving consistency when several observers are present and allowed to interact with one another (to 'consult' one another) as well as with other object-systems ~~_____~~

In order to develop a language for interpreting our pure wave mechanics for composite systems we shall find it useful to develop quantitative definitions for such notions as the nearness of a state function ψ to an eigenfunction of an operator A, i.e. the "sharpness" or definiteness of A, and the degree of correlation between subsystems of a composite system, or ~~between~~ a pair of operators ~~in~~ the subsystems, so that we can use these ~~notions~~ concepts in an unambiguous manner. The mathematical ~~~~ development of these notions will be carried out in the next chapter (II) using some concepts borrowed from 'Information Theory'. We shall develop there the general definitions of information and correlation, as well as some of their more important properties. While we shall use the language of probability theory in this chapter, to facilitate the exposition, we shall nevertheless subsequently apply the mathematical definitions directly to state functions by replacing probabilities by square amplitudes, <u>without</u>

however, making any reference to probability models.

We shall then investigate the Quantum formalism of composite systems (chap. IV), particularly the concept of relative state functions, and the meaning of the representation of subsystems by non-interfering mixtures of states characterized by density matrices. We shall also see that the existence of a strong correlation between an operator A in one subsystem and an operator B in the other implies the possibility of a representation of the composite system state as a superposition, for each element of which both A and B have nearly definite values, the degree of this definition improving with the degree of correlation. Finally we shall see that there is a characteristic correlation between the subsystems of a composite system which has some interesting properties.

This will be followed by an investigation of abstract measuring processes (chap. IX) considered simply as correlation inducing interactions between subsystems of a single isolated system, and the representation of the resulting state as a superposition of states for which the two subsystems have nearly definite values for the measured quantity. Ideal observers are then introduced and treated in similar fashion, and the validity of process 1 as a subjective phenomenon is deduced. The consistency of allowing several observers to interact is also shown.

The abstract treatment is then supplemented in chap. X, with a discussion of real processes. The existence and meaning of macroscopic objects

We do not wish to restrict ourselves to numerical
valued random variables, but to abstract valued
random variables — hence our general set considerations

of fairly well defined boundaries and shapes from the viewpoint of their atomic constitution and the wave mechanics of their constituent particles, ~~is~~ is discussed.

~~well as the justification of assigning shape ~~(~~)~~ the body of such objects is discussed.~~ In addition some other implications of the investigation are mentioned.

The final chapter summarizes the situation and describes further the advantages of the theory, as well as the difficulties of the other alternatives.

P

Insert 1:
" If we are to deny the possibility of B's use of a quantum mechanical description (wave function *obeying wave equation*) for A+S, then we must be supplied with some alternative description for systems which contain observers (or measuring apparatus). Furthermore we would have to have a criterion for telling precisely what type of systems would have the preferred positions of "measuring apparatus" or "observer" and be subject to the alternate description. Such a criterion is probably not capable of rigorous formulation.

Insert 2:
Or else B's state function, with its purely causal character, is an inadequate description of what is happening to A+S.

Insert 3:
which this equally orthodox quantum theorist believes to be a complete description of the room and its contents. If B's state function calculation tells beforehand exactly what is going to be in the notebook, then A is incorrect in his belief about the indeterminacy of the outcome of his measurement. We therefore assume that B's state function contains non-zero amplitudes over several of the notebook entries.

Insert
4

Alternative 3: ~~but to~~ To admit the validity of the state function description, deny the possibility that B could ever be in possession of the state function of A+S. Thus one might argue that a determination of the state of A would constitute such a drastic intervention that A would cease to function as an observer.

The first objection to this view is that no matter what the state of A+S is, there is in principle a complete set of commuting operators for which it is an eigenstate, so that, at least, the determination of these quantities will not affect the state nor in any way disrupt the operation of A. There are ~~no~~ fundamental restrictions in the usual theory about the knowability of any state functions, and the introduction of any such restrictions to avoid the paradox must therefore require extra postulates.

The second objection is that it is not particularly relevent whether or not B actually knows the precise state function of A+S. If he merely believes that the system is described by a state function, which he does not presume to know, then the difficulty still exists. He must then believe that this state function changed deterministically, and hence that there was no probability in A's determination.

<u>Alternative 5:</u> To assume the universal validity of the quantum description, by the complete abandonment of Process 1. The general validity of pure wave mechanics, <u>without any statistical assertions</u>, is assumed for <u>all</u> physical systems, including observers and measuring apparatus. Observation processes are to be described completely by the state function of the composite system which includes the observer and his object system, and which at all times obeys the wave equation (Process 2).

Insert 5

all of physics is presumed to follow from this function alone.

Insert 6

The present thesis will be devoted to showing that this concept of a universal wave mechanics, together with the necessary correlation machinery for its interpretation, forms a logically self consistent description of a universe in which several observers are at work, ~~and one which can be put into correspondence with our experience~~.

Insert 7

Insert 8

We shall be able to introduce into the theory systems which represent observers. Such systems can be conceived as automatically functioning machines (servomechanisms) possessing recording devices (memory) and which are capable of responding to their environment. The behavior of these observers shall always be treated within the framework of wave mechanics. Furthermore, we shall deduce the probabilistic assertions of Process 1 as subjective appearances to such observers, thus placing the theory in correspondence with experience.

Insert 9

In order to bring about this correspondence with experience for the pure wave mechanical theory, we shall exploit the correlations between subsystems of a composite system which is described by a state function.

Insert 10

It is, however, an inescapable consequence that if one regards an observer at a subsystem of the composite system that after the interaction has taken place there will not, in general, exist a single observer state. There will, however, be a superposition of the composite system states, each element of which contains a definite observer state and a definite relative object system state. Furthermore each of these relative object system states will be, approximately, the eigenstates of the observation corresponding to the value attained by the observer which is described by the same element of the superposition.

Handwritten Draft Conclusion to the Long Thesis

This is an early handwritten draft of the conclusion to the long thesis. Some of Everett's usual themes are described in novel ways here, and some themes treated here were left out of the thesis. Of the first sort is the characterization of classical laws as laws that approximately hold for relative subsystems over short times (compare with pgs. 136 and 158), his goal of deducing everything from the universal wave function alone, and the role of the principle of psychophysical parallelism in suggesting that one require there to be physical models for observers. Of the second sort is the explanation of why observer interactions between branches would not disturb the history of a branch and the sense in which having a deterministic formulation of quantum mechanics like pure wave mechanics that avoids probabilistic jumps scattered about space-time might be a virtue in formulating a quantum mechanical version of general relativity or relativistic field theory.

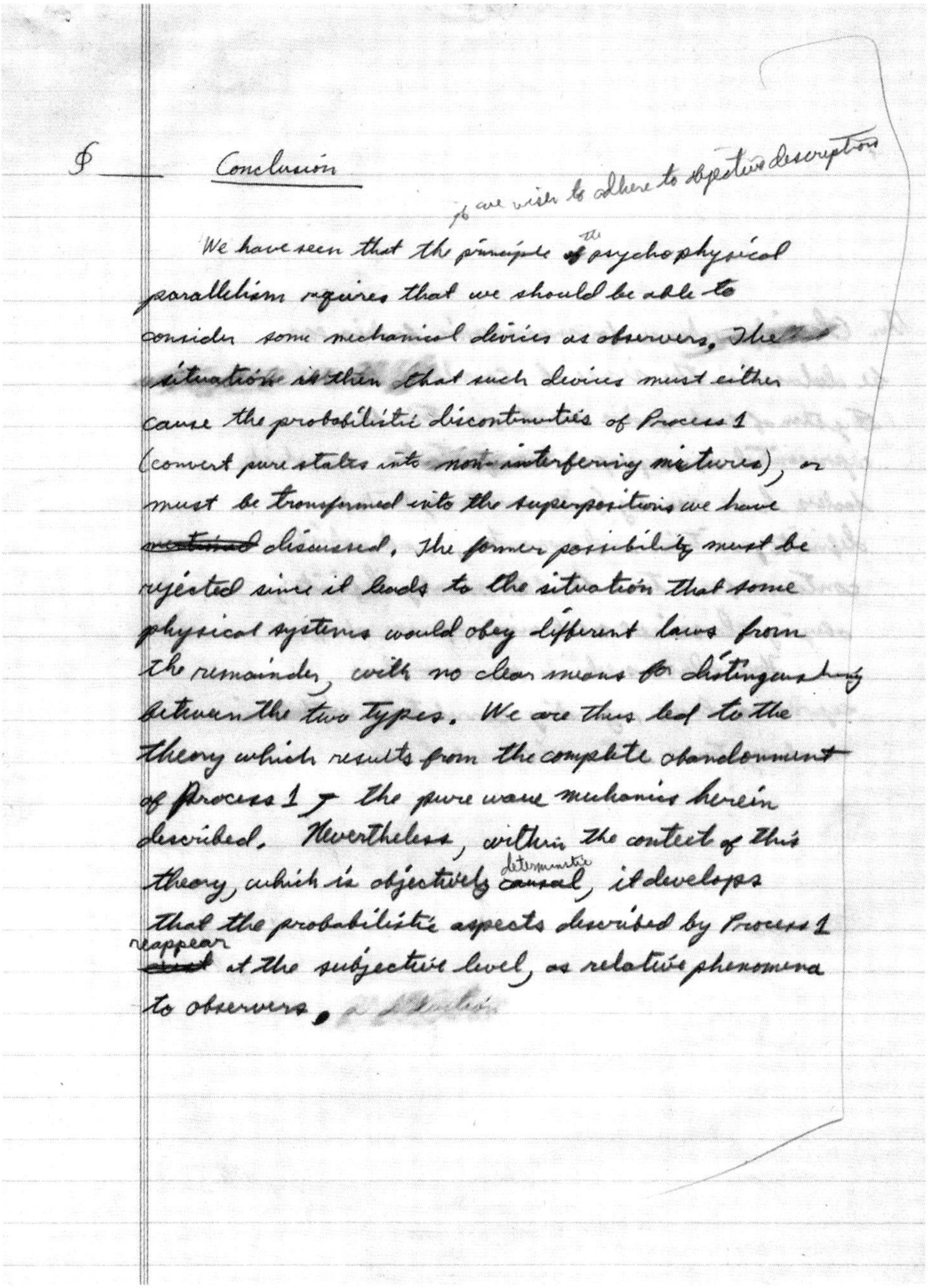

§ ___ Conclusion

We have seen that if we wish to adhere to objective description the principle of the psychophysical parallelism requires that we should be able to consider some mechanical devices as observers. The situation is then that such devices must either cause the probabilistic discontinuities of Process 1 (convert pure states into non-interfering mixtures), or must be transformed into the superpositions we have ~~mentioned~~ discussed. The former possibility must be rejected since it leads to the situation that some physical systems would obey different laws from the remainder, with no clear means for distinguishing between the two types. We are thus led to the theory which results from the complete abandonment of Process 1 — the pure wave mechanics herein described. Nevertheless, within the context of this theory, which is objectively ~~causal~~ deterministic, it develops that the probabilistic aspects described by Process 1 ~~exist~~ reappear at the subjective level, as relative phenomena to observers.

Figure E.1. Everett's handwritten draft conclusion to the Long Thesis.

thus Classical mechanics for macroscopic bodies can
be deduced in the sense of correlation laws —
System of such bodies can at one time be
represented as superposition of states in which
bodies have nearly (within uncertainty limits)
definite positions and momenta, and which
continue for a time to be nearly definite,
obeying classical mechanics approximately.

Thus class. mech. is ~~a a law~~
expresses laws regarding correlations ~~with~~ in
such systems between different times.

One thus begins with a theory, *is may free to of the universe* which postulates only the existence of a Universal wave function, which obeys a linear wave equation. One then *investigates* the internal correlations in this wave function with the aim of deducing laws of physics, which are simply statements of the form, *which condition The property A* such and *such a property* of a subsystem of the universe (subset of the total collection of coordinates for the wave function) is correlated with the property of **B** of another subsystem, with the manner of the correlation being specified. (For example, the classical mechanics of a system of massive particles becomes a law which states that there is a certain correlation between the positions and momenta (approximate) of the particles at one time and the positions and momenta at another time. (i.e. at any instant a state for the system can, for example, be represented as a superposition, each element of which describes the particles as independent gaussian wave packets, and each element of which goes into a state at a later *(but nearly)* time for which the packets have moved in a nearly classical manner.) All statements about subsystems then become relative statements, i.e. statements about one subsystem relative to a prescribed state for the remainder, and all laws are correlation laws.

shettons

It is a complete, causal theory of conceptual simplicity; which It maintains fully the principle of Psyco Physical Parallelism. (All of the correlation Paradoxes (Einstein. P.) find easy explanation. ~~The paradox Field probably do deal with any number of observers consistently.~~ Disappearance of Paradox of introduction, since all elements of a superposition equally valid (no need to bear that any present existence will be upset by a future observer, since he would merely correlate over whole superposition, in no way affect past history.

Since
— This viewpoint will be applicable to all forms of Quantum mechanics where the superposition principle holds.
Viewpoint Avoid ~~running to~~ considering anomalous probabilistic jumps scattered about space time, can assert that field equations valid everywhere and everywhen, then deduce any statistical assertions that are possible by the present method.

— We should also like to remark upon the fundamental nature of the correlation information, as defined here, (its invariance, etc) As a basic quantity characteristic of coupled systems, Also its relation to entropy, etc. And the possibility of deducing useful relations concerning it. (regard to precision of possible predictions, etc.

(Has such remarkable properties that one cannot escape the feeling that it has a fundamental significance.)

The theory also forms a framework for the discussion
of, in addition to ordinary phenomena, observation processes
themselves & in a logical, unambiguous fashion. While
this theory ~~justifies~~ the solipsist
position (Expressed in Alternative 4 of the introduction), since
that is in fact a deduction of this theory — the subjective
appearance to observers, it forms a broader frame in
which to ~~~~ understand the consistency of that
view. (which is maintained by correlations.) It transcends that position, however,
in its ability to deal logically with questions
of imperfect observations (approximate measurements).
It may ~~also~~ prove a fruitful framework
for the interpretation of new quantum formalisms
as they appear, ~~such as~~ Field theories, particularly
any which might be relativistic in the sense of
General relativity, since one is free to ~~use the~~ construct
formal (non-probabilistic) theories, and supply any
possible statistical interpretations later, ~~~~ (
~~deductions in the manner carried out here~~ By
focussing attention upon questions of correlations,
one may be able to deduce useful relations for
such interpretations. (Correlation laws). Quantized
Fields do not generally have pointwise independent field
values, the value at one place and time being correlated to
those at neighboring points in a manner, as is to
be expected, approximating the behavior of classical fields.)
If correlations are important in systems with a finite no
of degrees of freedom, how much more important must they
be in systems with infinitely many.

Finally, aside from any possible <u>all</u> <u>practical</u> advantages of the theory, it remains a matter of intellectual interest that the statistical assertions of the usual theory are not independent hypotheses, but are deducible (in the present sense) from the pure wave mechanics.

APPENDIX F

Handwritten Revisions to the Long Thesis for Inclusion in DeWitt and Graham (1973)

The following scanned pages come from Everett's personal copy of the original long version of the thesis. They contain examples of the modifications he made to the text in preparation for the inclusion of the long thesis in the DeWitt and Graham anthology (1973). The pages included here are those where Everett's changes to the text affect its meaning. Notable among these changes are that Everett consistently changes "observers" to "observer states" (pgs. 356 and 360) and that he consistently strengthens his claims concerning the status of his derivation of probability (pgs. 359 and 361). The latter changes are particularly salient in the context of Everett's disagreement at this point with DeWitt and Graham concerning the proper understanding of probability in the relative-state formulation. On the page taken from the second appendix (pg. 362), Everett substitutes the term "operationalist" for the term "positivist," a change that did not appear in the printed version of the anthology. Note also the "language difficulty" footnote (pg. 363) and Everett's "omit" notation in the left margin. This footnote did appear in the anthology.

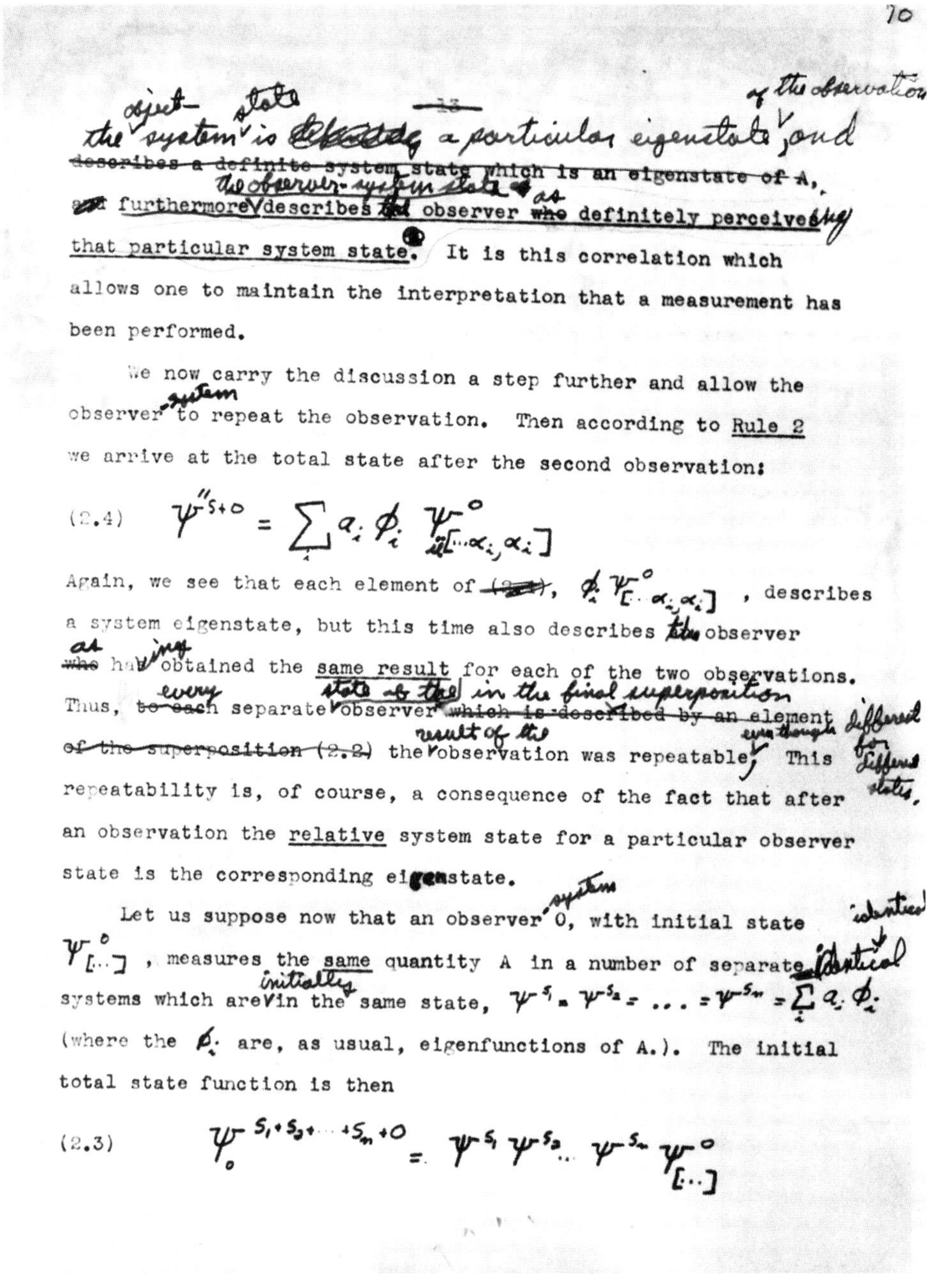

70

the ~~object~~ 'system' is ~~already~~ a particular eigenstate, and ~~of the observation~~
~~describes a definite system state which is an eigenstate of A,~~
~~and~~ furthermore describes ~~the~~ observer ~~who~~ definitely perceives ~~the observer-system state as~~
that particular system state. It is this correlation which

allows one to maintain the interpretation that a measurement has

been performed.

We now carry the discussion a step further and allow the
observer ~~system~~ to repeat the observation. Then according to Rule 2
we arrive at the total state after the second observation:

$$(2.4) \qquad \psi''_{S+O} = \sum_i a_i \phi_i \, \psi^o_{i[\dots\alpha_i,\alpha_i]}$$

Again, we see that each element of ~~(2.4)~~, $\phi_i \psi^o_{i[\dots\alpha_i,\alpha_i]}$, describes

a system eigenstate, but this time also describes ~~the~~ observer
~~who~~ ha~~d~~ving ~~as~~ obtained the same result for each of the two observations.

Thus, ~~to each~~ every separate ~~state of the~~ observer ~~which is described by an element~~ in the final superposition ~~different~~
~~of the superposition (2.2)~~ the result of the observation was repeatable. This ~~even though~~ ~~differs~~ ~~states~~

repeatability is, of course, a consequence of the fact that after ~~states~~

an observation the relative system state for a particular observer

state is the corresponding eigenstate. ~~system~~

Let us suppose now that an observer O, with initial state ~~identical~~ ~~identical~~
$\psi^o_{[\dots]}$, measures the same quantity A in a number of separate
systems which are initially in the same state, $\psi^{S_1} = \psi^{S_2} = \dots = \psi^{S_n} = \sum_i a_i \phi_i$

(where the ϕ_i are, as usual, eigenfunctions of A.). The initial

total state function is then

$$(2.3) \qquad \psi_o^{S_1+S_2+\dots+S_n+O} = \psi^{S_1} \psi^{S_2} \dots \psi^{S_n} \psi^o_{[\dots]}$$

Figure F.1. Everett's handwritten revisions to the Long Thesis.

71

- ~~14~~ -

We shall assume that the measurements are performed on the systems in the order S_1, S_2, ...S_m. Then the total state after the first measurement will be, by <u>Rule 1</u>,

$$(2.4) \qquad \psi_1^{S_1 + S_2 + \cdots S_m + O} = \sum_i a_i \phi_i^{S_1} \psi^{S_2} \cdots \psi^{S_m} \psi^O_{i[\cdots \alpha_i^1]}$$

(where α_i refers to the first system, S_1)

After the second measurement it will be, by <u>Rule 2</u>,

$$(2.5) \qquad \psi_2^{S_1 + S_2 + \cdots + O} = \sum_{i,j} a_i a_j \phi_i^{S_1} \phi_j^{S_2} \psi^{S_3} \cdots \psi^{S_m} \psi^O_{ij[\cdots \alpha_i^1, \alpha_j^2]}$$

and in general, after r measurements have taken place ($r \neq m$) <u>Rule 2</u> gives the result:

$$(2.6) \quad \psi_r = \sum_{ij\cdots k} a_i a_j \cdots a_k \phi_i^{S_1} \phi_j^{S_2} \cdots \phi_k^{S_r} \psi^{S_{r+1}} \cdots \psi^{S_m} \psi^O_{\substack{ij\cdots k \\ [\cdots \alpha_i^1, \alpha_j^2, \cdots \alpha_k^r]}}$$

We can give this state, ψ_r, the following interpretation. It consists of a superposition of states:

$$(2.7) \qquad \psi'_{ij\cdots k} = \phi_i^{S_1} \phi_j^{S_2} \cdots \phi_k^{S_r} \psi^{S_{r+1}} \cdots \psi^{S_m} \psi^O_{[\alpha_i^1, \alpha_j^2, \cdots, \alpha_k^r]}$$

each of which describes the observer with a definite memory sequence $[\alpha_i^1 \alpha_j^2 \cdots \alpha_k^r]$, and relative to whom the (observed) system states are the corresponding eigenfunctions $\phi_i^{S_1}, \phi_j^{S_2} \cdots \phi_k^{S_r}$, the remaining systems $S_{r+1}, \cdots S_m$, being unaltered.

In the language of subjective experience, the observer which is described by a typical element, $\psi'_{ij\cdots k}$, of the superposition has perceived an apparently random sequence of definite results for the observations. It is furthermore true ~~that if any of the observers described by such an element of the super-~~

Ins. 1

↳ since in each element the system has been left in an eigenstate of the measurement,

Ext. fige

- 15 -

~~position should repeat any one of the preceeding determinations,~~ /e
~~that observer would obtain the same result that was obtained~~
~~for the earlier observation, since the relative system states~~
~~are now eigenstates. That is,~~ *that* if at this stage a redetermination of an earlier system observation (S_ℓ) takes place, every
element of the resulting final superposition will describe ~~the~~
observer ~~whose~~ *with a* memory state ~~is~~ *configuration* of the form $\left[\alpha^1_{\cdot,}..\alpha^\ell_{j,}..\alpha^r_{\kappa,}\alpha^\ell_i\right]$
in which the earlier memory coincides with the later--i.e. the
memory states are correlated. It will thus appear to the observer which is described by a typical element of the superposition that each initial observation on a system caused the
system to "jump" into an eigenstate in a random fashion and
& subsequent measurements on the same system.
thereafter remain there. Therefore, qualitatively, at least,
the probabilistic assertions of Process 1 appear to be valid
to the observer described by a typical element of the final
superposition. *start new page here (material to be inserted)*

In order to establish quantitative results, we must put
some sort of measure (weighting) on the elements of a final
superposition. This is necessary to be able to make assertions
which will hold for almost all of the observers described by
elements of a superposition. In order to make quantitative
statements about the relative frequencies of the different possible results of observation which are recorded in the memory
of a typical observer we must have a method of selecting a
typical observer.

→ go directly to page 79 from here

(contd. from pg 82)

P Having deduce that there is a unique measure which will satisfy our requirements, the square-amplitude measure, we continue our deduction.

- 16 -

We choose for this measure the square amplitude of the coefficients of the superposition, a choice which we shall subsequently see is not as arbitrary as it appears. This ~~choice~~ measure then assigns to the $i, j, \ldots, \kappa \stackrel{\text{th}}{=}$ element of the superposition (S.6),

$$(2.8) \qquad \phi_i^{s_1} \phi_j^{s_2} \ldots \phi_\kappa^{s_r} \psi^{s_{r+1}} \psi^{s_m} \psi^0 \qquad [\alpha_{i,}^1 \alpha_{j,}^2 \ldots, \alpha_\kappa^r]$$

Put ( ) but no number

the measure (weight)

$$(2.9) \qquad M_{ij \ldots \kappa} = (a_i \, a_j \ldots a_\kappa)^* (a_i \, a_j \ldots a_\kappa)$$

so that the observer ~~whose~~ with memory ~~state~~ state configuration $[\alpha_{i,}^1 \alpha_{j,}^2 \ldots \alpha_\kappa^r]$ is assigned the measure $a_i^* a_i \, a_j^* a_j \ldots a_\kappa^* a_\kappa = M_{ij \ldots \kappa}$. We see immediately that this is a product measure, namely

$$(2.10) \qquad M_{ij \ldots \kappa} = M_i \, M_j \ldots M_\kappa$$

where $\qquad M_\ell = a_\ell^* a_\ell$

so that the measure assigned to a particular memory sequence $[\alpha_{i,}^1 \alpha_{j,}^2 \ldots \alpha_\kappa^r]$ is simply the product of the measures for the individual components of the memory sequence.

We notice now a direct correspondence of our measure structure to the probability theory of random sequences. Namely, <u>if we were to regard</u> the $M_{ij \ldots \kappa}$ as **probabilities** for the sequences $[\alpha_{i,}^1 \alpha_{j,}^2 \ldots \alpha_\kappa^r]$, then the sequences are equivalent to the random sequences which are generated by ascribing to each term the <u>independent</u> probabilities $M_\ell = a_\ell^* a_\ell$. Now probability theory is equivalent to measure theory mathematically,

- 21 -

Any observation of a quantity B, between two successive observations of quantity A (all on the same system) will destroy
the one-one correspondence between the earlier and later memory
states for the result of A. Thus for alternating observations
of different quantities there are fundamental limitations upon
the correlations between memory states for the same observed
quantity, these limitations expressing the content of the uncertainty principle.

In conclusion, we have described in this section processes
involving an idealized observer, processes which are entirely
deterministic and continuous from the over-all viewpoint (the
total state function is presumed to satisfy a wave equation at
all times) but whose result is a superposition, each element of
which describes the observer with a different memory state. We
have seen that in almost all of these observers states it appears to the observer that
the probabilistic aspects of the usual form of quantum theory
are valid. We have thus seen how pure wave mechanics, without
any initial probability assertions, can lead to these notions on
a subjective level, as appearances to observers.

(stop here)

Start here on continuation from Pg 72

- 2 -

§3.   Remarks on the choice of square amplitude measure:

While at first sight an artificial choice, and one which
seems to give rise to the danger of begging the question, a
little reflection shows that this choice of measure is not so
arbitrary as it appears, but is the only reasonable choice for
the purpose of making statistical deductions.

Let us ~~therefore~~ consider the search for a general scheme for assigning
a measure to the elements of a superposition of orthogonal states
$\sum a_i \phi_i$. We require then a positive function m of the complex
coefficients of the elements of the superposition, so that $m(a_i)$
shall be the measure assigned to the element ϕ_i. In order
that this general scheme shall be unambiguous we must first
require that the states themselves always be normalized, so that
we can distinguish the coefficients from the states. However,
we can still only determine the **coefficients,** in distinction to
the states, up to an arbitrary phase factor, and hence the function m must be a function of the amplitudes of the coefficients
alone, (i.e. $m(a_i) = m(\sqrt{a_i^* a_i})$), in order to avoid ambiguities.

If we now impose the additivity requirement that if we
regard a subset of the superposition, say $\overset{m}{\underset{i=1}{\sum}} a_i \phi_i$, as a single
element $\alpha \phi'$:

(3.1) $$\alpha \phi' = \sum_{i=1}^{n} a_i \phi_i$$

then the measure assigned to ϕ' shall be the sum of the measures
assigned to the ϕ_i (i from 1 to n):

It is necessary to say a few words about a view which is sometimes expressed, the idea that a physical theory should contain no
elements which do not correspond directly to observables. This position seems to be founded on the notion that the only purpose of a
theory is to serve as a summary of known data, and overlooks the
second major purpose, the discovery of totally new phenomena. The
major motivation of this viewpoint appears to be the desire to construct perfectly "safe" theories which will never be open to contradiction. Strict adherence to such a philosophy would probably seriously stifle the progress of physics.

The critical examination of just what quantities are observable
in a theory does, however, play a useful role, since it gives an insight into ways of modification of a theory when it becomes necessary.
A good example of this process is the development of Special Relativity.
Such successes of the ~~positivist~~ viewpoint, when used merely as a
However, operationalist
tool for deciding which modifications of a theory are possible, in
no way justify its universal adoption as a general principle which
all theories must satisfy.

In summary, a physical theory is a logical construct (model),
consisting of symbols and rules for their manipulation, some of whose
elements are associated with elements of the perceived world. The
fundamental requirements of a theory are logical consistency and
correctness. There is no reason why there cannot be any number of
different theories satisfying these requirements, and further criteria such as usefulness, simplicity, comprehensiveness, pictorability, etc., must be resorted to in such cases to further restrict the
number. Even so, it may be impossible to give a total ordering of
the theories according to "goodness", since different ones may rate
highest according to the different criteria, and it may be most advantageous to retain more than one.

As a final note, we might comment upon the concept of <u>causality</u>.
It should be clearly recognized that causality is a property of a

the observer as definitely perceiving that particular system state

¹ At this point we encounter a language difficulty.
Wheras before the observation we had a single observer
state afterwards there were a number of different states
for the observer, all occurring in a superposition. Each
of these separate states is a state for an observer, so
that we can speak of the different observers described
by the different states. On the other hand, the same
physical system is involved, and from this viewpoint it
is the same observer, which is in different states for
different elements of the superposition (i.e., has had
different experiences in the separate elements of the
superposition). In this situation we shall use the
singular when we wish to emphasize that a single physical
system is involved, and the plural when we wish to
emphasize the different experiences for the separate
elements of the superposition. (e.g., "The observer
performs an observation of the quantity A, after which
each of the observers of the resulting superposition has
perceived an eigenvalue.")

Handwritten Notes on Everett's Copy of DeWitt and Graham (1973)

The following two scanned pages come from Everett's personal copy of the DeWitt and Graham anthology (DeWitt and Graham, 1973). Bryce DeWitt and his student Neill Graham found Everett's derivation of probability in the long and short theses uncompelling. Everett, in turn, was frustrated by what he took as their lack of understanding concerning how his derivation worked. Everett's marginal notes on the following two pages indicate his sentiments. The first scanned page is from DeWitt's paper, "The Many-Universes Interpretation of Quantum Mechanics." The second is from Graham's paper, "The Measurement of Relative Frequency."

Then, making use of (4.13), we find

$$(4.16) \qquad \langle \chi_N^s | \chi_N^s \rangle = \langle \Psi | \Psi \rangle \sum_{\substack{s_1 \ldots \\ \delta(s_1 \ldots s_N) \geqslant s}} w_{s_1} w_{s_2} \ldots = \langle \Psi | \Psi \rangle \sum_{\substack{s_1 \ldots s_N \\ \delta(s_1 \ldots s_N) \geqslant s}} w_{s_1} \ldots w_{s_N}$$

$$< \frac{1}{\varepsilon} \langle \Psi | \Psi \rangle \sum_{s_1 \ldots s_N} \delta(s_1 \ldots s_N) w_{s_1} \ldots w_{s_N} = \frac{1}{N\varepsilon} \langle \Psi | \Psi \rangle \sum_{s} w_s (1 - w_s) < \frac{1}{N\varepsilon} \langle \Psi | \Psi \rangle .$$

From this it follows that no matter how small we choose ε we can always find an N big enough so that the norm of $|\chi_N^s\rangle$ becomes smaller than any positive number. This means that

$$(4.17) \qquad \qquad \lim_{N \to \infty} |\Psi_N^s\rangle = |\Psi\rangle .$$

It will be noted that, because of the orthogonality of the basis vectors $|s_1\rangle |s_2\rangle \ldots$, this result holds regardless of the quality of the measurements, *i.e.* independently of whether or not the condition

$$(4.18) \qquad \qquad \langle \Phi[s_1 \ldots s_N] | \Phi[s_1' \ldots s_N'] \rangle = \langle \Phi | \Phi \rangle \prod_{n=1}^{N} \delta_{s_n s_n'}$$

for good measurements is satisfied.

A similar result is obtained if $|\Psi_N^s\rangle$ is redefined by excluding, in addition, elements of the superposition (4.2) whose memory sequences fail to meet any finite combination of the infinity of other requirements for a random sequence. Moreover, no other choice for the w's but (2.4) will work. *The conventional statistical interpretation of quantum mechanics thus emerges from the formalism itself.* Nonrandom memory sequences in the superposition (4.8) are of *measure zero* in the Hilbert space, in the limit $N \to \infty$ (*). Each automaton (that is, apparatus *cum* memory sequence) in the superposition sees the world obey the familiar statistical quantum laws. This conclusion obviously admits of immediate extension to the world of cosmology. Its state vector is like a tree with an enormous number of branches. Each branch corresponds to a possible universe-as-we-actually-see-it.

The alert student may now object that the above argument contains an element of circularity. In order to derive the *physical* probability interpreta-

---

(*) Everett's original derivation of this result [1] invokes the formal equivalence of measure theory and probability theory, and is rather too brief to be entirely satisfying. The present derivation is essentially due to GRAHAM [7] (see also ref. [8]). A more rigorous treatment of the statistical interpretation question, which deals carefully with the problem of defining the Hilbert space in the limit $N \to \infty$, has been given by HARTLE [9].

only to you!

Figure G.1. Handwritten notes on Everett's copy of DeWitt and Graham.

If we again refer to Table 1.1, we see that the measure of those worlds with R successes does indeed have a peak at $R = NP = 36$. But this is of little comfort when we observe the degree to which these worlds are in a numerical minority.

In short, we criticize Everett's interpretation on the grounds of insufficient motivation. Everett gives no connection between his measure and the actual operations involved in determining a relative frequency, no way in which the value of his measure can actually influence the reading of, say, a particle counter. Furthermore, it is extremely difficult to see what significance such a measure can have when its implications are completely contradicted by a simple count of the worlds involved, worlds that Everett's own work assures us must all be on the same footing.

(To be sure Everett argues that the measure defined by (1.13) is unique. But remember that Gleason [8] has shown that the probabilities defined by the Born interpretation, considered as a measure on a Hilbert space, are themselves unique. Nevertheless, this (hopefully) does not deter anyone from inquiring into the connection between those probabilities and experiments that measure relative frequency.)

It thus appears that, in the "one step" measurement we have described, any attempt to show that the probability interpretation holds in the majority of the resulting Everett worlds is doomed to failure. As mentioned earlier, we shall attempt to improve matters by considering instead a "two step" measurement, in which a macroscopic apparatus mediates between a microscopic system and a macroscopic observer. To better motivate the technical work that follows, we now give a brief outline of this approach.

To begin, define the relative frequency of ℓ in the sequence $m_1, \ldots, m_N$ by

$$f_\ell(m_1, \ldots, m_N) = \frac{1}{N} \sum_{i=1}^{N} \delta_{m_i \ell} \, . \tag{1.15}$$

Since relative frequency is an observable, on the same footing with any other observable in quantum mechanics, we may associate with it a Hermitean operator. We define this operator, F_ℓ, in the obvious way:

Although Everett did not do any more work on his theory, the materials stored for so many years in Mark's basement shed a good deal of light on what Everett was thinking as he developed it—and how he viewed it empirically. As the reader can see from the presentations in this volume, Everett really did care about the fate of his bright idea.

Among his basement papers were notes he had made after the American Institute of Physics asked him (in May 1957) to prioritize his top five scientific capabilities according to a list of skills. At the bottom of the list, he put "servomechanisms." Above that was "operations research." Skill number three was "relativity and gravity." Two was "decision game theory." He added a new category to the list—"foundations of quantum mechanics"—but, upon reflection, he settled upon another term for pride of first place: "quantum mechanics."

Sadly, the gradual acceptance of the revolutionary theory as a viable interpretation of the quantum world came about largely after the untimely demise of the theorist even as his personal world was collapsing. At the time of his death at age 51, Everett's consulting firm was disintegrating due to cutbacks in military research budgets and his mismanagement of the business end. He was ruined financially and personally—his body wracked by compulsive smoking, drinking, and eating. His marriage had long been troubled, and his self-effacing wife, Nancy, was resigned to accepting his multiple affairs with employees and prostitutes (while engaging in a few of her own). Only Nancy knew of the black depression that fueled her husband's alcoholic self-destruction, and she felt powerless to intervene.

Ironically, Everett had devoted his life to making models of reality, but he was largely oblivious to the harm he caused those closest to him. He barely acknowledged the existence of his troubled children, Mark and Liz, who yearned for paternal attention. When Everett died suddenly of a heart attack on July 19, 1982, his teenage son, Mark, trying to revive his corpse, reflected that he could not recall ever having touched his father in life. After his father died, Mark moved to Los Angeles and became a successful songwriter known to his fans as E of the band Eels. Liz committed suicide in 1996, leaving behind a note saying that she was going to join her father in another universe. Nancy died of lung cancer in 1998, probably from her husband's secondhand smoke. Mark, the sole survivor, found solace in writing songs about his vanished family and in popularizing the story of his family through concerts, television, and books—helping to give his father, as he put it, "a day in the sun." And to that end, he authorized the publication of this volume: which we dedicate to the memory of Hugh, Nancy, and Elizabeth.

Peter Byrne
Petaluma, California
January 2011

Aharonov, Y., and D. Bohm (1959). "Significance of Electromagnetic Potentials in the Quantum Theory." *Physical Review, 115*(3), 485–491.

Albert, David (1986). "How to Take a Photograph of Another Everett World." *Annals of the New York Academy of Sciences, 480*, 498–502.

Albert, David (2010). "Probability in the Everett Picture." *Many Worlds?: Everett, Quantum Theory, and Reality*. Edited by Simon Saunders, Jonathan Barrett, Adrian Kent, and David Wallace. Oxford: Oxford University Press. 355–368.

Albert, David, and Barry Loewer (1988). "Interpreting the Many Worlds Interpretation." *Synthese, 77*(2), 195–213.

Albert, David Z. (1992). *Quantum Mechanics and Experience*. Cambridge, Mass. Harvard University Press.

Barrett, Jeffrey (1999). *The Quantum Mechanics of Minds and Worlds*. New York: Oxford University Press.

Barrett, Jeffrey A. (2008). "Everett's Relative-State Formulation of Quantum Mechanics." *The Stanford Encyclopedia of Philosophy*. Edited by Edward N. Zalta. Palo Alto, Calif. The Metaphysics Research Lab (http://plato.stanford.edu/archives/win2008/entries/qm-everett/).

Barrett, Jeffrey A. (2010). "A Structural Interpretation of Pure Wave Mechanics." *Humana.Mente*, (13).

Barrett, Jeffrey A. (2011a). "Everett's Pure Wave Mechanics and the Notion of Worlds." *European Journal for Philosophy of Science, 1* (2), 277–302.

Barrett, Jeffrey A. (2011b). "On the Faithful Interpretation of Pure Wave Mechanics." *British Journal for the Philosophy of Science, forthcoming*.

Bell, J. S. (1964). "On the Einstein-Podolsky-Rosen Paradox." *Physics, 1*, 195–200.

Bell, J. S. (1971). "On the Hypothesis that the Schroedinger Equation is Exact." Ref.TH.1424-CERN (27 October 1971). Everett's personal copy of Bell's paper contains Everett's unpublished, undated marginal comments. A copy of the original unpublished document is available at http://cdsweb.cern.ch/record/956196/files/CM-P00058456.pdf?version=1. A revised version was published as Bell (1987).

Bell, J. S. (1987). "Quantum Mechanics for Cosmologists." *Speakable and Unspeakable in Quantum Mechanics*. New York: Cambridge University Press.

Ben-Dov, Yoav (1990). "Everett's Theory and the 'Many-Worlds' Interpretation." *American Journal of Physics, 58*(9), 829.

Bohm, D. (1951). *Quantum Theory*. New York: Prentice-Hall.

Bohm, D. (1952). "A Suggested Interpretation of the Quantum Theory in Terms of Hidden Variables, I and II." *Physical Review, 85*, 166–193.

Bohr, Niels (1934). *Atomic Theory and the Description of Nature*. Cambridge, U.K.: Cambridge University Press.

Bohr, Niels (1949). "Discussions with Einstein on Epistemological Problems in Atomic Physics." *Albert Einstein: Philosopher-Scientist*. Edited by P. A. Schilpp. Evanston, Ill.: The Library of Living Philosophers, Inc.

Bohr, Niels, and Leon Rosenfeld (1933). "Zur frage der messbarkeit der elektromagnetischen feldgrössen [On the question of the measurability of electromagnetic field quantities]." *Kongelige Danske Videnskabernes Selskab, Matematisk-Fysiske Meddelelser [Royal Danish Academy of Science and Letters, Journal of Mathematical Physics]*, 12(8).

Bopp, Friedrich (1947). "Quantenmechanik und Korrelationsrechnung." *Zeitschrift für Naturforschung*, 2a(4), 202–208.

Bopp, Friedrich (1953a). "Für die Quantenmechanik bemerkenswerter Satz der Korrelationsrechnung." *Zeitschrift für Naturforschung*, 7a, 82–87.

Bopp, Friedrich (1953b). "Statische Untersuchung des Grundprozesses der Quantentheorie der Elementarteilchen." *Zeitschrift für Naturforschung*, 8a, 6–13.

Bub, Jeffrey, Robert Clifton, and Bradley Monton (1998). "The Bare Theory Has No Clothes." *Quantum Measurement: Beyond Paradox*. Edited by Richard Healey and Geoffrey Hellman. Volume 17 of Minnesota Studies in the Philosophy of Science. Minneapolis: University of Minnesota Press, 32–51.

Byrne, Peter (2010). *The Many Worlds of Hugh Everett III: Multiple Universes, Mutual Assured Destruction, and the Meltdown of a Nuclear Family*. New York: Oxford University Press.

Chateaubriand, François-Auguste-René de (1802). *Génie du Christianisme [The Genius of Christianity]*. Paris: Chez Migneret.

Cooper, Leon, and Deborah van Vechten (1969). "On the Interpretation of Measurement within the Quantum Theory." *American Journal of Physics*, 37(12), 1212–1220. Reprinted in DeWitt and Graham (1973).

Deutsch, David (1999). "Quantum Theory of Probability and Decisions." *Proceedings of the Royal Society of London A*, 456, 1759.

DeWitt, Bryce (1988). "Confidential Referee Report." This document is an unpublished referee report that DeWitt wrote on Ben-Dov (1990). Available at the UCIspace archive at http://hdl.handle.net/10575/1166.

DeWitt, Bryce, and R. Neill Graham (1973). *The Many-Worlds Interpretation of Quantum Mechanics*. Princeton, N.J.: Princeton University Press.

DeWitt, Bryce S. (1970). "Quantum Mechanics and Reality." *Physics Today*, 23, 30–35. Reprinted in DeWitt and Graham (1973).

DeWitt, Bryce S. (1971). "The Many Universes Interpretation of Quantum Mechanics." Preprint distributed by the Department of Physics, University of North Carolina at Chapel Hill. A version was later printed in DeWitt and Graham (1973). Everett's personal copy includes two significant marginal comments.

DeWitt, C. M. (1957). "Conference on the Role of Gravitation in Physics." Transcript. *National Technical Information Service*, AD118180.

Dirac, Paul A. M. (1930). *The Principles of Quantum Mechanics*. Oxford: Clarendon Press.

Doob, J. L. (1953). *Stochastic Processes*. New York: Wiley.

Dowker, F., and A. Kent (1996). "On the Consistent Histories Approach to Quantum Mechanics." *Journal of Statistical Physics*, 83(5–6), 1575–1646.

Einstein, Albert (1949). "Remarks to the Essays Appearing in this Collective Volume." *Albert Einstein: Philosopher-scientist*. Edited by P. A. Schilpp. Evanston, Ill.: The Library of Living Philosophers, Inc.

Einstein, Albert (1950). *The Meaning of Relativity*. Third edition. Princeton, N.J.: Princeton University Press.

Einstein, Albert, and Nathan Rosen (1935). "The Particle Problem in the General Theory of Relativity." *Physical Review*, 48(1), 73–77.

Einstein, Albert, Boris Podolsky, and Nathan Rosen (1935). "Can Quantum-Mechanical Description of Physical Reality Be Considered Complete?" *Physical Review*, 47(10), 777–780.

Everett, Hugh, III (1957a). *On the Foundations of Quantum Mechanics*, Ph.D. thesis, Princeton University, Department of Physics.

Everett, Hugh, III (1957b). " 'Relative State' Formulation of Quantum Mechanics." *Reviews of Modern Physics*, 29, 454–462.

Everett, Hugh, III (1959). "Letter from Hugh Everett, III to Boris Podolsky dated March 12, 1959." Available at the UCIspace archive (http://hdl.handle.net/10575/1166).

Everett, Hugh, III (1973). "The Theory of the Universal Wave Function," In DeWitt and Graham (1973), pgs. 3–140.

Feller, W. (1950). *An Introduction to Probability Theory and Its Applications*. New York: Wiley.

Frank, Philipp (1946). *Foundations of Physics*. Volume 1 (7) of International Encyclopedia of Unified Science. Chicago: University of Chicago Press.

Frank, Philipp (1950). *Relativity: A Richer Truth*. Boston: Beacon Press.

Frank, Philipp (1954). *The Validation of Scientific Theories*. Boston: Beacon Press.

Frank, Philipp (1955). *Modern Science and Its Philosophy*. New York: George Braziller.

Gell-Mann, Murray, and James B. Hartle (1990). "Quantum Mechanics in Light of Quantum Cosmology." *Complexity, Entropy, and the Physics of Information*. Edited by W. H. Zurek. Volume VIII of Proceedings of the Santa Fe Institute Studies in the Sciences of Complexity. Redwood City, Calif.: Addison-Wesley. 422–458.

Geroch, Robert (1984). "The Everett Interpretation." *Noûs*, *18*(4), 617–633.

Ghirardi, G. C., A. Rimini, and T. Weber (1986). "Unified dynamics for microscopic and macroscopic systems." *Phys. Rev. D*, *34*(2), 470.

Gosse, Philip (1857). *Omphalos: An Attempt to Untie the Geological Knot*. London: John Van Voorst.

Halmos, P. R. (1950). *Measure Theory*. New York: Van Nostrand.

Hardy, G. H., J. E. Littlewood, and G. Pólya (1952). *Inequalities*. New York: Cambridge University Press.

Heisenberg, Werner (1955). "The development of the interpretation of the quantum theory." *Niels Bohr and the Development of Physics*. Edited by Wolfgang Pauli. New York: McGraw-Hill.

Henry, O. (1909). "Roads of Destiny." *Roads of Destiny*. New York: Doubleday, Page & Co.

Jammer, Max (1966). *The Conceptual Development of Quantum Mechanics*. New York: McGraw-Hill.

Jammer, Max (1974). *The Philosophy of Quantum Mechanics: The Interpretations of Quantum Mechanics in Historical Perspective*. New York: John Wiley & Sons.

Kelley, J. (1955). *General Topology*. New York: Van Nostrand.

Kent, Adrian (2010). "One World Versus Many: The Inadequacy of Everettian Account of Evolution, Probability, and Scientific Confirmation." *Many Worlds?: Everett, Quantum Theory, and Reality*. Edited by Simon Saunders,

Jonathan Barrett, Adrian Kent, and David Wallace. Oxford: Oxford University Press, 307–354.

Khinchin, A. I. (1949). *Mathematical Foundations of Statistical Mechanics*. Translated by George Gamow. New York: Dover.

Kuhn, Harold W. (1997). *Classics in Game Theory*. Princeton, N.J.: Princeton University Press.

Landau, L., and E. Lifshitz (1951). *The Classical Theory of Fields*. Translated by M. Hamermesh. Cambridge, Mass.: Addison-Wesley Press.

Lévy-Leblond, Jean-Marc (1976). "Towards a Proper Quantum Theory (Hints for a recasting)." *Dialectica*, *30*(2/3), 162–196.

Lorentz, Hendrik (1892). *La Théorie électromagnetique de Maxwell et son application aux corps mouvants [Maxwell's electromagnetism and its application to moving bodies*. Leide: E. J. Brill.

Maudlin, Tim (2010). "Can the World Be Only Wavefunction." *Many Worlds?: Everett, Quantum Theory, and Reality*. Edited by Simon Saunders, Jonathan Barrett, Adrian Kent, and David Wallace. Oxford: Oxford University Press, 144–153.

Misner, Charles W. (1957). "Feynman Quantization of General Relativity." *Reviews of Modern Physics*, *29*(3), 497–509.

Monton, Brad (1998). "Quantum-Mechanical Self-Measurement." *The Modal Interpretation of Quantum Mechanics*. Edited by Dennis Dieks and Pieter E. Vermaas. Dordrecht: Kluwer Academic.

Pais, Abraham (1991). *Niels Bohr's times: In physics, philosophy, and polity*. Oxford: Clarendon Press.

Poincaré, Henri (1905). *Science and Hypothesis*. Translated by George Bruce Halsted. New York: The Science Press.

Poincaré, Henri (1921). "Science and Method." *The Foundations of Science*. New York: The Science Press.

Quine, Willard Van Orman (1951). "Two Dogmas of Empiricism." *Philosophical Review*, *60*, 20–43.

Saunders, Simon (1995). "Time, Quantum Mechanics, and Decoherence." *Synthese*, *102*, 235–266.

Saunders, Simon (1996). "Time, Quantum Mechanics, and Tense." *Synthese*, *107*, 19–53.

Saunders, Simon (1998). "Time, Quantum Mechanics, and Probability." *Synthese*, *114*, 373–404.

Saunders, Simon, Jonathan Barrett, Adrian Kent, and David Wallace (2010). *Many Worlds?: Everett, Quantum Theory, and Reality*. Oxford: Oxford University Press.

Schrödinger, Erwin (1952a). "Are There Quantum Jumps? Part I." *British Journal for the Philosophy of Science*, *3*(10), 109–123.

Schrödinger, Erwin (1952b). "Are There Quantum Jumps? Part II." *British Journal for the Philosophy of Science*, *3*(11), 233–242.

Schrödinger, Erwin (1983). "The Present Situation in Quantum Mechanics." *Quantum Theory and Measurement*. Edited by John Archibald Wheeler and Wojciech Hubert Żurek. Princeton, N.J.: Princeton University Press.

Schrödinger, Erwin (1995). *The interpretation of quantum mechanics: Dublin seminars (1949–1955) and other unpublished essays*. Woodbridge, CT: Ox Bow Press.

Shannon, Claude E., and Warren Weaver (1949). *The Mathematical Theory of Communication*. Champaign, IL: University of Illinois Press.

Tauber, G. E. (1979). *Albert Einstein's Theory of Relativity*. New York: Crown Publishers.

ter Haar, D. (1954). *Elements of Statistical Mechanics*. New York: Rinehart.

van Fraassen, Bas C. (2008). *Scientific Representation: Paradoxes of Perspective*. Oxford: Oxford University Press.

von Neumann, John (1932). *Mathematische Grundlagen der Quantenmechanik [Mathematical Foundations of Quantum Mechanics]*. Berlin, Germany: Springer Verlag.

von Neumann, John (1955). *Mathematical Foundations of Quantum Mechanics*. Trans. R. T. Breyer. Princeton, N.J.: Princeton University Press.

Wallace, David (2003). "Everettian rationality: defending Deutsch's approach to probability in the Everett interpretation." *Studies in the History and Philosophy of Modern Physics*, 34, 415–439.

Wallace, David (2010a). "Decoherence and Ontology." *Many Worlds?: Everett, Quantum Theory, and Reality*. Edited by Simon Saunders, Jonathan Barrett, Adrian Kent, and David Wallace. Oxford: Oxford University Press, 53–72.

Wallace, David (2010b). "How to Prove the Born Rule." *Many Worlds?: Everett, Quantum Theory, and Reality*. Edited by Simon Saunders, Jonathan Barrett, Adrian Kent, and David Wallace. Oxford: Oxford University Press, 227–263.

Wallace, David (2011). *The emergent multiverse: Quantum theory according to the Everett interpretation*. Oxford: Oxford University Press. Forthcoming.

Wheeler, John Archibald (1955). "Geons." *Physical Review*, 97, 511–536.

Wheeler, John Archibald (1971). "From Mendeleev's Atom to the Collapsing Star." *Transaction of the New York Academy of Sciences, Series 2*, 33, 745–779.

Wheeler, John Archibald, and Kenneth Ford (1998). *Geons, Black Holes, and Quantum Foam: A Life in Physics*. New York: W. W. Norton & Co.

Wiener, Norbert (1948). *Cybernetics, or Control and Communications in the Animal and the Machine*. Cambridge, Mass.: MIT University Press.

Wiener, Norbert, and Armand Siegel (1955). "The Differential Space Theory of Quantum Systems." *Nuovo Cimento Suppl.*, 2(4), 982–1003.

Wigner, Eugene P. (1961). "Remarks on the Mind-Body Problem." *The Scientist Speculates*. Edited by I. J. Good. London: Heinemann, 284–302.

Woodward, P. M. (1953). *Probability and Information Theory, with Applications to Radar*. New York: McGraw-Hill.

Zeh, H. Dieter (1970). "On the Interpretation of Measurement in Quantum Theory." *Foundations of Physics*, 1, 69–76.

Zurek, W. H. (1991). "Decoherence and the Transition from Quantum to Classical." *Physics Today*, 44(10), 36–44.